Protein Purification Protocols

Methods in Molecular Biology™
John M. Walker, SERIES EDITOR

60. **Protein NMR Protocols**, edited by *David G. Reid, 1996*
59. **Protein Purification Protocols**, edited by *Shawn Doonan, 1996*
58. **Basic DNA and RNA Protocols**, edited by *Adrian J. Harwood, 1996*
57. **In Vitro Mutagenesis Protocols**, edited by *Michael K. Trower, 1996*
56. **Crystallographic Methods and Protocols**, edited by *Christopher Jones, Barbara Mulloy, and Mark Sanderson, 1996*
55. **Plant Cell Electroporation and Electrofusion Protocols**, edited by *Jac A. Nickoloff, 1995*
54. **YAC Protocols**, edited by *David Markie, 1995*
53. **Yeast Protocols:** *Methods in Cell and Molecular Biology,* edited by *Ivor H. Evans, 1996*
52. **Capillary Electrophoresis:** *Principles, Instrumentation, and Applications,* edited by *Kevin D. Altria, 1996*
51. **Antibody Engineering Protocols**, edited by *Sudhir Paul, 1995*
50. **Species Diagnostics Protocols:** *PCR and Other Nucleic Acid Methods,* edited by *Justin P. Clapp, 1996*
49. **Plant Gene Transfer and Expression Protocols**, edited by *Heddwyn Jones, 1995*
48. **Animal Cell Electroporation and Electrofusion Protocols**, edited by *Jac A. Nickoloff, 1995*
47. **Electroporation Protocols for Microorganisms**, edited by *Jac A. Nickoloff, 1995*
46. **Diagnostic Bacteriology Protocols**, edited by *Jenny Howard and David M. Whitcombe, 1995*
45. **Monoclonal Antibody Protocols**, edited by *William C. Davis, 1995*
44. ***Agrobacterium* Protocols**, edited by *Kevan M. A. Gartland and Michael R. Davey, 1995*
43. **In Vitro Toxicity Testing Protocols**, edited by *Sheila O'Hare and Chris K. Atterwill, 1995*
42. **ELISA:** *Theory and Practice,* by *John R. Crowther, 1995*
41. **Signal Transduction Protocols**, edited by *David A. Kendall and Stephen J. Hill, 1995*
40. **Protein Stability and Folding:** *Theory and Practice,* edited by *Bret A. Shirley, 1995*
39. **Baculovirus Expression Protocols**, edited by *Christopher D. Richardson, 1995*
38. **Cryopreservation and Freeze-Drying Protocols**, edited by *John G. Day and Mark R. McLellan, 1995*
37. **In Vitro Transcription and Translation Protocols**, edited by *Martin J. Tymms, 1995*
36. **Peptide Analysis Protocols**, edited by *Ben M. Dunn and Michael W. Pennington, 1994*
35. **Peptide Synthesis Protocols**, edited by *Michael W. Pennington and Ben M. Dunn, 1994*
34. **Immunocytochemical Methods and Protocols**, edited by *Lorette C. Javois, 1994*
33. **In Situ Hybridization Protocols**, edited by *K. H. Andy Choo, 1994*
32. **Basic Protein and Peptide Protocols**, edited by *John M. Walker, 1994*
31. **Protocols for Gene Analysis**, edited by *Adrian J. Harwood, 1994*
30. **DNA–Protein Interactions**, edited by *G. Geoff Kneale, 1994*
29. **Chromosome Analysis Protocols**, edited by *John R. Gosden, 1994*
28. **Protocols for Nucleic Acid Analysis by Nonradioactive Probes**, edited by *Peter G. Isaac, 1994*
27. **Biomembrane Protocols:** *II. Architecture and Function,* edited by *John M. Graham and Joan A. Higgins, 1994*
26. **Protocols for Oligonucleotide Conjugates:** *Synthesis and Analytical Techniques,* edited by *Sudhir Agrawal, 1994*
25. **Computer Analysis of Sequence Data:** *Part II,* edited by *Annette M. Griffin and Hugh G. Griffin, 1994*
24. **Computer Analysis of Sequence Data:** *Part I,* edited by *Annette M. Griffin and Hugh G. Griffin, 1994*
23. **DNA Sequencing Protocols**, edited by *Hugh G. Griffin and Annette M. Griffin, 1993*
22. **Microscopy, Optical Spectroscopy, and Macroscopic Techniques**, edited by *Christopher Jones, Barbara Mulloy, and Adrian H. Thomas, 1993*
21. **Protocols in Molecular Parasitology**, edited by *John E. Hyde, 1993*
20. **Protocols for Oligonucleotides and Analogs:** *Synthesis and Properties,* edited by *Sudhir Agrawal, 1993*
19. **Biomembrane Protocols:** *I. Isolation and Analysis,* edited by *John M. Graham and Joan A. Higgins, 1993*
18. **Transgenesis Techniques:** *Principles and Protocols,* edited by *David Murphy and David A. Carter, 1993*
17. **Spectroscopic Methods and Analyses:** *NMR, Mass Spectrometry, and Metalloprotein Techniques,* edited by *Christopher Jones, Barbara Mulloy, and Adrian H. Thomas, 1993*
16. **Enzymes of Molecular Biology**, edited by *Michael M. Burrell, 1993*
15. **PCR Protocols:** *Current Methods and Applications,* edited by *Bruce A. White, 1993*
14. **Glycoprotein Analysis in Biomedicine**, edited by *Elizabeth F. Hounsell, 1993*
13. **Protocols in Molecular Neurobiology**, edited by *Alan Longstaff and Patricia Revest, 1992*
12. **Pulsed-Field Gel Electrophoresis:** *Protocols, Methods, and Theories,* edited by *Margit Burmeister and Levy Ulanovsky, 1992*
11. **Practical Protein Chromatography**, edited by *Andrew Kenney and Susan Fowell, 1992*
10. **Immunochemical Protocols**, edited by *Margaret M. Manson, 1992*
9. **Protocols in Human Molecular Genetics**, edited by *Christopher G. Mathew, 1991*
8. **Practical Molecular Virology:** *Viral Vectors for Gene Expression,* edited by *Mary K. L. Collins, 1991*
7. **Gene Transfer and Expression Protocols**, edited by *Edward J. Murray, 1991*
6. **Plant Cell and Tissue Culture**, edited by *Jeffrey W. Pollard and John M. Walker, 1990*
5. **Animal Cell Culture**, edited by *Jeffrey W. Pollard and John M. Walker, 1990*

Methods in Molecular Biology™ • 59

Protein Purification Protocols

Edited by

Shawn Doonan

*Department of Life Sciences,
University of East London, UK*

Humana Press ✻ Totowa, New Jersey

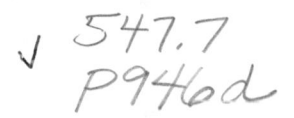

© 1996 Humana Press Inc.
999 Riverview Drive, Suite 208
Totowa, New Jersey 07512

All rights reserved. No part of this book may be reproduced, stored in a retrieval system, or transmitted in any form or by any means, electronic, mechanical, photocopying, microfilming, recording, or otherwise without written permission from the Publisher. Methods in Molecular Biology™ is a trademark of The Humana Press Inc.

All authored papers, comments, opinions, conclusions, or recommendations are those of the author(s), and do not necessarily reflect the views of the publisher.

This publication is printed on acid-free paper. ∞
ANSI Z39.48-1984 (American Standards Institute)
Permanence of Paper for Printed Library Materials.

Cover illustration: Fig. 1 from Chapter 16, "Affinity Chromatography," by Paul Cutler.

For additional copies, pricing for bulk purchases, and/or information about other humana titles, contact Humana at the above address or at any of the following numbers: Tel.: 201-256-1699; Fax: 201-256-8341; E-mail: humana@interramp.com

Photocopy Authorization Policy:
Authorization to photocopy items for internal or personal use, or the internal or personal use of specific clients, is granted by Humana Press Inc., provided that the base fee of US $5.00 per copy, plus US $00.25 per page, is paid directly to the Copyright Clearance Center at 222 Rosewood Drive, Danvers, MA 01923. For those organizations that have been granted a photocopy license from the CCC, a separate system of payment has been arranged and is acceptable to Humana Press Inc. The fee code for users of the Transactional Reporting Service is: [0-89603-336-8/96 $5.00 + $00.25].

Printed in the United States of America. 10 9 8 7 6 5 4 3

Library of Congress Cataloging in Publication Data

Main entry under title:

Methods in molecular biology™.

Protein purification protocols / edited by Shawn Doonan.
    p. cm. — (Methods in molecular biology™ ; 59)
    Includes index.
    ISBN 0-89603-336-8 (alk. paper)
    1. Proteins—Purification—Laboratory manuals. I. Doonan, Shawn. II. Series: Methods in molecular biology™ (Totowa, NJ) ; 59.
QP551.P69756   1996
547.7'5406—dc20
                                                       96-3250
                                                       CIP

Preface

Hans Neurath has written that this is the second golden era of enzymology (*Protein Science* [1994], vol. 3, pp. 1734–1739); he could with justice have been more general and referred to the second golden age of protein chemistry. The last two decades have seen enormous advances in our understanding of the structures and functions of proteins arising on the one hand from improvements and developments in analytical techniques (*see* the companion volume, *Basic Protein and Peptide Protocols,* in this series) and on the other hand from the technologies of molecular genetics. Far from turning the focus away from protein science, the ability to isolate, analyze, and express genes has increased interest in proteins as gene products. Hence, many laboratories are now getting involved in protein isolation for the first time, either as an essential adjunct to their work in molecular genetics or because of a curiosity to know more about the products of the genes that they have been studying.

Protein Purification Protocols is aimed mainly at these newcomers to protein purification, but it is hoped that it will also be of value to established practitioners who may find here techniques that they have not tried, but which might well be most applicable in their work. With the exception mainly of the first and last chapters, the format of the contributions to the present book conform to the established format of the *Methods in Molecular Biology* series. That is, they introduce the theoretical background of a method or group of related methods, provide a list of the reagents and equipment required for the procedure, follow with a detailed step-by-step description of how to carry out the protocol, and conclude with a set of Notes dealing with whatever problems are likely to arise and precisely how to deal with them. The aim has been to cover everything that is required to obtain a pure protein from initial extraction to drying and storage of the product. The success of a purification schedule is critically dependent on obtaining a good initial extract with which to work; thus much of Chapters 2–12 is concerned with procedures for the extraction of proteins from various sources, including subcellular fractions of plant and animal tissues, for concentrating extracts, and for maintaining the integrity of proteins in those extracts. The remainder of the book then deals with individual techniques for fractionation and purification, including the special approaches required in the case of membrane proteins.

The first and last chapters of *Protein Purification Protocols* are somewhat different. Many of the individual methods described in the main section of the book are forms of column chromatography, but the chapters deal with particular techniques and not with the chromatographic method *per se*. Hence we have considered it worthwhile to conclude with a general chapter on practical column chromatography to fill in some of the experimental details that are common to all its applications. The first chapter deals with strategies for protein purification and is intended to put the rest of the book in context. Individual techniques used in protein purification are of limited value in isolation; the appropriate ones have to be used in the correct sequence to construct a complete purification schedule. It is hoped that this first chapter will help the newcomer to protein purification, as well as the seasoned investigator, to decide what has to be done and that the rest of *Protein Purification Protocols* will enable him or her to do it.

Shawn Doonan

Contents

Preface v
Contributors ix

CH. 1. General Strategies,
          Shawn Doonan 1

CH. 2. Preparation of Extracts from Animal Tissues,
          Shawn Doonan 17

CH. 3. Protein Extraction from Plant Tissues,
          Peter R. Shewry and Roger J. Fido 23

CH. 4. Extraction of Recombinant Protein from Bacteria,
          D. Margaret Worrall 31

CH. 5. Protein Extraction from Fungi,
          Paul Bridge 39

CH. 6. Subcellular Fractionation of Animal Tissues,
          Norma M. Ryan 49

CH. 7. Subcellular Fractionation of Plant Tissues: *Isolation of Chloroplasts and Mitochondria from Leaves,*
          Alyson K. Tobin 57

CH. 8. The Extraction of Enzymes from Plant Tissues Rich in Phenolic Compounds,
          William S. Pierpoint 69

CH. 9. Avoidance of Proteolysis in Extracts,
          Robert J. Beynon and Simon Oliver 81

CH. 10. Concentration of Extracts,
          Shawn Doonan 95

CH. 11. Making and Changing Buffers,
          Shawn Doonan 103

CH. 12. Purification and Concentration by Ultrafiltration,
          Paul Schratter 115

CH. 13. Bulk Purification by Fractional Precipitation,
          Shawn Doonan 135

CH. 14. Ion-Exchange Chromatography,
          David Sheehan and Richard FitzGerald 145

CH. 15. Hydrophobic Interaction Chromatography,
          Paul O'Farrell 151

Сн. 16.	Affinity Chromatography, Paul Cutler 157
Сн. 17.	Dye-Ligand Affinity Chromatography, D. Margaret Worrall 169
Сн. 18.	Lectin Affinity Chromatography, Iris West and Owen Goldring 177
Сн. 19.	Immunoaffinity Chromatography, George W. Jack and David J. Beer 187
Сн. 20.	Immobilized Metal Ion Affinity Chromatography, Tai-Tung Yip and T. William Hutchens 197
Сн. 21.	Chromatography on Hydroxyapatite, Shawn Doonan 211
Сн. 22.	Affinity Precipitation Methods, Jane A. Irwin and Keith F. Tipton 217
Сн. 23.	Isoelectric Focusing, Reiner Westermeier 239
Сн. 24.	Chromatofocusing, Timothy J. Mantle and Patricia Noone 249
Сн. 25.	Size-Exclusion Chromatography, Paul Cutler 255
Сн. 26.	Fast Protein Liquid Chromatography (FPLC) Methods, David Sheehan 269
Сн. 27.	Reversed-Phase Chromatography of Proteins, Bill Neville 277
Сн. 28.	Extraction of Membrane Proteins, Kay Ohlendieck 293
Сн. 29.	Removal of Detergent from Protein Fractions, Kay Ohlendieck 305
Сн. 30.	Purification of Membrane Proteins, Kay Ohlendieck 313
Сн. 31.	Lyophilization of Proteins, Ciarán Ó Fágáin 323
Сн. 32.	Storage of Pure Proteins, Ciarán Ó Fágáin 339
Сн. 33.	Electroelution of Proteins from Polyacrylamide Gels, Michael J. Dunn 357
Сн. 34.	Electroblotting of Proteins from Polyacrylamide Gels, Michael J. Dunn 363
Сн. 35.	High-Performance Electrophoresis Chromatography, Serge Desnoyers, Sylvie Bourassa, and Guy G. Poirier 371
Сн. 36.	Practical Column Chromatography, Shawn Doonan 381
Index	 397

Contributors

DAVID J. BEER • *Centre for Applied Microbiology and Research, Porton Down, Salisbury, UK*
ROBERT J. BEYNON • *Department of Biochemistry and Applied Molecular Biology, UMIST, Manchester, UK*
SYLVIE BOURASSA • *Eastern Quebec Peptide Sequencing Facility, CHUL Research Centre and Laval University, Ste. Foy, Quebec, Canada*
PAUL BRIDGE • *International Mycological Institute, Egham, Surrey, UK*
PAUL CUTLER • *SmithKline Beecham Pharmaceuticals, Welwyn, UK*
SERGE DESNOYERS • *Demerec Laboratory, Cold Spring Harbor, NY*
SHAWN DOONAN • *Department of Life Sciences, University of East London, UK*
MICHAEL J. DUNN • *Department of Cardiothoracic Surgery, Heart Science Centre, National Heart and Lung Institute, Harefield, UK*
ROGER J. FIDO • *Institute of Arable Crops Research-Long Ashton Research Station, Long Ashton, Bristol, UK*
RICHARD FITZGERALD • *Agricultural Institute, Moorepark Research Centre, Fermoy, Ireland*
OWEN GOLDRING • *North East Surrey College of Technology, Ewell, Surrey, UK*
T. WILLIAM HUTCHENS • *Department of Food Science and Technology, University of California, Davis, CA*
JANE A. IRWIN • *Department of Biochemistry, Trinity College, Dublin, Ireland*
GEORGE W. JACK • *Centre for Applied Microbiology and Research, Porton Down, Salisbury, UK*
TIMOTHY J. MANTLE • *Department of Biochemistry, Tinity College, Dublin, Ireland*
BILL NEVILLE • *SmithKline Beecham Pharmaceuticals, Welwyn, UK*

PATRICIA NOONE • *Department of Biochemistry, Trinity College, Dublin, Ireland*
CIARÁN Ó FÁGÁIN • *School of Biological Sciences, Dublin City University, Dublin, Ireland*
PAUL O'FARRELL • *Division of Biosciences, University of Hertfordshire, Hatfield, UK*
KAY OHLENDIECK • *Department of Pharmacology, University College, Dublin, Ireland*
SIMON OLIVER • *Department of Biochemistry and Applied Molecular Biology, UMIST, Manchester, UK*
WILLIAM S. PIERPOINT • *Institute of Arable Crops Research-Rothamsted, Harpenden, UK*
GUY G. POIRIER • *Eastern Quebec Peptide Sequencing Facility, CHUL Research Centre and Laval University, Ste. Foy, Quebec, Canada*
NORMA M. RYAN • *Department of Biochemistry, University College, Cork, Ireland*
PAUL SCHRATTER • *Amicon Inc., Beverly, MA*
DAVID SHEEHAN • *Department of Biochemistry, University College, Cork, Ireland*
PETER R. SHEWRY • *Institute of Arable Crops Research-Long Ashton Research Station, Long Ashton, Bristol, UK*
KEITH F. TIPTON • *Department of Biochemistry, Trinity College, Dublin, Ireland*
ALYSON K. TOBIN • *Plant Sciences Laboratory, School of Biological and Medical Sciences, University of St. Andrews, UK*
IRIS WEST • *North East Surrey College of Technology, Ewell, Surrey, UK*
REINER WESTERMEIER • *ETC Elektrophorese-Technik, Kirchentellinsfurt, Germany*
D. MARGARET WORRALL • *Department of Biochemistry, University College, Dublin, Ireland*
TAI-TUNG YIP • *Department of Food Science and Technology, University of California, Davis, CA*

CHAPTER 1

General Strategies

Shawn Doonan

1. Defining the Problem

The chapters that follow in this volume give detailed instructions on how to use the various methods that are available for purification of proteins. The question arises, however, of which of these methods to use and in which order to use them to achieve purification in any particular case. That is, the purification problem must be clearly defined. What follows outlines the sorts of questions that need to be asked as part of that definition and how the answers affect the approach that might be taken to developing a purification schedule. It should be noted here that the discussion does not touch on the special cases of purification of proteins at industrial scale or for therapeutic applications; these raise very specific problems that are outside the scope of this chapter (*see* refs. *1* and *2,* respectively, for a coverage of these topics).

1.1. How Much Do I Need?

The answer to this question depends on the purpose for which the protein is required. For example, to carry out a full chemical and physical analysis of a protein may require several hundreds of milligrams of purified material while a kinetic analysis of the reaction catalyzed by an enzyme could perhaps be done with a few milligrams and <1 mg would be required to raise a polyclonal antibody. At the extreme end of the scale, if the objective is to obtain limited sequence information from the N-terminus of a protein as a preliminary to design of an oligonucleotide probe for clone screening, then using modern microsequencing techniques, a few micrograms will be sufficient. These different requirements for quantity may well dictate the source of the protein chosen (*see* Section 1.4.) and will certainly influence the approach to purification. Purification of large quantities of protein requires use of techniques, at least in the early stages, which have high capacity but low resolving power, such as fractional precipitation with salt or organic solvents (Chapter 13). Only when the volume and protein content of the extract has been reduced to manageable levels can methods of medium resolution and capacity, such as ion-exchange chromatography (Chapter 14) be used leading on, if necessary, to high-resolution but generally lower capacity techniques, such as affinity chromatography (Chapter 16) and isoelectric focusing (Chapter 23). On the other hand for isolation of small to medium amounts of proteins, it will usually be possible to move directly to the more refined methods of purification without the need for initial use of bulk methods. This is, of course, important because the fewer the steps that have to be used, the higher the final yield of the protein will be and the less time it will take to purify it.

1.2. Do I Want to Retain Biological Activity?

If the answer to this is positive then it restricts to some extent the range of techniques that can be employed and the conditions under which they can be performed. Most proteins retain activity when handled in neutral aqueous buffers at low temperature (although there are exceptions and these exceptions lend themselves to somewhat different approaches to purification). This consideration then rules out use of those techniques in which the conditions are likely to deviate substantially from the above. For example, immunoaffinity chromatography is a very powerful method but the conditions required to elute bound proteins are often rather severe, for example, the use of buffers of low pH, because of the tightness of binding between antibodies and antigens (*see* Chapters 16 and 19 for a discussion of this problem). Similarly, reversed-phase chromatography (Chapter 27) requires the use of organic solvents to elute proteins and rarely will be compatible with recovering an active species. Ion-exchange chromatography provides the most general method for isolation of proteins with retention of activity unless the protein has special characteristics that offer alternative strategies (*see* Section 2.4.). With labile molecules it is important to plan the purification schedule to contain as few steps as possible and with minimum requirement for changing buffers (Chapter 11), since this will reduce losses of activity.

General Strategies

In some cases, retention of biological activity is not required. This would be the case, for example, if the protein is needed for sequence analysis or perhaps for raising an antiserum. There is then no restriction on the methods that can be used and, indeed, the very powerful separation method of polyacrylamide gel electrophoresis in the presence of sodium dodecyl sulphate (SDS-PAGE) followed by blotting or elution from the gel can be used to isolate small amounts of pure protein either from partially purified extracts or even from crude extracts (Chapters 33–35). It is important in this context to differentiate between loss of biological activity arising from loss of three-dimensional structure, which will not be of concern in the applications outlined above, from loss of activity owing to modification of the chemical structure of the protein, which certainly would be a major concern. The most important route to chemical modification is proteolytic cleavage and ways in which this can be detected and avoided are discussed in Chapter 9.

1.3. Do I Need a Completely Pure Protein?

The concept of purity as applied to proteins is not entirely straightforward. It ought to mean that the protein sample contains, in addition to water and things like buffer ions that have been purposefully added, only one population of molecules all with identical covalent and three-dimensional structures. This is an unattainable goal and indeed an unnecessary one. What is required is a sample of protein that does not contain any species which will interfere with the experiments for which the protein is intended. This is not simply an academic point since it will usually become more and more difficult to remove residual contaminants from a protein sample as purification progresses. Extra purification steps will be required which take time (effectively an increase in cost of the product) and will inevitably lead to decreasing yields. What is required is an operational definition of purity for the particular project in hand because this will not only define the approach to the purification problem but may also govern its feasibility. It may not be possible to obtain a highly purified sample of a labile protein but it may be possible to obtain it in a sufficient state of purity for the purposes of a particular investigation.

The usual criterion of purity used for proteins is that a few micrograms of the sample produces a single band after electrophoresis on SDS-PAGE when stained with a reagent such as Coomasie blue or some similar nonspecific stain (*see* ref. *3* for practical details of this procedure and other chapters in the same volume for many other basic protein protocols). This simple criterion begs several questions. The most important of these is that SDS-PAGE separates proteins effectively on the basis of size and it may be that the sample contains two or more components that are sufficiently similar not to be resolved; the answer here is to subject the sample to an additional procedure, such as nondenaturing PAGE *(4)* since it is unlikely that two proteins will migrate identically in both systems. It must always be born in mind, however, that even if a single band is observed in two such systems, minor contaminants will inevitably become visible if the gel is more heavily loaded or if staining is carried out using a more sensitive procedure, such as silver staining *(5)*.

The major question is: Does it matter if the protein is 50, 90, or 99% pure? The answer is that it depends on the purpose of the purification. For example, a 50% pure protein may be entirely acceptable for use in raising a monoclonal antibody but a 95% pure protein may be entirely unacceptable for raising a monospecific polyclonal antibody particularly if the contaminants are highly immunogenic. Similarly, a relatively impure preparation of an enzyme may be acceptable for kinetic studies provided that it does not contain any competing activities; an affinity chromatography method might provide a rapid way of obtaining such a preparation. As a final example, a 95% pure protein sample is perfectly adequate for amino acid sequence analysis and indeed a lower state of purity is acceptable if proper quantitation is carried out to ensure that a particular sequence does not arise from a contaminant.

The message here is that preparation of a sample of protein approaching homogeneity is difficult and may not always be necessary so long as one knows what else is there. By taking account of the purpose for which the protein is required, it may be possible to decide on an acceptable level of contaminants and consideration of the nature of acceptable contaminants may suggest a purification strategy to be adopted.

1.4. What Source Should I Use?

The answer to this question may be partly or entirely dictated by the problem in hand. Clearly if the objective is to study the enzyme ribulose bisphosphate carboxylase, then there is no choice but to isolate it from a plant, but the plant can be chosen for its ready availability, high content of the enzyme, ease of extraction of proteins (Chapter 3), and low content of interfering polyphenolic compounds (Chapter 8). Of course, if one is interested in, for example, comparative biochemistry or molecular evolution, then not only the desired protein but also its source may be completely constrained.

In general, however, plants will not be the source of choice for isolation of a protein of general occurrence and where species differences are not of interest. Microbial or fungal sources may be a better choice since they can usually be grown under defined conditions thus assuring the consistency of the starting material and, in some cases, allowing for manipulation of levels of desired proteins by control of growth media and conditions (Chapters 4 and 5). They have the disadvantage, however, of possesing tough cell walls that are difficult to break and, consequently, microorganisms are not ideal for large scale work unless the laboratory has specialized equipment needed for their disruption.

The most convenient source of proteins in most cases is animal tissue, such as heart and liver and, except for relatively small scale work, the tissues will normally be obtained from a commercial abattoir. Laboratory animals provide an alternative for smaller scale purifications. Content of a particular protein is likely to be tissue specific in which case the most abundant source will probably be the best choice. It is worth noting, however, that it is easier to isolate proteins from tissues, such as heart, than from liver (the reasons for which are outlined in Chapter 2) and hence the heart may be the better bet even if the levels of the protein are lower than in liver.

A different sort of question arises if the protein of interest exists in soluble form in a subcellular organelle, such as the mitochondrion or chloroplast. Once the source organism has been chosen, there remains the decision as to whether to carry out a total disruption of the tissue under conditions where the organelles will lyse or whether to homogenize under conditions where the organelles remain intact and can be isolated by methods such as those described in Chapters 6 and 7. The latter approach will, of course, result in a very significant initial enrichment of the protein and subsequent purification will be easier because the range and amount of contaminating proteins will be much decreased. In the case of animal tissues the decision will probably depend on the scale at which it is intended to work (assuming, of course, that access to the necessary preparative high-speed centrifuges is available). Subcellular fractionation of a few hundred grams of tissue is a realistic objective, but if it is intended to work with larger amounts, then the time required for organelle isolation probably will be prohibitive and is unlikely to compensate for the extra work which will be involved in purification from a total cellular extract. Subcellular fractionation of plants is a much more difficult operation in most cases (*see* Chapter 7). Hence, except in the most favorable cases and for small scale work, purification from a total cellular extract will probably be the only realistic option.

In the case of membrane proteins, there again will be a considerable advantage in isolating as pure a sample of the membrane as possible before attempting purification. The ease with which this can be done depends on the organism and membrane system in question. Chapters 6 and 30 give some approaches to this problem for specific cases, but if it is intended to isolate a membrane protein from other sources, then a survey of the extensive literature on membrane purification is recommended (*see* ref. 6).

For proteins which are present in only very small quantities or which are found only in inconvenient sources, gene cloning and expression in a suitable host now provide an alternative route to purification (for a review of methods *see* ref. 7). This is, of course, a major undertaking and is likely to be used only when conventional methods are not successful. Sufffice it to say that once the protein is expressed and extracted from the host cell (*see* Chapter 4 for a method of extracting recombinant proteins from bacteria), the methods of purification are the same as those for proteins from conventional sources.

1.5. Has it Been Done Before?

It is quite common to need to purify a protein whose purification has been reported previously, perhaps to use it as an analytical tool or perhaps to carry out some novel investigations on it. In this case the first approach will be to repeat the previously described procedure. The chances are, however, that it will not work exactly as described since small variations in starting material, experimental conditions, and techniques—which are inevitable between different laboratories—can have a significant effect on the behavior of a protein during purification. This should not matter too much since adjustments to the procedures should be relatively easy to make once a little experience has been gained of the behavior of the protein. One pitfall to watch out for is the conviction that there ought to be a better way of doing it. It is possible to spend a great deal of time trying to improve on a published procedure often to little avail.

Even if the particular protein of interest has not been isolated previously, it may be that a related molecule has been, for example, the same

General Strategies

protein but from a different organism or a member of a closely related class of proteins. In the former case, particularly if the organisms are closely related, then the properties of the proteins should be quite similar and only minor variations in procedures, for example, the pH used for an ion-exchange step, might be required. Even if the family relationships are more distant, significant clues might still be available, such as the fact that the target is a glycoprotein, which will provide valuable approaches to purification (*see* Section 2.4.). Much time and wasted effort can be saved by using information in the literature rather than trying to re-invent the wheel.

2. Exploiting Differences

Protein purification involves the separation of one species from perhaps a thousand or more species of essentially the same general characteristics (they are all proteins!) in a mixture of which it may constitute a small fraction of 1% of the total. It is, therefore, necessary to exploit to the full those properties in which proteins differ from one another in devising a purification schedule. The following lists the most important of those properties and outlines the techniques that make use of them with comments on their practical application. More details on each technique will be found in the chapters that follow.

2.1. Solubility

Proteins differ in the balance of charged, polar, and hydrophobic amino acids that they display on their surfaces and hence in their solubilities under a particular set of conditions. In particular, they tend to precipitate differentially from solution on addition of species such as neutral salts or organic solvents and this provides a route to purification (*see* Chapter 13). It is however, a rather gross procedure since precipitation will occur over a range of solute concentrations and those ranges necessarily overlap for different proteins. It is not to be expected, therefore, that a high degree of purification can be achieved by such methods (perhaps two- to threefold in most circumstances), but the yield should be high and, most importantly, fractional precipitation can be carried out easily on a large scale provided only that a suitable centrifuge is available. It is, therefore, very common for this technique to be used at the stage immediately following extraction when working on a moderate to large scale. An important added advantage is that a substantial degree of concentration of the extract can be obtained at the same time which, considering that water is the major single contaminant in a protein solution, is a considerable added benefit.

2.2. Charge

Proteins differ from one another in the proportions of the charged amino acids (aspartic and glutamic acids, lysine, arginine, and histidine) that they contain. Hence they will differ in net charge at a particular pH or, another manifestation of them same difference, in the pH at which the net charge is zero (the isoelectric point). The first of these differences is exploited in ion-exchange chromatography which is perhaps the single most powerful weapon in the protein purifier's armory (Chapter 14). This makes use of the binding of proteins carrying a net charge of one sign onto a solid supporting material bearing charged groups of the opposite sign; the strength of binding will depend on the magnitude of the charge on the particular protein. Proteins may then be eluted from the matrix in exchange for ions of the opposite charge with the concentration of the ionic species required being determined by the magnitude of the charge on the protein.

Ion-exchange chromatography is a technique of moderate to high resolution depending on the way in which it is implemented. For large-scale work (around 100 g of protein), use is generally made of fibrous cellulose-based resins that give good flow rates with large bed volumes but not particularly high resolution; this would normally be done at an early stage in a purification. Better resolution is available with the more advanced Sepharose-based materials but generally on a smaller scale. For small quantities (≤ 10 mg), the technique of fast protein liquid chromatography (Chapter 26) is available which makes use of packing materials with very small diameters and correspondingly high resolving power; this, however, requires specialized equipment that may not be available in all laboratories. Because of the small scale, this method would usually be used at a late stage for final clean-up of the product. It should be born in mind that two proteins which carry the same charge at a particular pH might well differ in charge at a different pH. Hence it is quite common for a purification procedure to contain two or more ion-exchange steps either using the same resin at different pH values or perhaps using two resins of opposite charge characteristics (e.g., one carrying the negatively charged carboxymethyl [CM] group and the other the positively charged diethylaminoethyl [DEAE] group).

General Strategies

There are two main ways of exploiting differences in isoelectric points between proteins. Chromatofocusing is essentially an ion-exchange technique in which the proteins are bound to an anion exchanger and then eluted by a continuous decrease of the buffer pH so that proteins elute in order of their isoelectric points (*see* Chapter 24). It is a method of moderately high resolving power and capacity and is hence best used to further separate partially purified mixtures. The other technique is isoelectric focusing (Chapter 23) in which proteins are caused to migrate in an electric field through a system containing a stable pH gradient. At the pH at which a particular protein has no net charge (the isoelectric point), it will cease to move; if it diffuses away from that point, then it will regain a charge and migrate back again. This method, although of low capacity, is capable of very high resolution and is frequently used to separate mixtures of proteins which are otherwise difficult to fractionate.

2.3. Size

This property is exploited directly in the techniques of size exclusion chromatography (Chapter 25) and ultrafiltration (Chapter 12). In the former, the protein solution is passed through a column of porous beads, the pore sizes being such that large proteins do not have access to the internal space, small proteins have free access to it, and intermediate sized proteins have partial access; a range of these materials with different pore sizes is available. Clearly, large proteins will pass through the column most rapidly and small proteins most slowly with a range of behavior in between. The method is of limited resolving power but is useful in some circumstances particularly when the protein of interest is at one of the extremes of size. The capacity is low because of the need to keep the volume of solution applied to the column as small as possible.

In ultrafiltration, liquid is forced through a membrane with pores of a controlled size such that small solutes can pass through but larger ones cannot. It, therefore, can be used to obtain a separation between large and small protein molecules and also has the advantage that it is not limited by scale. Use of the method for protein fractionation is, however, restricted to a few special cases (*see* Chapter 12) and the principal value of the technique is for concentration of protein solutions.

A completely different approach to the use of size differences to effect protein separation is SDS-PAGE. In this method, the protein molecules are denatured and coated with the detergent so that they carry a large negative charge (the inherent charge is swamped by the charge of the detergent). The proteins then migrate in gel electrophoresis on the basis of size; small proteins migrate most rapidly and large ones slowly because of the sieving effect of the gel. The method has enormously high resolving power and its use in various forms for analytical purposes is one of the most important techniques in analytical protein chemistry (3). The development of methods for recovery of the protein bands from the gel after electrophoresis (*see* Chapters 33 and 34) has enabled this resolving power to be exploited for purification purposes. Obviously the scale of separation is small and the product is obtained in a denatured state, but a sufficient amount often can be obtained from very complex mixtures for the purposes of further invesigation (*see* Section 1.2.).

2.4. Specific Binding

Most proteins exert their biological functions by binding to some other component in the living system. For example, enzymes bind to substrates and sometimes to activators or inhibitors, hormones bind to receptors, antibodies bind to antigens, and so on. These binding phenomena can be exploited to effect purification of proteins usually by attaching the ligand to a solid support and using this as a chromatographic medium. An extract or partially purified sample containing the target protein is then pased through this column to which the protein binds by virtue of its affinity for the ligand. Elution is achieved by varying the solvent conditions or introducing a solute that binds strongly either to the ligand or to the protein itself.

Various types of affinity chromatography, as the method is called, are described in detail in Chapters 16–20. Immunoaffinity chromatography in particular is capable of very high selectivity because of the extreme specificity of antibody–antigen interactions. As mentioned above and dealt with in more detail in Chapters 16 and 19, the most common problem with this technique is to effect elution of the target protein under conditions that retain biological activity. Lectin-affinity chromatography (Chapter 18) exploits the selective binding between members of this class of plant proteins and particular carbohydrates. It, therefore, has found widespread use both in the isolation of glycoproteins and in removal of glycoprotein contaminants from other proteins, and is also capable of high specificity.

Affinity methods that rely on interactions of the target protein with low-mol-wt compounds (e.g., enzymes with substrates or substrate analogs) are frequently less specific because the ligand may bind to several proteins in a mixture. For example, immobilized NAD^+ will bind to many dehydrogenases and benzamidine will bind to most serine proteases; thus a group of related enzymes rather than individual species may be isolated using these ligands. A novel application of affinity methods is provided by the use of bifunctional NAD^+ derivatives to selectively precipitate dehydrogenases from solution (Chapter 22).

The use of organic dyes as affinity ligands (Chapter 17) is interesting because these molecules seem to bind fairly specifically to nucleotide-binding enzymes, although from their structures it is not at all clear why they should do so; it is likely that hydrophobic interactions between the dye and protein also contribute to binding. Use of the latter interaction has led to development of a specific form of chromatography which uses hydrophobic stationary phases (Chapter 15); this method has elements of biospecificity in that some proteins have binding sites for natural hydrophobic ligands, but in the general case it relies on the fact that all proteins have hydrophobic surface regions to a greater or lesser extent.

Finally, many proteins are known that bind metal ions with varying degrees of specificity and this forms the basis of immobilized metal ion affinity chromatography (Chapter 20). Specific affinity of proteins for calcium ions may also be the basis, in part, for binding to hydroxyapatite but ion-exchange effects are probably also involved (Chapter 21).

In summary there is a variety of affinity methods available, ranging from medium to very high selectivity, and in favorable cases affinity chromatography can be used to obtain a single-step purification of a protein from an initial extract. Generally, however, the capacities of affinity media are not high and the materials can be very expensive thus rendering their use on a large scale unrealistic. For these reasons affinity methods are usually used at a late stage in a purification schedule.

2.5. Special Properties

In a sense the specific binding properties discussed in the previous section are "special," but that is not what is meant here. Some proteins have, for example, the property of greater than normal heat stability and in those circumstances it may be possible to obtain substantial purification by heating a crude extract at a temperature where the target protein is stable but contaminants are denatured and precipitate from solution (*see* ref. *8* for an example of the use of this method). It is not likely, of course, that this approach will be useful in purification of proteins from thermophilic organisms since all or most of the proteins present would be expected to share the property of thermostability. Another possibility is that the protein of interest may be particularly stable at one or other of the extremes of pH; in this case incubation of an extract at low or high pH might well lead to selective precipitation of contaminants. It is always worthwhile carrying out some preliminary experiments with an unknown protein to see if it possesses special properties of this kind that would assist in its purification.

Finally mention should be made of the fact that it is now feasible, if the need is sufficiently great, to engineer special properties into proteins to assist in their purification. Typical examples include addition of polyarginine or polylysine tails to improve behavior on ion-exchange chromatography, or of polyhistidine tails to introduce affinity on immobilized metal affinity chromatography *(9)*. It is, however, likely that these techniques would be used as a last resort if all other attempts to purify the protein failed unless recombinant DNA technology had been selected as the route to protein production and purification in the first place (*see* Section 1.4.).

3. Documenting the Purification

It is vitally important to keep an inventory at each stage of a purification of volumes of fractions, total protein content, and content of the protein of interest. The last of these is particularly important since otherwise it is very easy to end up with a vanishingly small yield of target protein and not to know at which step the protein was lost. If the protein has a measurable activity then it is equally important to monitor this since it is also possible to end up with a protein sample which is inactive if one or more steps in the purification involves conditions under which the protein is unstable.

Measurement of total protein content of fractions presents no problems. At early stages of a purification it is usually sufficient to determine the absorbance of the solution at 280 nm (making sure that it is optically clear to avoid errors owing to light scattering) and to use the rough approximation that $A^{1\%}_{280\,nm} = 10$. At later stages one of the more accurate methods, such as the Bradford procedure *(10)* or the bicinchoninic acid assay *(11)*, should be used unless the absorbance/dry weight correlation for the target protein happens to be known.

Measurement of the amount and/or activity of the protein of interest may or may not be straightforward. For example, many enzymes can be assayed using simple and rapid spectrophotometric methods. For other proteins the assay may be more difficult and time consuming, such as bioassay or immunoassay (it should also be recognized that these are not necessarily the same thing; immunoassay frequently will not distinguish between inactive and active molecules so care must be taken in the interpretation of results using this method). In other situations the protein of interest may have no measurable biological activity; in such cases immunoassay can be used or, more commonly, quantitation of the appropriate band after separation of the protein on polyacrylamide gels *(12)*. Indeed it may be that the target protein will only have been identified as a spot on two-dimensional polyacrylamide gels *(13)* and purification is being attempted as a preliminary to determining its biological activity.

Obviously it is not possible to be presciptive here about what methods of analysis and quantitation to use in any specific case. What must be said, however, is that it is very unwise to embark on an attempted purification without first devising a method for quantitation of the protein of interest. Not to do so is courting failure.

4. An Example

To give the newcomer to protein purification a "feel" for what the process might look like in practice, Table 1 shows the fully documented results of the isolation of a particular enzyme starting from 5 kg of pig liver. All techniques used are described in detail in subsequent chapters and are only summarized here.

The strategy was to start by totally homogenizing the tissue in 10 L of buffer and, after removal of cell debris by centrifugation, to carry out an initial crude purification by fractional precipitation with ammonium sulphate. This had the added advantages of removing residual insoluble material from the extract (this precipitated in the first ammonium sulphate fraction) and achieving a very large reduction in volume of the active fraction. Ammonium sulphate was removed from the active fraction by dialysis.

Because of the large amount of protein remaining in the active fraction, the next step was a relatively crude ion-exchange separation using a large column (7 × 50 cm) of CM-cellulose CM23 (this has a high capacity and good flow rates but is of only moderate resolving power). Conditions were chosen so that the enzyme was absorbed onto the column and then, after washing off unbound contaminants, it was eluted with a single stepwise increase in ionic strength to $0.1 M$ using sodium chloride.

Table 1
Example Protein Purification Schedule

Fraction	Volume, mL	Protein concentration, mg/mL	Total protein, mg	Activity,[a] U/mL	Total activity, U	Specific activity, U/mg	Purification factor[b]	Overall yield,[c] %
Homogenate	8500	40	340,000	1.8	15,300	0.045	1	100
45–70% (NH$_4$)$_2$SO$_4$	530	194	103,000	23.3	12,350	0.12	2.7	81
CM-cellulose	420	19.5	8190	25	10,500	1.28	28.4	69
Affinity chromatography	48	2.2	105.6	198	9500	88.4	1964	62
DEAE-Sepharose	12	2.3	27.6	633	7600	275	6110	50

[a]The unit of enzyme activity is defined as that amount which produces 1 µmol of product per minute under standard assay conditions.
[b]Defined as: purification factor = (specific activity of fraction/specific activity of homogenate).
[c]Defined as: overall yield = (total activity of fraction/total activity of homogenate).

Previous trial experiments had shown that the enzyme bound to an affinity matrix in a buffer at the same pH and salt content as that with which it was eluted from CM-cellulose, and so affinity chromatography was used for the next step without changing the buffer and without prior concentration. The enzyme was eluted by applying a linear salt gradient up to a concentration of $1M$.

At this stage, electrophoresis of the active fraction under nondenaturing conditions showed the presence of two major contaminants, both of them more basic than the protein of interest. Hence, the sample was applied to a column of DEAE-Sepharose under conitions where the target protein was absorbed but the majority of the contaminating protein was not; the sample was equilibrated in starting buffer by dialysis before application to the column. The target protein was eluted from the column using a linear salt gradient and was found to be homogeneous by the usual techniques (*see* Section 1.3.).

The results in Table 1 show that the pruification procedure was quite successful in that a high yield (50% overall) of enzyme activity was obtained; this was achieved by using a small number of steps each of which gave a good step yield. There will inevitably be losses on any purification step and the important point is that these and the number of steps should be kept as low as possible (a 5-step schedule in which the yield from each step is 50% will give an overall yield of 3%; a 10-step schedule with 80% step yield will give a final yield of 11%). It can also be seen from the final purification factor that the amount of this particular enzyme in the liver was low (about 0.016% of soluble protein) and hence a relatively large amount of tissue had to be used to obtain the required amount of product. This was an important factor in deciding the first two steps in the scedule (*see* Section 1.1.).

The purification in its final form can be completed in 5–6 working days. It must be born in mind, however, that each step has been optimized and that development of the procedure took several months of work. This is common when working out a new purification schedule and it is always necessary to be conscious of the time commitment when deciding to embark on purifying a protein.

References

1. Asenjo, J. A. and Patrick, I. (1990) Large-scale protein purification, in *Protein Purification Applications: A Practical Approach* (Harris, E. L. V. and Angal, S., eds.), IRL, Oxford, UK, pp. 1–28.

2. Bristow, A. F. (1990) Purification of proteins for therapeutic use, in *Protein Purification Applications: A Practical Approach* (Harris, E. L. V. and Angal, S., eds.), IRL, Oxford, UK, pp. 29–44.
3. Smith, B. J. (1994) SDS polyacrylamide gel electrophoresis of proteins, in *Methods in Molecular Biology, Vol. 32: Basic Protein and Peptide Protocols* (Walker, J. M., ed.), Humana, Totowa, NJ, pp. 23–34.
4. Walker, J. M. (1994) Nondenaturing polyacrylamide gel electrophoresis of proteins, in *Methods in Molecular Biology, Vol. 32: Basic Protein and Peptide Protocols* (Walker, J. M., ed.), Humana, Totowa, NJ, pp. 17–22.
5. Dunn, M. J. and Crisp, S. J. (1994) Detection of proteins in polyacrylamide gels using an ultrasensitive silver staining technique, in *Methods in Molecular Biology, Vol. 32: Basic Protein and Peptide Protocols* (Walker, J. M., ed.), Humana, Totowa, NJ, pp. 113–118.
6. Graham, J. M. and Higgins, J. A. (eds.) (1993) *Methods in Molecular Biology, Vol. 19: Biomembrane Protocols: I. Isolation and Analysis,* Humana, Totowa, NJ.
7. Murray, E. J. (ed.) (1991) *Methods in Molecular Biology, Vol. 7: Gene Transfer and Expression Protocols,* Humana, Totowa, NJ.
8. Banks, B. E. C., Doonan, S., Lawrence, A. J., and Vernon, C. A. (1968) The molecular weight and other properties of aspartate aminotransferase from pig heart muscle. *Eur. J. Biochem.* **5,** 528–539.
9. Brewer, S. J. and Sassenfeld, H. M. (1990) Engineering proteins for purification, in *Protein Purification Applications: A Practical Approach* (Harris, E. L. V. and Angal, S., eds.), IRL, Oxford, UK, pp. 91–111.
10. Kruger, N. J. (1994) The Bradford method for protein quantitation, in *Methods in Molecular Biology, Vol. 32: Basic Protein and Peptide Protocols* (Walker, J. M., ed.), Humana, Totowa, NJ, pp. 9–15.
11. Walker, J. M. (1994) The Bicinchoninic acid (BCA) assay for protein quantitation, in *Methods in Molecular Biology, Vol. 32: Basic Protein and Peptide Protocols* (Walker, J. M., ed.), Humana, Totowa, NJ, pp. 5–8.
12. Smith, B. J. (1994) Quantification of proteins on polyacrylamide gels (nonradioactive), in *Methods in Molecular Biology, Vol. 32: Basic Protein and Peptide Protocols* (Walker, J. M., ed.), Humana, Totowa, NJ, pp. 107–111.
13. Pollard, J. W. (1994) Two-dimentional polyacrylamide gel electrophoresis of proteins, in *Methods in Molecular Biology, Vol. 32: Basic Protein and Peptide Protocols* (Walker, J. M., ed.), Humana, Totowa, NJ, pp. 73–85.

CHAPTER 2

Preparation of Extracts from Animal Tissues

Shawn Doonan

1. Introduction

Extraction of soluble proteins from most animal tissues is relatively straightforward, since the cell membranes are weak and easily ruptured by a combination of mechanical and osmotic forces. The tissue is disrupted in a suitable buffer using a homogenizer and the homogenate then clarified by centrifugation. Extraction of membrane proteins requires more specialized methods, and these are dealt with in Chapter 28.

Thought needs to be given to the tissue to be used and the source of that tissue. These will sometimes be dictated by the particular project in hand. For example, the protein of interest may be restricted to a particular organ, and the objective may be to compare the structure of that protein as expressed in the rat to that expressed in some other animal. In this case the parameters are obviously fixed. In other cases, the protein may be of general occurrence and the animal of origin of no particular interest, e.g., if the objective is to study the mechanism of action of an enzyme. In such cases, availability of the tissue, amount of soluble protein in the tissue, and the ease of obtaining clarified extracts become the important criteria.

The cheapest and most convenient source of large quantities of animal tissue (pig, sheep, cattle) will normally be a commercial abattoir. These companies will usually allow personal collection of tissues, which can be packed in ice for transport back to the laboratory; tissues may be stored frozen at $-20°C$ for later use, but this is not to be generally recommended.

Alternatively, smaller animals (rats, mice, rabbits, and so forth) bred in animal facilities can be used, but this is a more expensive option and will require a large number of animals to be sacrificed for a large-scale preparation. The choice of tissue is also important. Heart is ideal if the protein is present in a reasonable quantity since it is very easy to obtain clarified homogenates from this tissue. Liver is less desirable, since it contains large quantities of nucleic acids and of fats, which make clarification of extracts more difficult. It also contains much larger quantities of soluble protein than does heart, and this increases the problems of subsequent purification.

Clearly, the tissue distributions of proteins vary enormously, and it is always advisable to carry out preliminary experiments to measure the relative content of the protein of interest in the various tissues. The final choice of starting material will then be determined by the balance between content, ease of obtaining the tissue, and the various technical issues described.

A final consideration is the subcellular distribution of the protein. If it is localized to a specific organelle, such as the mitochondrion, then it may be advantageous to carry out a subcellular fractionation (*see* Chapter 6) and extract the protein from the isolated organelles. This will clearly simplify subsequent purification, but will necessarily restrict the amount of protein obtainable since subcellular fractionation can only be carried out on a small scale. If large amounts of protein are required, then the only practical approach is to homogenize the tissue in the conventional manner, under which conditions subcellular organelles will lyse, and to purify the protein from the total extract. An example of a large-scale purification of a mitochondrial enzyme using this approach is given in ref. *1*.

The method of extraction described herein is of general applicability, but there are other specialized procedures that may be of use in particular cases. Comprehensive coverage of these procedures is given in ref. *2*, which, though now old, is still a valuable survey of methods.

2. Materials

In addition to normal laboratory glassware and equipment, the following are necessary:

1. Waring blender: These are available from general laboratory equipment suppliers and come with vessels ranging from about 10 mL to 4 L in capacity. The vessels may be of glass or stainless steel; the latter is to be pre-

ferred, since they may be cooled more readily. An alternative for large-scale work is a domestic food blender.
2. Refrigerated centrifuge: Again, several models are available, for example, the MSE Mistral or the Sorvall RC2-B. The choice of rotors will depend on the scale of purification work envisaged. Generally useful is a six-place angle rotor for 250- or 300-mL tubes. For larger scale work, a six-place, 1000-mL swing-out rotor is useful.
3. Polypropylene screw-cap centrifuge tubes for the rotors to be used.

3. Methods

1. Trim fat and connective tissue from the tissue and cut into pieces of a few grams each.
2. Place the tissue in the precooled blender vessel (*see* Note 1) and add cold extraction buffer using 2 vol of buffer by weight of tissue (*see* Note 2). Use a blender vessel that has a capacity approximately that of the volume of buffer plus tissue so that the air space is minimized; this will reduce aerosol formation.
3. Homogenize at full speed for 1–3 min depending on the toughness of the tissue (or until the tissue is fully disrupted as judged by eye). For long periods of homogenization, it is best to blend in 1-min bursts with a few minutes in between to avoid excessive heating.
4. Pour the homogenate into a glass beaker, place on ice, and stir for 15–30 min to ensure full extraction.
5. Remove cell debris and other particulate matter from the homogenate by centrifugation at 4°C. For large-scale work use a 6 × 1000-mL swing-out rotor operated at about 2000–3000g for 30 min. For smaller scale work (up to 3 L of homogenate) a 6 × 250-mL angle rotor operated at about 5000g is to be prefered (*see* Note 3).
6. Pour off the supernatant carefully to avoid disturbing sedimented material. Any fatty material that has floated to the top of the tube should be removed by filtering the extract through cheesecloth or through a plug of glass wool in a filter funnel. The pellet may be re-extracted with more buffer to increase the yield (*see* Note 4).

   The extract obtained at this stage may require further treatment to remove insoluble material before fractionation, depending mainly on the tissue that has been used and on the planned next stage in the purification (*see* Note 5).

4. Notes

1. Such tissues as liver, kidney, brain, and heart are easily homogenized in a blender. Such tissues as skeletal muscle, lung, and so on, are tougher, and in these cases, it is advisable to grind them using a domestic meat mincer before homogenization. Very fibrous tissues (e.g., mammary glands) should

be frozen before homogenization to facilitate disruption. Small amounts of soft tissues (1–5 g) and animal cells obtained from culture may be more conveniently homogenized using a handheld Dounce homogenizer *(3)*.
2. The choice of buffer may be dictated by the particular problem in hand, but generally a buffer of moderate ionic strength at a neutral pH (e.g., $0.1M$ phosphate, pH 7.0) is used. Some proteins tend to attach electrostatically to tissue debris and/or to membrane fragments in which case addition of $0.1M$ NaCl to the extraction buffer will increase yield; this can easily be established in trial experiments. A major consideration is whether to add protease inhibitors to the extraction buffer (*see* Chapter 9 for a discussion of the problem of proteolysis). In general, this is not essential because the protein content of the extract will be high and the bulk protein lends protection to the particular protein of interest. Some proteins are, however, more susceptible to proteases than others, and some tissues (e.g., liver) have much higher protease levels than others (e.g., heart), so proteolysis may be a problem depending on the protein of interest and the source tissue. Use of protease inhibitors, particularly at the stage of extraction when volumes are large, is expensive and hence best avoided if possible. Ideally, a trial experiment should be carried out in which the level of the protein of interest is measured in an extract over a period of a few hours to see if measurable losses occur; if so, proteolysis should be suspected and steps taken to avoid it.

Some proteins are susceptible to oxidation, in which case addition of dithiothreitol (1 mM) or β-mercaptoethanol ($0.1M$) to the extraction buffer is advised. Similarly, if the protein of interest is inhibited by heavy metals, then EDTA ($0.1M$) should be added. In cases where the protein of interest has a known cofactor, it is sometimes worth including the cofactor in the homogenization buffer to prevent dissociation of the cofactor with possible consequent decrease in stability of the protein. In all these situations, trial experiments should be done to see if any advantage is obtained by potential additives.
3. Great care must be taken when using centrifuges and, in particular, to ensure that the tubes are balanced across the central axis (i.e., in a six-place rotor, tubes 1 and 4, 2 and 5, and 3 and 6 contain the same weight to within 1 g of suspensions of the same density). Hence, if only 250 mL of homogenate are available, this should not be balanced against a tube containing an equal weight of water—the densities will be different and the centers of gravity of the two tubes will differ. In this case, the homogenate should be split between two tubes. It is also unwise to fill the tubes completely when using an angle rotor. This is because the neck of the tube may distort slightly at high rotor speeds, allowing leakage of liquid; in addition to being potentially

dangerous, this could lead to damage to the centrifuge rotor. Ideally, the tubes should be filled to such an extent that when placed in the rotor, the liquid level is just below the screw-cap at its lowest point.

Conditions for centrifugation are always (or should always be) quoted in terms of the average centrifugal field, i.e., the field at the midpoint of the tube, rather than as revolutions per min (rpm). Technical data supplied with centrifuge rotors will invariably include a chart for conversion between the two. It is, however, easy to calculate the g-value from the dimensions of the rotor and the rpm using the relationship:

$$\text{Relative centrifugal field} = 1.118 \times 10^{-5} \times n^2 \times r$$

where n is the number of revolutions per minute and r is the distance in cm from the center of rotation to the choosen point in the centrifuge tube.

4. Using a buffer-to-tissue ratio of 2:1, a quarter to a third of the tube after centrifugation will be occupied by the pellet of cell debris, depending on the particular tissue used. This pellet will, of course, contain a substantial amount of buffer, and hence, of soluble protein. Yields can be increased by resuspending the pellet in buffer and centrifuging again, but this will obviously result in dilution of the final extract, which is undesirable since water is a major contaminant that has to be disposed of at some stage. In general, it is better to sacrifice some yield of protein in the interest of keeping the volume down, unless the tissue being used is scarce, in which case re-extraction should be done.

5. With tissues such as heart the procedure given will result in an extract of sufficient clarity to proceed to the next step in purification. With such tissues as liver, kidney, and brain, this is not so, and further steps need to be taken to remove particulate matter before proceeding, partcularly if the next step is to be column chromatography where partculate matter can cause serious clogging problems.

An obvious approach is to use much higher centrifugal fields for longer periods. With a 6 × 250-mL rotor, values of up to $6000g$ can be obtained, but this does not always solve the problem; use of high centrifugal fields is also impractical when large volumes are to be processed. The problem arises mainly from the release of large quantities of nucleic acids and nucleoprotein into the homogenate; this increases the viscosity, and hence, decreases the rate of sedimentation. The best way of overcoming this difficulty is to add a polyamine, such as protamine sulfate, to the partially clarified homogenate, stir for some time, and then recentrifuge. The polyamine causes aggregation of the nucleic acids and nucleoproteins, which now sediment much more easily. A concentration of 0.1% (w/v) is usually adequate to produce a clear supernatant, but trial experiments may be required to find optimal conditions. It is also necessary, of course, to confirm that the protein of interest is not precipitated or inactivated by this procedure.

A very effective solution to the problem, but one that is useful in only a limited number of cases is heat precipitation. This requires that the protein of interest be stable to heating at 70°C for 10–15 min. Under these conditions, many proteins are denatured and form readily sedimentable aggregates along with other suspended matter. Not only is it then easy to clarify the extract, but then there will also be a substantial purification of the protein of interest. An example where this method has been successfully used is given in ref. *4*. In practice, it is necessary to heat the clarified extract rapidly to the desired temperature (usually 70°C) by putting it in a lage round-bottomed glass flask, immersing the flask in a water bath at least 10°C hotter than the target temperature, and swirling constantly until the target temperature is reached. After the required time at this temperature, the flask and contents are rapidly cooled under running tap water and precipitated material removed by centrifugation.

In practice, it may be possible to ignore the problem if the first step in the proposed purification scheme is to be fractional precipitation by salt or by organic solvents (*see* Chapter 13) and if the protein of interest does not precipitate in the first fraction taken. In the case of precipitation with organic solvents, addition of the solvent decreases both the density and the viscosity of the solution, and hence, sedimentation of particulate matter will occur readily in the first fraction taken. During salt fractionation, particulate material tends to aggregate and again will usually sediment in the first fraction. Clearly, if the protein of interest precipitates early, the particulate matter will be carried with it and the problem compounded.

References

1. Barra, D., Bossa, F., Doonan, S., Fahmy, H. M. A., Martini, F., and Huges, G. J. (1976) Large-scale purification and some properties of the mitochondrial aspartate aminotransferase from pig heart. *Eur. J. Biochem.* **64,** 519–526.
2. Morton, R. K. (1955) Methods of extraction of enzymes from animal tissues. *Methods Enzymol.* **1,** 25–51.
3. Dignam, J. D. (1990) Preparation of extracts from higher eukaryotes. *Methods Enzymol.* **182,** 194–203.
4. Banks, B. E. C., Doonan, S., Lawrence, A. J., and Vernon, C. A. (1968) The molecular weight and other properties of aspartate aminotransferase from pig heart muscle. *Eur. J. Biochem.* **5,** 528–539.

CHAPTER 3

Protein Extraction from Plant Tissues

Peter R. Shewry and Roger J. Fido

1. Introduction

Plant tissues contain a wide range of proteins, which vary greatly in their properties, and require specific conditions for their extraction and purification. It is therefore not possible to recommend a single protocol for extraction of all plant proteins. Plant tissues do pose specific problems that must be taken into account when developing protocols for extraction. The first is the presence of a rigid cellulosic cell wall, which must be sheared to release the cell contents. The second is the presence of specific contaminating compounds that may result in protein degradation or modification and, where the protein of interest is an enzyme, subsequent loss of catalytic activity. Such compounds include phenolics and a range of proteases. Although it is sometimes possible to avoid these problems or partially control them by using a specific tissue or plant species, in other cases (for example, enzymes involved in secondary product synthesis), this is not possible and the biochemist must find ways to remove or inactivate active contaminants. The removal of phenolics is dealt with elsewhere in this volume (Chapter 8). Because many plant proteases are of the serine type, it is often convenient to include the serine protease inhibitor phenylmethylsulfonylfluoride (PMSF) in extraction buffers on a routine basis (*see* Chapter 9 for a general discussion of protease inhibition).

Animals have many highly specialized tissues (e.g., liver, muscle, brain) that are rich sources of specific enzymes, thus facilitating their purification. This is not usually the case with plant enzymes, which may be present at low levels in highly complex protein mixtures. An exception to this is storage organs, such as seeds, tubers, and tap roots. These organs contain high levels of specific proteins whose role is to act as a store of nitrogen, sulfur, and carbon. These storage proteins are among the most widely studied proteins of plant origin, because of their abundance and ease of purification, and their economic and nutritional importance as food, feed for livestock, and raw material in the food and other industries. Indeed, seed proteins were among the earliest of all proteins to be studied in detail, with wheat gluten having been isolated in 1745 *(1)*, the Brazil nut globulin, edestin, crystallized in 1859 *(2)* and a range of globulin storage proteins being subjected to ultracentrifugation analysis by Danielsson in 1949 *(3)*.

Comparative studies of the extraction and solubility of plant proteins also formed the basis for the first systematic attempt to classify proteins. T. B. Osborne, working at the Carnegie Institute of Washington between about 1880 and 1930, compared and characterized proteins from a range of plant sources, including the major storage proteins of cereal and legume seeds *(4)*. He defined four groups that were extracted sequentially in water (albumins), dilute salt solutions (globulins), alcohol/water mixtures (prolamins), and dilute acid or alkali (glutelins). These "Osborne groups" still form the basis for studies of seed storage proteins, whereas the terms albumin and globulin have become accepted into the general vocabulary of protein chemists.

Three detailed protein extraction protocols are given. The first two are for the extraction of the enzymically active proteins, ribulose 1,5-bisphosphate carboxylase/oxygenase (Rubisco) (EC 4.1.1.39) and nitrate reductase (EC 1.6.6.1), from vegetative tissues. Rubisco is a hexadecameric protein (eight subunits of approx 50–60 kDa and eight subunits of 12–20 kDa) with an M_r of 500,000, which catalyzes the fixation of carbon in the chloroplast stroma. It often represents more than 50% of the total chloroplast protein and is recognized as the most abundant protein in the world. In contrast, the complex enzyme nitrate reductase (NR), which has a M_r of approx 200,000, is present in plant tissues at <5 mg/kg fresh weight *(5)*. This low abundance, combined with susceptibility to proteolysis and loss of functional prosthetic groups during extraction and purification, often leads to a very low recovery of the enzyme. The third protocol is a specialized procedure for the extraction of seed proteins from cereals, based on the classical Osborne fractionation.

Finally, two rapid methods are described for the extraction of leaf and seed proteins for SDS-PAGE analysis. These are suitable for monitoring the expression of transgenes in engineered plants.

2. Materials

1. Buffer A (Rubisco): 20 mM Tris-HCl, pH 8.0, 10 mM NaHCO$_3$, 10 mM MgCl$_2$, 1 mM EDTA, 5 mM DTT, 0.002% Hibitane, and 1% insoluble PVP.
2. Buffer B (Nitrate Reductase [NR]): 0.5M Tris-HCl, pH 8.6, 1 mM EDTA, 5 µM Na$_2$MoO$_4$, 25 µM FAD, 5 mM PMSF, 5 µg/mL pepstatin, 10 µM antipain, and 3% BSA.
3. Buffer C: 0.0625M Tris-HCl, pH 6.8, 2% (w/v) SDS, 5% (v/v) 2-mercaptoethanol or 1.5% (w/v) dithiothreitol, 10% (w/v) glycerol, 0.002% (w/v) bromophenol blue.
4. Buffer D: 0.1M Tris-HCl, pH 8.0, 0.01M MgCl$_2$, 18% (w/v) sucrose, 40 mM 2-mercaptoethanol.

3. Methods

3.1. Extraction of Enzymically Active Preparations from Leaf Tissues

All procedures are carried out at 0–4°C with precooled reagents and apparatus. Tissue can be used fresh, or after rapid freezing using liquid nitrogen, and stored at –20 to –80°C or under liquid nitrogen. Tissue homogenization can be a made in pestle and mortar or a ground glass homogenizer (for small volumes), or a Waring blender or Polytron for larger initial weights.

The method for the extraction of Rubisco from wheat leaves is taken from Keys and Parry (6). It is reported that the extraction procedure and extraction buffers used are important in affecting the initial rate and total activities of the enzyme (see Note 1). It is also important for initial activity measurements to maintain the extract at a temperature of 2°C.

1. Cut 3-wk-old wheat leaves into 1-cm lengths and homogenize in an ice-cold buffer using a ratio of 6:1.
2. Filter the homogenate through four layers of muslin, and then add sufficient solid (NH$_4$)$_2$SO$_4$ to give 35% saturation (see Chapter 13).
3. After 20 min, centrifuge the suspension at 20,000g for 15 min. Discard the pellet.
4. Add further solid (NH$_4$)$_2$SO$_4$ to give 55% saturation. After centrifugation dissolve the pellet in 20 mM Tris-HCl containing 1 mM DTT, 1mM MgCl$_2$, and 0.002% Hibitane (see Note 2) at pH 8.0. After clarification, the Rubisco can then be fractionated by sucrose-density centrifugation.

The method for NR extraction, using a complex extraction buffer (*see* buffer B), is taken from Somers et al. *(7),* who attempted to identify whether barley NR was regulated by enzyme synthesis and degradation or by an activation–inactivation mechanism.

1. Both root and shoot tissues were excised at different ages (days), weighed, frozen in liquid nitrogen and stored at –80°C.
2. Pulverize the frozen tissue in a pestle and mortar under liquid nitrogen. Extract with 1 mL/g fresh weight of buffer B (*see* Note 3).
3. Filter the homogenate through two layers of cheesecloth, and centrifuge at 30,000g to clarify. The supernatant can be used directly for enzyme activity measurements (*see* Note 4).

3.2. Extraction of Cereal Seed Proteins Using a Modified Osborne Procedure

The procedure is based on Shewry et al. *(8).* Air-dry grain (≈14% water) is milled to pass a 0.5-mm mesh sieve. The meal is then extracted by stirring (*see* Note 5) with the following series of solvents: 10 mL of solvent is used/g meal and each extraction is for 1 h. Extractions are carried out at 20°C and repeated as stated.

1. Water-saturated 1-butanol (twice) to remove lipids.
2. 0.5M NaCl to extract salt-soluble proteins (albumins and globulins) and nonprotein components (twice) (*see* Note 6).
3. Distilled water to remove residual NaCl.
4. 50% (v/v) 1-Propanol containing 2% (v/v) 2-mercaptoethanol (or 1% [w/v] dithiothreitol) and 1% (v/v) acetic acid (three times) to extract prolamins (*see* Note 7).
5. 0.05M borate buffer, pH 10, containing 1% (v/v) 2-mercaptoethanol and 1% (w/v) sodium dodecyl sulfate to extract residual proteins (glutelins) (*see* Note 8).

The supernatants are separated by centrifugation (20 min at 10,000g) and treated as follows:

6. Supernatants 2 and 3 are combined and dialyzed against several changes of distilled water at 4°C over 48 h. Centrifugation removes the globulins, allowing the soluble albumins to be recovered by lyophilization.
7. Supernatants from 4 are combined, and the prolamins recovered after precipitation, either by dialysis against distilled water or addition of 2 vol of 1.5M NaCl, followed by standing overnight at 4°C.
8. Supernatants from 5 are combined and the glutelins recovered by dialysis against distilled water at 4°C followed by lyophilization (*see* Note 9). SDS can be removed from the protein using standard procedures.

3.3. Extraction of Proteins for SDS-PAGE Analysis

The methods described in Sections 3.1. and 3.2. are suitable for the bulk extraction of proteins for purification of individual components. However, in some situations, for example, analysis of transgenic plants or studies of seed protein genetics, it is advantageous to extract total proteins for direct analysis by SDS-PAGE. The following methods are specially designed for this purpose.

3.3.1. Extraction of Leaf Tissues

The method, based on Nelson et al. *(9)*, gives good results with chlorophyllous tissues.

1. Freeze the tissue in liquid N_2.
2. Grind for about 30 s in a mortar with 3 mL of buffer D/g of tissue (*see* Note 10).
3. Filter through muslin, and centrifuge for 15 min in a microfuge.
4. Dilute to about 2 mg protein/mL, ensuring that the final solution contains about 2% (w/v) SDS, 0.002% (w/v) bromophenol blue, and at least 6% (w/v) sucrose (*see* Note 11).
5. Separate aliquots by SDS-PAGE.

3.3.2. Extraction of Seed Proteins

1. Grind in a mortar with 25 µL of buffer C/mg meal.
2. Transfer to an Eppendorf tube, and allow to stand for 2 h.
3. Suspend in a boiling water bath for 2 min.
4. Allow to cool, and then spin in a microfuge.
5. Separate 10–20-µL aliqouts by SDS-PAGE.

4. Notes

4.1. Extraction of an Enzymically Active Preparation from Leaf Tissue

1. A wide range of buffers can be used, depending on the pH range required for optimal enzyme activity and the preference (or prejudice) of the operator. However, Tris is very widely used.
2. A range of specific additions can be made in order to help preserve the activity of the enzyme under consideration. For example, with NR, it is advantageous to add a flavin compound (i.e., FAD) in order to maintain the endogenous levels needed for catalytic activity. The inclusion of both CO_2 and Mg^{2+} ions in the extraction buffer of Rubisco has been reported to be necessary *(10)*. There is no single simple method to guarantee activity—the operator should consult published protocols for the extraction of related enzymes and be prepared to carry out exploratory extractions using different buffer compositions.

a. 1 mM Dithiothreitol or 10 mM 2-mercaptoethanol to preserve sulphydryl groups.
b. 1 mM EDTA to chelate metals, especially with phosphate buffers that commonly contain inhibitory concentrations of ferrous ions.
c. 50 mM Sodium fluoride to inhibit phosphatases that inactivate phosphoenzymes.
d. 25 g/kg Fresh weight of PVP. This is an insoluble compound that binds phenolic compounds. It forms a slurry and can be removed by centrifugation.
e. 0.1 mM PMSF to inhibit serine proteinases. This is readily dissolved in a small volume of 1-propanol prior to mixing with the buffer. **It is highly toxic.**
f. Glycerol (up to 30%) or other organic alcohols (ethylene glycol, mannitol) may help to stabilize some highly labile enzymes.
g. The addition of exogenous proteins, e.g., casein and BSA, has been used to stabilize enzymes by preventing hydrolysis owing to protease activity.
h. Antibacterial agents, such as Hibitane, can also be added.
3. Similar methods can be used to extract enzymes from seed tissues, either by direct homogenization or after milling. Lipid-rich tissues can either be defatted with cold (4°C) acetone or an acetone powder can be made *(11)*. **Extreme care should be taken due to the low flash point of acetone: Operations should be carried out in a fume cupboard and electrical sparks avoided.**
4. The supernatant may be concentrated by precipitation with $(NH_4)_2SO_4$ *(12)* and assayed directly after desalting on a column of Sephadex G25.

4.2. Extraction of Cereal Seed Proteins Using a Modified Osborne Procedure

5. Extraction may be carried out by stirring magnetically or with a paddle—the mechanical grinding that occurs may assist extraction.
6. It may be advantageous to extract the salt-soluble proteins at 4°C and include 1.0 mM PMSF (*see* Note 2) to minimize proteolysis.
7. It is sometimes of interest to extract the prolamins in two fractions. Extraction twice with 50% (v/v) 1-propanol gives monomeric prolamins and alcohol-soluble disulfide-stabilized polymers, whereas subsequent extraction twice with 50% (v/v) 1-propanol with 2% (v/v) 2-mercaptoethanol and 1% (v/v) acetic acid gives reduced subunits derived from alcohol-insoluble disulfide-bonded polymers.
8. It is usual to determine the amounts of extracted proteins by Kjeldahl N analysis of aliquots removed from the supernatants. The values can then be multiplied by a factor of 5.7 for prolamins or 6.25 for other fractions to give the amount of protein.
9. SDS-PAGE is used to monitor the compositions of the fractions.

4.3. Extraction of Proteins for SDS-PAGE Analysis

10. Addition of 2% (w/v) SDS to buffer D allows the extraction of membrane and other insoluble proteins.
11. If required, soluble and insoluble proteins can be extracted in two sequential fractions. Soluble proteins are initially extracted in buffer D (3 mL/g) and insoluble proteins by re-extracting the pellet with 0.05 vol (relative to the original homogenate) of 2% (w/v) SDS, 6% (w/v) sucrose, and 40 mM 2-mercaptoethanol.

References

1. Beccari, J. B. (1745) *De Frumento. De Bononiensi Scientiarum et Artium atque Academia Commentarii, Tomi Secundi.* Bononia.
2. Matschke, O. (1859) Ueber den Bau und die Bestandtheile der Kleberbläschen in Bertholletia, deren Entwickelung in Ricinus, nebst einigen Bemerkunger über Amylonbläschen. *Botanische Zeitung* **17,** 409–447.
3. Danielsson, C. E. (1949) Seed globulins of the Gramineae and Leguminosae. *Biochem. J.* **44,** 387–400.
4. Osborne, T. B. (1924) *The Vegetable Proteins.* Longmans, Green, London.
5. Wray, J. L. and Fido, R. J. (1990) Nitrate and nitrite reductase, in *Methods in Plant Biochemistry* vol. 3 (Lea, P. J., ed.), Academic, New York, pp. 241–256.
6. Keys, A. J. and Parry, M. A. J. (1990) Ribulose bisphosphate carboxylase/oxygenase and carbonic anhydrase, in *Methods in Plant Biochemistry,* vol. 3 (Lea, P. J., ed.), Academic, New York, pp. 1–14.
7. Somers, D. A., Kuo, T.-M., Kleinhofs, A., Warner, R. L., and Oaks, A. (1983) Synthesis and degradation of barley nitrate reductase. *Plant Physiol.* **72,** 949–952.
8. Shewry, P. R., Franklin, J., Parmar, S., Smith, S. J., and Miflin, B. J. (1983) The effects of sulfur starvation on the amino acid and protein compositions of barley grain. *J. Cer. Sci.* **1,** 21–31.
9. Nelson, T., Harpster, M. H., Mayfield, S. P., and Taylor, W. C. (1984) Light regulated gene-expression during maize leaf development. *J. Cell. Biol.* **98,** 558–564.
10. Servaites, J. C., Parry, M. A. J., Gutteridge, S., and Keys, A. J. (1986) Species variation in the predawn inhibition of ribulose-1,5-bisphosphate carboxylase/oxygenase. *Plant Physiol.* **82,** 1161–1163.
11. Nason, A. (1955) Extraction of soluble enzymes from higher plants. *Methods Enzymol.* **1,** 62,63.
12. Green, A. A. and Hughes, W. L. (1955) Protein fractionation on the basis of solubility in aqueous solutions of salts and organic solvents. *Methods Enzymol.* **1,** 67–90.

CHAPTER 4

Extraction of Recombinant Protein from Bacteria

D. Margaret Worrall

1. Introduction

The advent of recombinant DNA technology and the overexpression of heterologous proteins in bacteria have posed some unique problems not previously encountered in extraction of bacterial proteins. Some of this technology has enabled secretion of recombinant proteins by bacteria into the media, thereby eliminating the need to lyze the cells. However, most situations still require lysis of the bacterial cell wall in order to extract the recombinant protein product. A number of methods based on enzymatic and mechanical means are available for breaking open the bacterial cell wall, and the choice will depend on scale of process *(1)*. Enzymatic methods use the activity of lysozyme, which cleaves the glucosidic linkages in the bacterial cell-wall polysaccharide. The inner cytoplasmic membrane can then be disrupted easily by detergents, osmotic pressure, or mechanical methods.

Overexpression of recombinant proteins from strong promoters on multiple-copy plasmids can result in expression levels of up to 40% of the total cell protein. However, in most cases, this results in the formation of insoluble protein aggregates known as inclusion bodies, which require further extraction *(2)*. Inclusion bodies are cytoplasmic granules seen as phase bright under the light microscope, and can contain most or all of the protein of interest. Scanning electron micrographs of *E. coli* containing inclusion bodies and isolated inclusions are shown in Fig. 1.

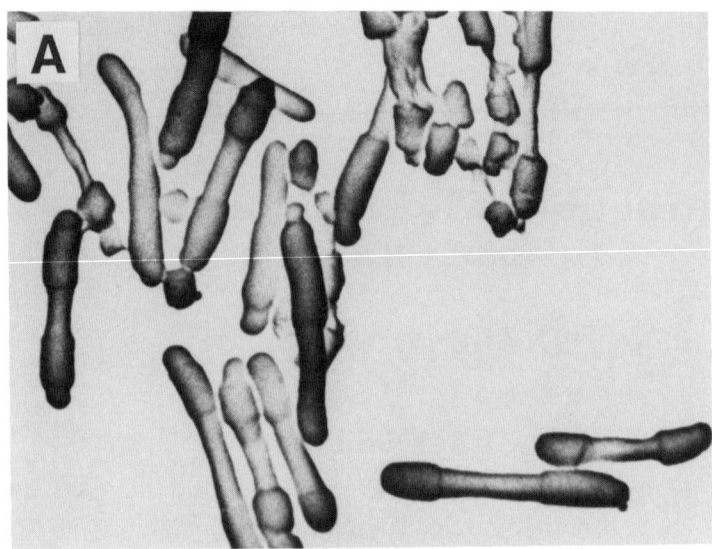

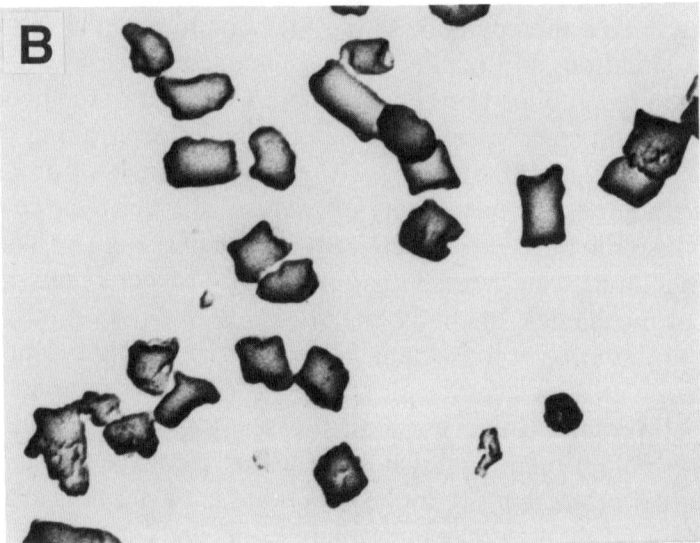

Fig. 1. **(A)** Scanning electron micrograph of *E. coli* containing inclusion bodies. The preparative techniques for electron micrographs have caused the cells to shrink. However, regions of the cells containing rigid inclusion bodies have not been able to shrink, showing clearly the outline of the inclusion bodies. **(B)** Scanning EM of isolated washed inclusion bodies. Again, it can be seen that the inclusions retain a rigid cylindrical shape.

Extraction of Recombinant Protein

Inclusion bodies were first reported by Williams et al. *(3)* on overexpression of proinsulin in *E. coli*. It is not known exactly how they are formed, but it is thought that the protein is partially or incorrectly folded. Formation of inclusion bodies is not only found on overexpression of foreign eukaryotic proteins, but is also seen on overexpression of bacterial proteins that are normally soluble *(4)*.

A number of methods have been described for increasing the soluble fraction of the protein. Growth temperature, media composition, and host strain have all been found to affect the partitioning of the overexpressed protein between the cytosol and inclusion body fractions *(5)*. Fusion proteins with a highly soluble protein, such as glutathione-S-transferase or thioredoxin, can also increase solubility of the protein of interest *(6,7)*.

The advantage of inclusion bodies is that they generally allow greater levels of expression, and they can be easily separated from a large proportion of bacterial cytoplasmic proteins by centrifugation giving an effective purification step. Some contaminating proteins are always present that may be associated with inclusion body formation.

The major disadvantage of inclusion bodies is that extraction of the protein of interest generally requires the use of denaturing agents. This can cause problems where native folded protein is required, since refolding methods are not always 100% effective and may be difficult to scale-up.

Some inclusion bodies can be solubilized by extremes of pH and temperature, but most require strong denaturing agents. Certain proteins, such as DNase 1, can be refolded after solubilization with SDS, but detergents are difficult to remove from most proteins and can interfere with subsequent refolding. The most commonly used solubilizing agents are water-soluble chaotropic agents, such as urea and guanidinium hydrochloride, which are more compatible with protein refolding. Most inclusions will be soluble in $8M$ urea, and a reducing agent, such as DTT, is generally required in order to prevent the formation of disulfide bonds between aggregates or denatured polypeptide chains.

2. Materials

All reagents are analaytical grade.

1. Prepare a stock solution of 100 mM PMSF in isopropanol, and store at $-20°C$. Add PMSF to buffers just before use. (Note—PMSF is a hazardous chemical and should be treated with caution.)

2. Lysis buffer: 50 mM Tris-HCl, pH 8.0, 1 mM EDTA, 50 mM NaCl, 1 mM PMSF (*see* Note 1).
3. Hen egg lysozyme (Sigma, Poole, Dorset, UK): 10 mg/mL stock solution.
4. DNase 1 (Boehringer Mannheim, Lewes, East Sussex, UK): 1 mg/mL stock solution (*see* Note 2).
5. Sodium deoxycholate (Sigma): 20 mg/mL stock solution.
6. Solubilization buffer: 50 mM Tris-HCl, pH 8.0, 8M urea, 1 mM DTT.

Urea solutions should be used within 1 wk of preparation and stored at 4°C, in order to reduce the formation of cyanate ions, which can react with protein amino groups forming carbamylated derivatives.

3. Methods
3.1. Enzymatic Lysis of E. coli

1. Harvest the bacterial cells by centrifugation at 1000g for 15 min at 4°C, and pour off the supernatant. Weigh the wet pellet. This is easiest to do if you have preweighed the centrifuge tubes which can then be deducted from the total weight.
2. Add approx 3 mL of lysis buffer for each wet gram of bacterial cell pellet and resuspend. Add lysozyme to a concentration of 300 µg/mL and stir the suspension for 30 min at 4°C (*see* Notes 3 and 4).
3. Add deoxycholate to a concentration of 1 mg/mL while stirring.
4. Place at room temperature, and add DNase1 to a concentration of 10 mg/mL and MgCl$_2$ to 10 mM. Stir suspension for a further 15 min to remove the viscous nucleic acid (*see* Note 2).
5. Centrifuge the suspension at 10,000g for 15 min at 4°C. Resuspend the pellet in lysis buffer to the same volume as the supernatant, and analyze aliquots of both for the protein of interest on SDS-PAGE. If the bulk of a normally soluble protein is found in the insoluble pellet fraction, then inclusion bodies are likely to have formed.

3.2. Washing of Inclusion Bodies

Washing of inclusion bodies prior to solubilization can remove further contaminant proteins, and using solutions other than water or buffer can increase the purification obtained (8). It is advisable to carry out a small scale trial to optimise the buffer and to ensure that the protein of interest is not solubilized.

1. Centrifuge 200-µL aliquots of the resuspended cell pellet in microfuge tubes at 12,000g for 10 min at 4°C. Resuspend the pellets in a range of test solutions. Lysis buffer containing 1, 2, 3, and 4M urea and 0.5% Triton X-100 are suggested. Mix and incubate for 10 min at room temperature. Centrifuge as before in a microfuge, and resuspend pellets in 200 µL H$_2$O.

Extraction of Recombinant Protein

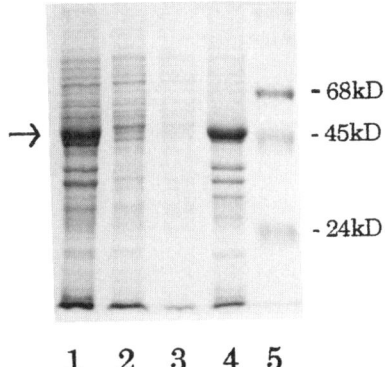

Fig. 2. Washing and solubilization of inclusion bodies containing recombinant plasminogen activator inhibitor-2 (45-kDa protein indicated). Lane 1: insoluble *E. coli* fraction; lane 2: first $2M$ urea wash; lane 3: second $2M$ urea wash; lane 4: inclusion body pellet solubilized in $8M$ urea.

2. Take equal volumes of the supernatant and the resuspended pellet and add to SDS boiling buffer. Analyze samples for the protein of interest on SDS-PAGE. The best washing buffer will contain the most contaminant proteins and little or none of the protein of interest.
3. Scale this procedure up, and wash the inclusion bodies twice with the optimum buffer. An example of the purification achieved on washing of inclusion bodies of plasminogen activator inhibitor-2 (PAI-2) is shown in Fig. 2.

3.3. Solubilization of Recombinant Protein from Inclusion Bodies

It is also important to optimize the solubilization solution, since a number of factors will affect solubility depending on the nature of the protein of interest. These include the nature and strength of the solubilization agent, the temperature and time taken to obtain efficient solubilization, protein concentration, and the presence or absence of reducing agents.

1. Resuspend the washed inclusion bodies in the solubilization buffer. Stir this suspension for 1 h at room temperature to ensure complete solubilization.
2. Centrifuge the solution at 20,000g for 20 min at 4°C to remove any remaining insoluble material. Check this pellet for the protein of interest on SDS-PAGE. If a substantial porportion of the protein has remained insoluble, then increase the incubation time with the solubilization buffer or try a different agent to solubilize the inclusions.

3. The extracted recombinant protein can be refolded from the urea at this stage (*see* Note 6), or may be purified under denaturing conditions (*see* Notes 5 and 7).

4. Notes

1. The lysis buffer for enzymatic digestion can be critical. Hen egg lysozyme has a pH optimum of between 7.0 and 8.6, and also works best in ionic strength of $0.05M$.
2. Bacterial extracts roughly consist of protein (40–70%), nucleic acid (10–30%), polysaccharide (2–10%), and lipid (10–15%). The nucleic acid fraction can often cause high viscosity. In addition to DNase treatment detailed in Section 3.1., step 4, this nucleic acid can also be removed from soluble protein solutions by precipitation with positively charged compounds, such as polyethyleneamine *(9)*. Precipitation methods should not be used with inclusion body preparations, since the precipitate will cocentrifuge with the inclusions.
3. If extracting soluble proteins from bacteria, it may be desirable to use more gentle methods of enzymatic lysis where sucrose buffers are used *(1)*. Inclusion of extra protease inhibitors, such as aprotinin or leupeptin, in the lysis buffer may also be advantageous if proteolysis of the target protein is occuring. This is generally not necessary when the protein is packaged in inclusion bodies, but inhibitors may be required in the solubilization and refolding buffers, since proteins in semifolded states are more susceptible to proteolysis.
4. Mechanical lysis is also effective in disrupting the bacterial outer cell wall, and a number of mechanical methods are available. Sonication is suitable for smaller scale purifications, but generation of heat during sonication can be difficult to control and may cause denaturation of proteins. For larger scale processing, the French press and the Mantin Gaulin press are most commonly used. These devices lyze the cells by applying pressure to the cell suspension followed by a release of pressure, which causes a liquid shear and thus cell disruption. Multiple passes of the cells through the presses are generally necessary to obtain adequate lysis *(1)*.
5. It is often desirable to carry out some purification of the protein under denaturing conditions before refolding. Urea solutions are compatible with ion-exchange chromatography, metal ion-affinity chromatography, and gel filtration. Guanidium hydrochloride is compatible with gel filtration chromatography but owing to its charge, it cannot be used in ion-exchange purification steps.
6. Many methods have been described for refolding of proteins from urea (for reviews, *see* refs. *10* and *11*). The most common techniques use dilution and dialysis to reduce the urea or guanidinium concentration gradually,

and the presence of thiol reagents is generally important for correct formation of disulfide bonds.
7. Proteins used for animal immunization purposes often do not require refolding unless antibodies to tertiary epitopes are required. It is also possible to use inclusion bodies directly injected, since particulate antigens are highly immunogenic. Sonication of the inclusions into smaller particles is recommended prior to injection *(12)*.

References

1. Cull, M. and McHenry, C. S. (1990) Preparation of extracts from prokaryotes. *Methods Enzymol.* **182,** 147–153.
2. Kane, J. F. and Hartley, D. L. (1989) Formation of recombinant protein inclusions in *Escherichia coli. Tibtech* **6,** 95–101.
3. Williams, D. C., Van Frank, R. M., Muth, W. L., and Burnett, J. P. (1982) Cytoplasmic inclusion bodies in *E. coli* producing biosynthetic human insulin proteins. *Science* **215,** 687–689.
4. Gribskov, M. and Burgess, R. R. (1983) Overexpression and purification of the sigma subunit of *Escherichia coli* RNA polymerase. *Gene* **26,** 109–118.
5. Schein, C. H. and Noteborn, M. H. M. (1988) Formation of soluble recombinant proteins is favoured by lower growth temperature. *BioTechnology* **6,** 291–294.
6. Smith, D. B. and Johnson, K. S. (1988) Single step purification of polypeptides expressed in *E. coli* as fusions with glutathione-S-transferase. *Gene* **67,** 31–40.
7. LaVallie, E. R., DiBlasio, E. A., Kovacic, S., Grant, K. L., Schendel, P. F., and McCoy, J. M. (1993) A thioredoxin gene fusion expresssion system that circumvents inclusion body formation in the *E.coli* cytoplasm. *BioTechnology* **11,** 187–193.
8. Schoner, R. G., Ellis, L. F., and Schoner, B. E. (1985) Isolation and purification of protein granules from *Escherichia coli*, cells overproducing bovine growth hormone. *BioTechnology* **3,** 151–154.
9. Burgess, R. and Jendrisak, J. (1975) A procedure for the rapid large-scale purification of *E. coli* DNA dependant RNA polymerase involving Polymin P precipitation and DNA cellulose chromatography. *Biochemistry* **14,** 4634–4645.
10. Marston, F. A. O. (1986) The purification of eukaryotic polypeptides synthesized in *Escherichia coli. Biochem. J.* **240,** 1–12.
11. Kohno, T., Carmichael, D. F., Sommer, A., and Thompson, R. C. (1990) Refolding of recombinant proteins. *Methods Enzymol.* **185,** 187–195.
12. Harlow, E. and Lane, D. (1988) *Antibodies: A Laboratory Manual.* Cold Spring Harbor Laboratory, Cold Spring Harbor, New York. pp. 88–91.

CHAPTER 5

Protein Extraction from Fungi

Paul Bridge

1. Introduction

In order to study proteins from yeasts and filamentous fungi, it is important to consider a number of basic features of the organisms. First, filamentous fungi undergo a growth cycle that includes differentiation and compartmentalization. In addition, both the filamentous fungi and yeasts will age during growth, and older cultures will undergo autolysis. As a result, particular proteins may only be associated with one part of the growth cycle, such as sporulation or autolysis, and this must be taken into account in determining growth conditions and sampling times.

Second, many of the enzymes produced during the growth period are sequential and may either be subject to significant repression or require induction by a substrate or substrate component. Examples of this include the requirement for chitin or chitin-like components to induce chitinases *(1)*, and the repression of some fungal proteases by glucose *(2)*.

Third, fungi possess rigid cell walls and complex cell-wall/membrane systems *(3)*. It is therefore important to ascertain the potential location of proteins prior to their extraction, since cell-wall-associated and extracellular proteins will be lost during intracellular extractions. An example of this is the utilization of many of the traditional fungal nutrients, such as cellulose and lignin, where significant levels of extracellular enzymes will be produced. Most of these extracellular enzymes can be produced in sufficient concentrations for them to be purified and characterized directly from the spent growth medium *(4,5)*.

A simple growth and extraction procedure is described here. This is a standard regime that will allow the extraction of intracellular proteins

From: *Methods in Molecular Biology, Vol. 59: Protein Purification Protocols*
Edited by: S. Doonan Humana Press Inc., Totowa, NJ

from a wide range of filamentous fungi, and has been used successfully with many fungal genera, including *Fusarium, Ganoderma, Aspergillus, Colletotrichum, Beauveria, Phoma, Verticillium,* and *Metarhizium (6).* The method has not been optimized toward any particular fungal group, and has proven suitable for filamentous ascomycetes and basidiomycetes as well as yeasts *(7–9).* The major variation that will be needed for different fungal groups is the growth medium and the length of the growth period (*see* Notes 1 and 2). Although a crude method, extracts produced in this way retain sufficient integrity and activity for enzyme assays and isoenzyme electrophoresis. An additional feature of this method is that the spent culture fluid may be retained for the detection of extracellular enzymes. Initially, this will only contain a small number of glucose-independent enzymes, but as the culture grows and the free glucose concentration decreases, further enzymes can be detected or extracted *(7,10).*

2. Materials

Fungal growth media and buffers should be sterilized prior to use. Growth media and buffers can routinely be sterilized at 10 psi for 10 min in a benchtop autoclave. Although the materials listed here are unaffected, is should be remembered that in complex media and buffers, individual components may break down or react during autoclaving, and so may need to be individually filter sterilized.

1. Malt extract agar (MEA): 20 g Malt extract (Oxoid, Basingstoke, UK), 1 g peptone (Oxoid; Bacteriological), 20 g glucose, 15 g agar, 1 L distilled water *(11).*
2. Glucose yeast medium (GYM): 1g $NH_4H_2PO_4$, 0.2 g KCl, 0.2 g $MgSO_4 \cdot 7H_2O$, 10 g glucose, 1 mL 0.5% aqueous $CuSO_4 \cdot 5H_2O$, 1 mL 1% aqueous $ZnSO_4 \cdot 7H_2O$, distilled water to 1 L *(7).*
3. Pectin broth: 0.9 g $NH_4H_2PO_4$, 2g $(NH_4)_2HPO_4$, 0.1 g $MgSO_4 \cdot 7H_2O$, 0.5 g KCl, 10 g citrus pectin (Sigma, Poole, UK), 1 L distilled water.
4. Tris-glycine buffer: 3 g Trizma (Sigma, Poole, UK), 14.4 g glycine, 1 L deionized water, pH 8.3.
5. Pectinase gel: 0.2 g Citrus pectin, 10 g acrylamide, 0.25 g methylene-bis-acrylamide, 0.1 mL TEMED (*N,N,N',N'*-tetramethylethylenediamine), 0.1 g ammonium persulfate, 100 mL gel buffer.
6. Gel buffer: 0.525 g Citric acid monohydrate, 4.598 g Tris, 1 L deionized water.
7. Electrode buffer: 7.22 g Boric acid, 15.75 g sodium tetraborate decahydrate, 1 L deionized water.
8. $0.1M$ Malic acid.

Protein Extraction from Fungi

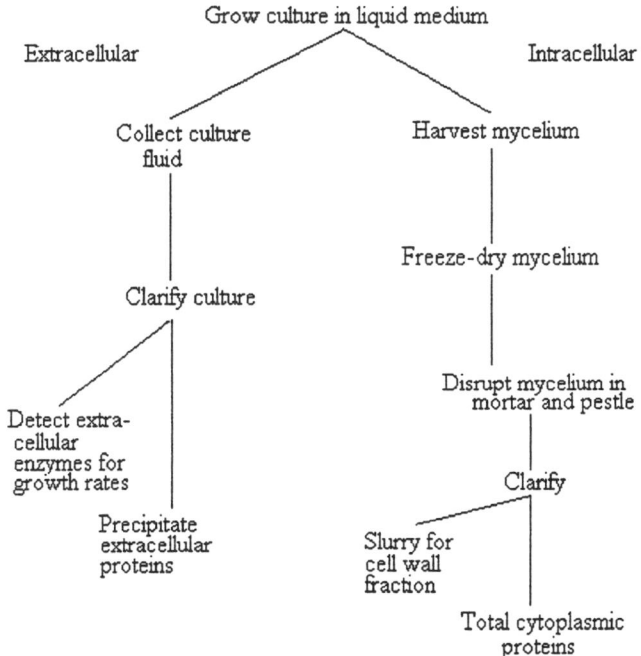

Fig. 1. Schematic diagram of the complete protocol.

9. 0.015% Aqueous ruthenium red.
10. Sterile deionized water.

3. Methods

The method presented here will enable the extraction of intracellular proteins and electrophoresis of extracellular enzymes from filamentous fungi and yeasts (Fig. 1). The Notes section details further considerations that may be needed for specific organisms or extractions

3.1. Extraction of Intracellular Proteins from **Metarhizium**

The following protocol describes the extraction of aqueous intracellular proteins from the filamentous fungus *Metarhizium anisopliae*. Further details regarding growth and extraction conditions for other filamentous fungi are given in the Notes 2–7.

1. Grow *Metarhizium* culture on malt extract agar for 7 d at 25–28°C.
2. Remove a plug (approx 0.5-cm diameter) of culture from the agar plate with a flamed cork borer or scalpel. Cut the plug into at least 10 smaller

pieces with a flamed scalpel and inoculate into 10-mL sterile GYM in a 28-mL Universal bottle or small flask (*see* Note 3).
3. The 10-mL culture is a starter culture for the subsequent main growth period and is required to ensure an actively growing inoculum. Incubate the starter culture at 25–28°C in an incubator, with shaking, for 60–72 h.
4. Aseptically transfer the starter culture into a 250-mL conical flask containing 60 mL of fresh sterile GYM. Replace the flask in the incubator and continue incubation with shaking for a further 60–72 h (*see* Note 4).
5. Harvest the mycelium from the flask by vacuum-assisted filtration. A Buchner funnel containing Whatman No. 3 filter paper is suitable for this purpose (*see* Note 5). Take care to ensure that any aerosols that may be formed from the filtration or the vacuum pump are minimized and contained.
6. Wash the mycelium once in the Buchner funnel with sterile deionized water, and transfer the harvested mycelium from the filter paper to a plastic Petri dish with a flamed or alcohol-sterilized spatula.
7. Freeze the mycelium at –20°C for storage prior to extraction.
8. Prepare the mycelium for extraction by freeze-drying for 24 h. This is one of several possible methods of obtaining cell breakage (*see* Notes 6 and 7).
9. Disrupt the freeze-dried mycelium by briefly grinding it in a mortar and pestle (*see* Note 6). The ground mycelium should be collected in sterile 1.5-mL microcentrifuge tubes in approx 500-mg amounts. This is roughly equivalent to the conical portion of the tube. These tubes of ground mycelium can be stored at –20°C for at least 3 mo.
10. Rehydrate 500 mg of ground mycelium in 1 mL of Tris-glycine buffer (*see* Note 8). This will need mixing with a micropipet tip or Pasteur pipet to form an even slurry.
11. Clarify the slurry by centrifugation at 12,500g for 40 min at 4°C. After centrifugation, collect the supernatant into another sterile microcentrifuge tube (*see* Note 9).
12. The collected supernatant will contain the total cytoplasmic proteins. The samples as prepared here typically contain 15–100 mg/mL protein (*see* Note 10). The extracts can be used directly in gel electrophoresis or column chromatography. Alternatively, standard precipitation techniques can be used to purify specific proteins further (*see* Note 9).

3.2. Electrophoresis of Extracellular Enzymes

The following method describes the induction and electrophoresis of extracellular pectinases. Pectin is used, both in the production medium and in the polyacrylamide gel, to induce the enzymes and to avoid the necessity for subsequent overlays for staining. The method given here was originally described by Cruickshank and Wade *(4),* and is given here as in

Paterson and Bridge *(6)*. The method has successfully been used without modifications with *Fusarium, Penicillium, Colletotrichum, Ganoderma,* and *Rhizoctonia*. A horizontal gel system is required, and self-adhesive vinyl tape (Dymo) is used to produce wells in a simple gel mold. Sephadex is used in the sample wells to make more well-defined enzyme patterns (*see* Note 11). Bromophenol blue is not used to measure the buffer front, since it can be seen adequately without the dye. This type of pectin/acrylamide gel gives enzyme patterns that are asymmetrical, unlike conventional isozyme bands. This is probably because of more definite chromatographic adsorption effects from charged pectin molecules, which are dissolved in the gel, i.e., the protein molecules are interacting with the gel.

1. Inoculate cultures into 2 mL pectin broth (PB) in a 5 mL Bijoux bottle.
2. When the cultures have grown sufficiently (usually 7 d), remove 200 µL of the culture fluid, and suspend in superfine (G-25) Sephadex at a concentration of 20 mg/200 µL.
3. Prepare a pectin-acrylamide gel for horizontal gel electrophoresis (*see* Note 11). Use three pieces of Dymo tape (2.5 × 5.0 mm) to form a well with a volume of 20 µL in the center of a gel mold. Similar pieces of tape are placed 1.25 cm apart across the mold to form the other wells.
4. Set up the horizontal electrophoresis with anode and cathode buffer compartments containing gel buffer. Use filter paper wicks to form the "contact" between the buffer and the gel.
5. Place the negative electrode 3.5 cm below the sample wells to allow adequate migration of any positively charged pectinase component.
6. After electrophoresis, submerge gels in $0.1 M$ malic acid for 1 h, rinse in distilled water, and stain overnight in 0.015% (w/v) ruthenium red at 4°C. Rinse the gel for 1 h with three changes of distilled water.
7. Enzyme patterns are recorded by:
   a. Tracing the gel and enzyme patterns onto vinyl transparencies and photocopying the tracing;
   b. Photographic recording; and/or
   c. Direct comparison from dried gels.
8. Polygalacturonase (PG) activity gives transparent zones, pectin esterase (PE) activity gives dark zones, and pectin lyase (PL) gives yellow zones.

4. Notes

1. The growth medium selected will vary depending on the requirements of the fungus under study. The GYM medium given in the Section 2. is a general-purpose medium suitable for the growth of many filamentous ascomycetes and basidiomycetes. For organisms that may be more fastidious, a

richer medium, such as peptone yeast extract glucose *(12)*, can be very useful (*see* ref. *6*). Oomycetes, such as *Pythium* and *Phytophora*, may require either a source of sterol or a more complex organic medium, such as V8 medium *(11)*. Organic components in growth media can vary between suppliers, and this variation may affect the biochemical properties of the fungus *(13)*. It is therefore important to standardize organic components, such as yeast extract, by only using a single source of supply.
2. Growth conditions are important in obtaining a reliable and constant source of protein, and it is usually desirable to extract from cultures consisting of material of a constant age. This is obtained by first inoculating fungal cultures into small starter cultures of the main growth medium. These starter cultures are incubated, usually as shaken liquids, and provide the inoculum for the main growth phase. Although the exact conditions will vary depending on the organism studied, a typical starter culture would consist of 10 mL of GYM medium in a 25–30 mL vessel, incubated on an orbital shaker at 25°C for 60 h. The entire starter culture is then used as inoculum for 60 mL of GYM medium in 250-mL conical flasks, which is then incubated at 25°C on the shaker for 3–5 d. Growth periods and temperatures will depend on the rate of growth and differentiation of the fungus studied. Most yeast cultures will produce sufficient suitable biomass after incubation at 30°C under a growth regime of 24 h in a starter culture and 24 h in a shaken flask. Many filamentous fungi however require 60 h in a starter culture followed by 3–5 d in shaken flasks at 25°C. In order to maintain reproducibility, incubation temperatures should be kept constant.
3. The inoculum used for the initial starter cultures is important and should contain a large number of potential "growing" points. That is, the inoculum should ideally be particulate so that mycelial or yeast growth can occur at many different points. This presents no problems if the culture is naturally particulate, such as yeast cells, or if conidial suspensions can be used (e.g., ref. *5*). However, if the fungus grows only as a mycelial mat, an inoculum made from broken mycelia will give a more homogenous and greater biomass than one derived from plugs or "lumps" of culture.
4. In many cases, the proteins of interest will be associated with active growth of the fungus. It is therefore often necessary to establish some form of growth curve for the organism prior to harvesting the growth. The growth rate of yeasts, and fungi growing in a yeast-like phase, can be estimated by sampling and measuring the turbidity of the growth medium over a period of time *(14)*. However, with filamentous organisms, this becomes impractical. One alternative is to assay the culture medium for the extracellular enzymes β-glucosidase, β-galactosidase, and diacetyl-chitobiosidase. In many fungi, these enzymes are each associated with particular features of the growth

cycle when grown in glucose-containing media. β-glucosidase is generally produced in the early phases of growth, β-galactosidase is not produced until the glucose concentration has been reduced, and diacetyl-chitobiosidase is generally produced during autolysis. These activities can be readily screened directly from culture fluids with the substrates 4-methylumbeliferyl β-D-glucoside, 4MU β-D-galactoside, and 4MU β-D-N,N'-diacetyl-chitobioside *(6,10)*. The substrates can be kept as 50-mM solutions in dimethylformamide, and for use, the stock solutions are diluted 0.15:9.85 mL in 0.05M sodium acetate, pH 5.4. Activity is determined by adding 50 µL of culture fluid to 50 µL of each substrate and incubating these for 4 h at 37°C. Positive enzyme activity can then be detected as fluoresence under UV light after the addition of 50 mL of saturated aqueous sodium bicarbonate.

5. When the required phase of growth has been reached, the fungal biomass must be harvested from the growth medium. Again methods differ depending on the organism, but in general, yeasts can be harvested by centrifugation at 3–7000g, whereas filamentous fungi are better harvested by vacuum filtration onto Whatman No. 3 filter papers in a Buchner funnel. In both cases, the resulting biomass should be washed in sterile distilled water to remove any residues from the media. If extracellular proteins, such as extracellular enzymes, are required, these can be extracted directly from the culture fluid after the mycelium has been harvested. Typically, the culture fluid is clarified by centrifugation and proteins extracted by one of the general protein extraction techniques, such as precipitation with acetone or methanol/ammonium or by immunoprecipitation *(15,16)*.

6. Efficient cell breakage is necessary to ensure a good recovery of proteins from filamentous fungi. A wide variety of cell breakage methods have been reported in the literature, varying from enzymatic digestion of the cell wall to physical disruption procedures, such as grinding mycelium in a mortar and pestle *(6)* or homogenizing thick cell pastes *(17)*. Enzymatic digestion has been used to generate protoplasts, which can be harvested from the remaining cell debris. The protoplast preparations can then be lysed to release intracellular proteins *(18,19)*. Many fungi produce active intra- and extracellular proteinases, and these are usually active under the conditions used for enzymatic digestion of cell walls. As a result, physical disruption of an "inert" sample is generally preferred. Protease activity has been supressed by phenylmethylsulfonylfluoride (PMSF) in some cases *(20)*. Fungal cell walls can be disrupted by briefly grinding freeze-dried mycelium in a mortar and pestle. Additional abrasives are not usually required, although carborundum may be added if required. This procedure should not be performed on an open bench owing to the possible hazardous nature of the dust produced. A single flask of 60 mL actively growing culture will

give about 500–2000 mg of freeze-dried material. The method of grinding a freeze-dried culture as described here is quick and reliable, but requires access to freeze-drying equipment. A commonly used alternative is to grind the harvested mycelium in liquid nitrogen *(15)*. However, appropriate safety measures should be considered to avoid any potential contact or splashing from the liquid nitrogen.

7. Yeast cells can be disrupted by passage through a pressure cell, such as a French Press *(21)*, although three or more passages may be necessary to achieve 70% breakage. Yeast cells can also be broken by shaking frozen cultures with glass beads. The glass beads used are generally larger than the Ballotini beads used for bacteria and 2–2.5-mm beads can produce adequate disruption after 10–15 min of shaking. This method can generate sufficient local heating to denature some proteins, so effective cooling of the system is required.
8. The ground mycelium is rehydrated in the method at 500 mg in 1 mL of Tris-glycine buffer. This will need mixing to form an even slurry. In some cases, detergent-containing buffers have been shown to give higher yields of total protein *(22,23)*, but the simple Tris-glycine buffer used here is adequate for general-purposes.
9. The total aqueous extract will contain the cytoplasmic proteins. Normally, the pellet of unbroken cells and debris is discarded although this can be saved if the cell-wall fraction is required. This extract will, however, contain many other components, including polysaccharides. The degree of further purification necessary will depend on the protein and level of activity required, as well as the level of polysaccharide contamination. As mentioned, the supernatant contains the total cytoplasmic proteins and may be used directly. Proteins in these crude extracts can be readily separated by polyacrylamide gel electrophoresis *(7,8)*, and this technique can be used to provide cell-free total protein patterns or, with specific stains, to demonstrate particular enzymes. Alternatively, it may be necessary to clarify the supernatant further by a second centrifugation or filtration through a 0.45-µm filter. Specific proteins can then be purified through standard precipitation techniques as mentioned in Note 5. Further clarification and separation, generally by differential centrifugation in sucrose density gradients will be required to extract the cell-wall fraction *(23)*.
10. The protein content of the supernatant should be estimated by one of the standard protein determination methods, such as Lowry's determination *(24)*.
11. Under the conditions used, pectinase isoenzymes may be both positively and negatively charged. It is therefore very important to undertake the electrophoresis in a horizontal gel system that allows for the construction of central sample wells.

References

1. St. Leger, R. J., Cooper, R. M., and Charnley, A. K. (1986) Cuticle-degrading enzymes of entomopathogenic fungi: regulation of production of chitinolytic enzymes. *J. Gen. Microbiol.* **132,** 1509–1517.
2. Clarkson, J. H. (1992) Molecular biology of filamentous fungi used for biological control, in *Applied Molecular Genetics of Filamentous Fungi* (Kinghorn, J. R. and Turner, G., eds.), Blackie Academic and Professional, Glasgow, Scotland, pp. 175–190.
3. Peberdy, J. F. (1990) Fungal cell walls—a review, in *Biochemistry of Cell Walls and Membranes in Fungi* (Kuhn, P. J., Trinci, A. P. J., Jung, M. J., Goosey, M. W., and Copping, L. G., eds.), Springer-Verlag, Berlin, Germany, pp. 5–30.
4. Cruickshank, R. H. and Wade, G. C. (1980) Detection of pectin enzymes in pectin acrylamide gels. *Analyt. Biochem.* **107,** 177–181.
5. Elad, Y., Chet, I., and Henis, Y. (1982) Degradation of plant pathogenic fungi by *Trichoderma harzianum*. *Can. J. Microbiol.* **28,** 719–725.
6. Paterson, R. R. M. and Bridge, P. D. (1994) *Biochemical Techniques for Filamentous Fungi.* CAB International, Wallingford, UK.
7. Mugnai, L., Bridge, P. D., and Evans, H. C. (1989) A chemotaxonomic evaluation of the genus *Beauveria*. *Mycological Res.* **92,** 199–209.
8. Jun, Y., Bridge, P. D., and Evans, H. C. (1991) An integrated approach to the taxonomy of the genus *Verticillium*. *J. Gen. Microbiol.* **137,** 1437–1444.
9. Monte, E., Bridge, P. D., and Sutton, B. C. (1990) Physiological and biochemical studies in *Coelomycetes*. *Phoma. Studies Mycology* **32,** 21–28.
10. Barth, M. G. and Bridge, P. D. (1989) 4-Methylumbelliferyl substituted compounds as fluorogenic substrates for fungal extracellular enzymes. *Lett. Appl. Microbiol.* **9,** 177–179.
11. Smith, D. and Onions, A. H. S. (1994) *The Preservation and Maintenance of Living Fungi,* 2nd ed. CAB International, Wallingford, UK.
12. Conti, S. F. and Naylor, H. B. (1959) Electron microscopy of ultrathin sections of *Schizosaccharomyces octosporus*. I. Cell division. *J. Bacteriol.* **78,** 868–877.
13. Filtenborg, O., Frisvad, J. C., and Thrane, U. (1990) The significance of yeast extract composition on secondary metabolite production in *Penicillium,* in *Modern Concepts in* Penicillium *and* Aspergillus *Classification* (Samson, R. A. and Pitt, J. I., eds.), Plenum, New York, pp. 433–441.
14. Barnett, J. A., Payne, R. W., and Yarrow, D. (1990) *Yeasts: Characteristics and Identification,* 2nd ed. Cambridge University Press, Cambridge, UK.
15. St. Leger, R. J., Staples, R. C., and Roberts, D. W. (1991) Changes in translatable mRNA species associated with nutrient deprivation and protease synthesis in *Metarhizium anisopliae*. *J. Gen. Microbiol.* **137,** 807–815.
16. Kim, K. K., Fravel, D. R., and Papavizas, G. C. (1990) Production, purification and properties of glucose oxidase from the biocontrol fungus *Talaromyces flavus*. *Can. J. Microbiol.* **36,** 199–205.
17. Hien, N. H. and Fleet, G. H. (1983) Separation and characterization of six (1→3)-β-glucanases from *Saccharomyces cerevisiae*. *J. Bacteriol.* **156,** 1204–1213.
18. Sambrook, J., Fritsch, E. F., and Maniatis, T. (1989) *Molecular Cloning: A Laboratory Manual,* vol. 3. Cold Spring Harbor Laboratory, Cold Spring Harbor, New York.

19. Messner, R. and Kubicek, C. P. (1990) Synthesis of cell wall glucan, chitin and protein by regenerating protoplasts and mycelia of *Trichoderma reesei*. *Can. J. Microbiol.* **36,** 211–217.
20. Kim, W. K. and Howes, N. K. (1987) Localization of glycopeptides and race-variable polypeptides in uredosporling walls of *Puccinia graminis tritici;* affinity to concalvin A, soybean agglutinin, and *Lotus* lectin. *Can. J. Botany* **65,** 1785–1791.
21. Schnaitman, C. A. (1981) Cell fractionation, in *Manual of Methods for General Bacteriology* (Gerhardt, P., Murray, R. G. E., Costilow, R. N., Nester, E. W., Wood, W. A., Krieg, N. R., and Briggs Phillips, G., eds.), American Society for Microbiology, Washington, DC, pp. 52–61.
22. Kim, W. K., Rohringer, R., and Chong, J. (1982) Sugar and amino acid composition of macromolecular constituents released from walls of uredosporlings of *Puccinia graminis triticii*. *Can. J. Plant Pathol.* **4,** 317–327.
23. Fèvre, M. (1979) Glucanase, glucan synthases and wall growth in *Saprolegnia monoica,* in *Fungal Walls and Hyphal Growth,* British Mycological Society Symposium 2 (Burnett, J. H. and Trinci, A. P. J., eds.), Cambridge University Press, Cambridge, UK, pp. 225–263.
24. Lowry, O. H., Rosebrough, N. J., Farr, A. L., and Randall, R. J. (1951) Protein measurement with the folin phenol reagent. *J. Biol. Chem.* **193,** 265–275.

CHAPTER 6

Subcellular Fractionation of Animal Tissues

Norma M. Ryan

1. Introduction

A number of techniques exist which exploit various physical parameters or biological properties of cells as means of investigating the complexity of cellular organelles and membranes. Subcellular fractionation using the centrifuge is the basis of traditional methods for separating cellular components. Even when other methods are used to effect the separation, it is often the case that better results will be obtained if the material is first purified or partially purified by centrifugal methods (*see* refs. *1* and *2* for a recently published discussion of the theory and applications of the use of the centrifuge in preparative procedures). In this chapter some of the basic centrifugal techniques used to fractionate a typical animal cell are described.

Subcellular fractionation involves three successive steps (for a detailed discussion of the principles involved *see* refs. *3* and *4*). The first step converts a tissue or cell suspension into a homogenate. The second step reintroduces a new kind of order into the system by grouping together, in separate fractions, those components of the homogenate of which certain physical properties, such as density or sedimentation coefficient, fall between certain limits set by the investigator. The third step consists of the analysis of the isolated fractions. Interpretation of the results involves a retracing of these three steps. It is essential to perform all steps of this process, regardless of whether the objective of the work is to study the spatial organization of components within a cell (i.e., subcellular fractionation itself) or to use the procedures as a means of obtaining a partially purified component which then can be more readily purified using other procedures.

Tissue disruption or homogenization may be accomplished by a variety of means such as grinding, ultrasonic vibrations, and by making use of the osmotic properties of cells. The grinding of tissues to form homogenates is usually accomplished with the Potter–Elvejhem homogenizer, which is highly effective with soft tissues, such as liver and kidney. Tough tissues, such as muscle, generally require the use of a mincer and blender of which a common example is the Waring Blender. In recent years the Chaikoff Press, which entails the forcing of the material through fine holes under high pressure, has found much use. Osmotic methods of tissue disruption are most useful when employed in connection with red blood cells and reticulocytes. The use of ultrasonic vibrations is useful for the disruption of bacterial cells and some animal cells. In such systems cooling tends to cause a problem because very high temperatures are generated at the point of disintegration. Homogenization should be carried out so as to leave the subcellular organelles virtually intact, although somewhat distorted, and yet shear the plasma membrane, endoplasmic reticulum, and other endomembranous systems into fragments which form spherical vesicles.

The suspension so prepared contains whole cells, partially broken cells, nuclei, mitochondria, and so on. Separation of the various components, in theory, can be achieved by the application of any method that exploits the differences between the physical and/or chemical properties of the constituents. Methods such as electrophoresis, counter-current distribution, and centrifugation are but a few of the methods available. The method most routinely used is centrifugation in one form or another, both because of the wide availability of the equipment necessary to carry out the procedures and also the facility of fractionating relatively large amounts of material with easy recovery of sample when fractionation is complete.

The behavior of particles in a centrifugal field and the various factors that affect their rate of sedimentation have been studied in great detail (5,6). The rate of movement of an ideal (spherical) particle in a centrifugal field is described by the following equation:

$$V = dr/dt = \{[a^2(Dp - Dm)]/18\eta\} \times \omega^2 r \tag{1}$$

where V is the velocity of particle in cm/s $= dr/dt$; a is the particle diameter; Dp is the particle density; Dm is the density of the medium; ω is the

Subcellular Fractionation

angular velocity in radian/s, which is equal to the number of revolutions per second × 2; r is the distance between the particle and the center of rotation in centimeters; and η is the viscosity of the medium.

The centrifugal force (CF) on a particle in a spinning rotor is given by:

$$CF = m\omega^2 r \qquad (2)$$

where m is the mass of the particle, r is the distance between the particle and the center of rotation in centimeters, and ω is the angular velocity in radian/s.

When centrifugation conditions are reported it is a common practice to list the relative centrifugal force (RCF). RCF is the force on a particle at some r value (generally for a point midway down the centrifuge tube, r_{av}) divided by the force on that particle in the earth's gravitational field:

$$RCF = \omega^2 r/g \qquad (3)$$

where g is the gravitational constant, 980 cm/s.

Thus:

$$RCF = \omega^2 r/980 \qquad (4)$$

Since temperature affects both the density and viscosity of the medium, it thus affects the rate of sedimentation of the particles.

In this chapter the subcellular fractionation of rat liver tissue is described as a fairly typical example of a fractionation procedure; ref. 7 should be consulted for information on variations in procedures appropriate for other tissues. The separation of the various constituents of an homogenate formed from rat liver tissue, using differential centrifugation, is described. The method described here does not yield an absolute purification of the individual components within a rat liver cell, but rather a partial purification of the cellular components results which will enable alternative/subsequent purification procedures to be carried out with a far higher chance of success.

2. Materials

All reagents are of Analar grade.

1. Homogenization buffer medium: 0.25*M* sucrose, 5 m*M* imidazole-HCl, pH 7.4 (*see* Notes 1 and 2).
2. 0.9% Saline for perfusion of the liver.
3. Resuspension buffer medium: 0.25*M* sucrose, 5 m*M* imidazole-HCl, pH 7.4 (*see* Note 3).

4. Refrigerated laboratory centrifuge, such as Sorvall RC5B fitted with an 8 × 50 mL rotor (*see* Note 2).
5. Potter–Elvejhem homogenizer.
6. Dounce homogenizer.
7. Beckman L65 ultracentrifuge fitted with an 8 × 35 mL rotor or equivalent (*see* Note 2).

3. Methods

3.1. Preparation of the Homogenate

1. Remove the liver from a rat large enough to yield approx 10 g liver (wet weight), having first perfused the liver with 0.9% saline. The perfusion should be carried out immediately after the sacrifice of the rat by appropriate means (*see* Note 4).
2. Wash the liver free of blood, hairs, and so on, by suspending it in ice-cold homogenizing medium.
3. Blot the tissue lightly with filter paper to dry it.
4. Mince the liver finely in a preweighed, chilled beaker and weigh again (*see* Note 5).
5. Homogenize the liver in homogenizing medium to form a 25% (w/v) homogenate, using six passes of a Potter–Elvejhem homogenizer (*see* Note 6).
6. Dilute homogenate to 12.5% (w/v).

3.2. Subcellular Fractionation (see Note 7)

1. Retain a small portion of the 12.5% (w/v) homogenate for analysis and centrifuge the remainder at 4°C at a speed of 600g for 10 min in a refrigerated laboratory centrifuge (*see* Note 8).
2. Resuspend the pellet in the same volume of homogenization medium as previously and centrifuge again at 600g for an additional 10 min (*see* Note 9).
3. Combine the post-600 g-minute supernatants and centrifuge at 15,000g for 10 min in order to prepare a fraction rich in mitochondria and lysosomes.
4. Resuspend the resulting pellet in the same volume of homogenization medium as previously and centrifuge again at 15,000g for another 10 min (*see* Note 10).
5. At the end of the washing steps pool all the supernatants and use in subsequent procedures (*see* Note 11). If the material retained in the pellet is required for further studies, resuspend it in homogenizing medium or other appropriate medium (*see* Note 3).
6. Pool the post-15 × 10^4 g-minute supernatants and centrifuge at 100,000g for 60 min in an ultracentrifuge at 4°C. The resulting complex micrososmal pellet contains vesicles derived from the plasma membrane, endoplasmic reticulum—some containing ribosomes and some ribosome-free—peroxisomes, polysomes, endosomes, Golgi stacks, and other such membranous systems from within the cell, while the supernatant will contain the remain-

Subcellular Fractionation

der of the cellular components, that is, the soluble components and smaller elements, such as free ribosomes (*see* Note 12).

7. Resuspend each of the pellets in a volume of $0.25M$ sucrose, 5 mM imidazole-HCl, pH 7.4 using a loose Dounce homogenizer. It may be more appropriate to use a different resuspension medium depending on the ultimate objectives (*see* Note 3).
8. Store all fractions in a deep freeze (*see* Note 13) until required for analysis (*see* Note 14) or further purification (*see* Note 15).

4. Notes

1. Homogenization is the first essential step in any fractionation procedure. It involves the disruption of an ordered system and results in a loss of some morphological information, but homogenization is necessary in order to apply the techniques of biochemistry to a study of subcellular components. The choice of an adequate homogenization medium is a critical one. The homogenization medium most suited to the biological material involved can only be elucidated by a process of trial and error. Sucrose is very widely used, but such details as concentration of sucrose, pH of the buffer, traces of specific cations, and so on, vary considerably. The best homogenate is the one which lends itself most successfully to fractionation and will result in a satisfactory resolution of the disrupted components of the homogenate. It is also the one which enables the next experiment to be carried out.
2. For best results the rat liver and every preparation made from the tissue must be kept cold (0–4°C) from the moment the organ is removed from the rat. Thus, all solutions must be prechilled to 0°C before addition to tissue, as should all glassware which comes in contact with the preparations. Centrifugation must be carried out in refrigerated centrifuges at 0–4°C, ensuring that the rotors are prechilled.
3. Depending on the final objectives of the purification procedures and the restrictions on the techniques following the differential centrifugation, it may be more appropriate to use a different resuspension medium than the one described here. Sucrose interferes with many assays and is not always easy to eliminate from preparations.
4. The method chosen for sacrifice of the rat will depend on the aims of the experiments. Most often this will be cervical dislocation. Anesthetics are occasionally used but these could have undesirable side effects on the tissue of interest, and therefore should only be employed if it is certain that the side effects will not affect the subcellular component/protein of interest.
5. Heavy metal ions are powerful enzyme inhibitors, so avoid sticking scissors, forceps, spatulas, and so on, into the tissue preparations. Manipulations, such as stirring of solutions or resuspension of pellets, should be carried out using glass rods.

6. This step is critical and the clearance of the homogenizer should be considered with care. Once again it is a matter of trial and error until precisely the right conditions are determined to suit the purposes of the experiment.
7. Subcellular fractionation is accomplished by the stepwise process of differential centrifugation, which separates particles from a supernatant in the form of a pellet. Differential pelleting is the simplest method for obtaining a crude separation which exploits the mass of the major organelles and membrane systems. All steps are carried out in the temperature range 0–4°C. A measured portion of the original homogenate is retained for further analysis. It is extremely important to keep an accurate record of all volumes and weights used in preparing the fractions. Otherwise it will be impossible to interpret the results in a meaningful way.
8. This first centrifugation step is designed to remove all nuclei, whole cells, partially intact cells, and plasma membrane sheets, and is very effective in this.
9. The objective of this washing step is to reduce contamination of the fraction by membrane components.
10. This washing step can be repeated two or three times, depending on how critical the degree of contamination by other membranous components is to the objectives of the experiment.
11. This fraction is not a pure preparation of mitochondria and lysosomes, but rather it is enriched in mitochondria and lysosomes with a reduction in the level of contaminating membranous components.
12. The ultimate location of the Golgi Apparatus and vesicles derived from the Golgi stacks and associated "trafficking" vesicles will be determined by the extent to which the homogenization procedures disrupt the networks of the Golgi. Sometimes these will be found in the post-6×10^3 g-minute supernatant, but more often they are detected primarily in the post-15×10^4 supernatant. The ribosomes may also be pelleted by subjecting the supernatant to sedimentation at 100,000g at 4°C in an ultracentrifuge for a minimum of 3 h.
13. In normal procedures, aging studies to determine the lability of the enzyme markers to be studied should be carried out. Freezing may damage some particles resulting in a loss of activity or of a constituent. The interference of components of the media used in the fractionation process with the assay of the markers should also be thoroughly assessed before proceeding with the analysis and interpretation.
14. After fractionation of the tissue each fraction should be assayed, in addition to the original homogenate, for selected markers for the purpose of following each constituent throughout the fractionation procedure. de Duve *(2,8)* laid down two criteria for the selection of an enzyme as a marker which can also be applied to a chemical constituent:

a. The enzyme must have a specific location, that is, be present only in one type of particle; and
b. That all subunits of a given population have the same enzyme content as related to their mass or total protein.

The second criterion is not absolutely essential for a marker. Known markers for specific organelles are used to appreciate the efficiency of the fractionation procedure and thereby the quality of the homogenate. If two markers show the same distribution pattern it is an indication, but not proof, that they may be associated with the same particle. By a comparison of the specific activity of the markers in the different fractions, an estimation of the contamination or degree of purity of the particles may be obtained.

The markers selected are usually biochemical (chemical constituents, enzyme activities, immunological), but morphological and sedimentation coefficient analysis can also prove very informative. A list of some of the classical enzymic markers normally employed in fractionation studies is provided in ref. 7. The amount of biochemical constituents, such as DNA, RNA, cholesterol, total protein, and lipids should be assayed and calculated in each of the fractions. The absolute activities and specific activities of each of the relevant enzymes should be assayed. The concentration of the components or enzyme activities should be expressed as a percentage of the total constituent or enzyme activity found in the homogenate, that is, the percent recovery of each marker is calculated for every fractionation experiment carried out. Quantitative recovery of each marker is important and before any interpretation of the data is made, quantitative recovery of any enzyme activity/constituent must be established. The results of the analysis are presented as distribution patterns or as frequency histograms if possible.

It is essential to prepare a balance sheet of the constituents and enzyme activities in the differing fractions compared with those in the original homogenate. This is achieved by:
a. Retaining a portion of the original homogenate;
b. Recording accurately the volume of the homogenate used for centrifugation and also the volumes of buffered medium in which the particles are resuspended; and
c. Carrying out all analyses on the homogenate as well as on the prepared fraction. Concentration of a constituent or enzyme activity should be expressed as a percentage of that which was present in the original homogenate.

A particular enzyme activity may be recovered in the fractions at values considerably less than or more than 100%. In the former case this may be caused by the handling of the materials or the removal of an influencing factor (i.e., an activator or a cofactor) which may have been separated into

another fraction. Recoveries in excess of 100% would tend to indicate the removal of some inhibiting substance.
15. The subcellular fractionation that has been described here applies principally to soft tissues, especially liver tissue. However, the properties described do apply to most animal cells, although different methods of homogenization, in particular, may be necessary to effect the desired subcellular fractionation. What has been described is the crude preparation of crude subcellular fractions which are partially purified and which may form the basis for purification of a specific membrane fraction or a particular protein by either techniques, such as density gradient centrifugation or other protein purification procedures as described in Chapters 14–30.

References

1. Rickwood, D. (ed.) (1992) *Preparative Centrifugation: A Practical Approach,* IRL, Oxford, UK.
2. de Duve, C. (1967) General principles, in *Enzyme Cytology* (Roodyn, D. B., ed.), Academic, London, pp. 1–26.
3. Birnie, G. D. and Rickwood, D. (eds.) (1978) *Centrifugal Separations in Molecular and Cell Biology,* Butterworths, London.
4. de Duve, C. and Beaufay, H. (1981) A short history of tissue fractionation. *J. Cell Biol.* **91,** 293s–299s.
5. Schachman, H. K. (1959) *Ultracentrifugation in Biochemistry,* Academic, London.
6. Svedberg, T. and Pederson, K. O. (1940) *The Ultracentrifuge,* Clarendon, Oxford (Johnson Reprint Corporation, New York).
7. Evans, W. H. (1992) Isolation and characterisation of membranes and cell organelles, in *Preparative Centrifugation: A Practical Approach* (Rickwood, D., ed.), IRL, Oxford, UK, pp. 233–270.
8. de Duve, C. (1971) Tissue fractionation—past and present. *J. Cell Biol.* **50,** 20d–55d.

CHAPTER 7

Subcellular Fractionation of Plant Tissues

Isolation of Chloroplasts and Mitochondria from Leaves

Alyson K. Tobin

1. Introduction

The successful isolation of intact, functional organelles from plant tissue is fraught with difficulties. Leaves are often covered in waxy cuticles, and frequently contain silica (as in grasses) and toxic components, such as phenolics, proteolytic enzymes, and high concentrations of acids and salts in the vacuole. In addition, all higher plants have one major barrier in common, the presence of a rigid, cellulose cell wall, which has to be broken in order to release the organelles. Mechanical isolation methods, i.e., where leaf material is macerated in a mechanical homogenizer, are likely to succeed only for a limited number of species, e.g., pea and spinach, where the leaves do not contain large amounts of tough, thickened tissue. Otherwise, the prolonged homogenization required to release significant numbers of organelles from tough leaves, e.g., of grasses, such as wheat or barley, results in most of them being broken and inactive. For this reason, the only viable method for obtaining good-quality organelles from species such as these is to isolate them from protoplasts. Although chloroplasts can be successfully isolated from protoplasts, the yield of mitochondria is so small that, unless very large-scale protoplast preparations are employed, the technique is generally inappropriate for mitochondrial isolation. This means that there are many plant species from whose leaves it has proven impossible to isolate good-quality mitochondria using existing techniques.

From: *Methods in Molecular Biology, Vol. 59. Protein Purification Protocols*
Edited by: S. Doonan Humana Press Inc., Totowa, NJ

A major factor in the success, or otherwise, of all of these methods is the quality of the plant material. Even the best technician cannot make good chloroplasts or mitochondria from poor-quality plants, and it is essential both to optimize and standardize the growing conditions if reproducible results are to be obtained.

Finally, isolated organelles are easily damaged and must be handled gently. Detergents and volatile solvents are extremely harmful, since they disrupt the lipid-rich membranes. Detergents should never be used to wash any apparatus used for organelle isolation. It is also important to work well away from anyone using volatile solvents (particularly phenol) in the laboratory, since even the vapor can be disruptive to organelles.

Clearly, there is no universally applicable technique, and it is necessary to optimize the conditions for each species. In this chapter, two different methods for isolating chloroplasts are described: from protoplasts and using a mechanical method. It is advisable always to try the mechanical method first, because this is much quicker and simpler, and to use protoplasts only if results are unsatisfactory. Two methods of protoplast and chloroplast isolation are described to illustrate the different requirements of different species, in this case wheat (*Triticum aestivum* L.) and barley (*Hordeum vulgare* L.).

2. Materials

Media 1–15 will keep for a maximum of 2 wk at 4°C providing that BSA, sodium pyrophosphate, $MnSO_4$, and Percoll are omitted, and added to the stocks, where required, immediately before use.

2.1. Solutions for Wheat Protoplast and Chloroplast Preparation

1. Medium 1: $0.5M$ sucrose, 5 mM MES, pH 6.0, 1 mM $CaCl_2$.
2. Medium 2: $0.4M$ sucrose, $0.1M$ sorbitol, 5 mM MES, pH 6.0, 1 mM $CaCl_2$.
3. Medium 3: $0.5M$ sorbitol, 5 mM MES, pH 6.0, 1 mM $CaCl_2$.
4. Medium 4: $0.4M$ sorbitol, 10 mM EDTA, 25 mM Tricine, pH 8.4.
5. Wheat digestion medium: 0.6 g Cellulysin (Sigma, St. Louis, MO), 6 mg Pectolyase (Sigma) in 30 mL of medium 3.

Wheat and barley digestion media are made immediately before use by adding cellulysin, pectolyase, and cellulase, as appropriate, to the surface of the medium. This is left to stand, without stirring, until the dry powders have been fully absorbed into the medium, preventing the formation of lumps of dry powder that are difficult to disperse.

2.2. Solutions for Barley Protoplast and Chloroplast Preparation

1. Medium 5: 0.4M sorbitol, 10 mM MES, pH 5.5, 1 mM CaCl$_2$, 1 mM MgSO$_4$.
2. Barley digestion medium: 1.5% (w/v) cellulase (Onozuka R10) in medium 5.
3. Medium 6: 35% (v/v) Percoll in medium 8.
4. Medium 7: 25% (v/v) Percoll in medium 8.
5. Medium 8: 0.4M sorbitol, 25 mM Tricine, pH 7.2, 1 mM CaCl$_2$, 1 mM MgSO$_4$.
6. Medium 9: 0.33M sorbitol, 50 mM Tricine, pH 7.8, 2 mM EDTA, 1 mM MgSO$_4$, 1 mM MnSO$_4$.

2.3. Solutions for Pea Chloroplast Preparation

1. Medium 10: 0.33M sorbitol, 50 mM Tricine, pH 7.9, 2 mM EDTA, 1 mM MgCl$_2$, 0.1% (w/v) BSA. BSA is added to solutions in the same way as described in Section 2.1. for addition of enzymes to digestion media.
2. Medium 11: 40% (v/v) Percoll, 0.33M sorbitol, 50 mM Tricine, pH 7.9, 0.1% (w/v) BSA.

2.4. Solutions for Chloroplast Assays

1. Medium 12: 0.33M sorbitol, 20 mM HEPES, pH 7.6, 10 mM EDTA, 0.2 mM KH$_2$PO$_4$, 30 mM MgCl$_2$.
2. Medium 13: double-strength medium 12.
3. Medium 14: 0.33M sorbitol, 50 mM Tricine, pH 8.2, 2 mM EDTA, 1 mM MgSO$_4$, 1 mM MnSO$_4$.
4. Medium 15: 0.33M sorbitol, 50 mM Tricine, pH 8.2, 10 mM KCl, 5 mM sodium pyrophosphate, 2 mM EDTA, 2 mM ATP.
5. 0.1M K$_3$Fe(CN)$_6$. Make in water, and store in the dark at $-20°C$; protect from bright light.
6. 1.0M NH$_4$Cl.
7. 1.0M KHCO$_3$ freshly made in the assay buffers.
8. 1.0M NaHCO$_3$ freshly made in the assay buffers.

2.5. Solutions for Mitochondrial Isolation

The mitochondrial isolation medium and wash medium will both keep at 4°C for up to 2 wk if stored in the absence of PVP, BSA, and cysteine which should be added immediately before use, where required. Solution I (minus BSA) will keep for 2 wk providing BSA is added fresh on the day, and solutions II and III should be freshly made.

1. Mitochondrial isolation medium: 0.3M sucrose, 50 mM MOPS-KOH, pH 7.5, 2 mM Na$_2$EDTA, 1 mM MgCl$_2$, 1% (w/v) PVP 40 (soluble), 0.4% (w/v) BSA, 4 mM cysteine.

2. Mitochondrial wash medium: $0.3 M$ sucrose, 20 mM MOPS-KOH, pH 7.5, 0.1% (w/v) BSA.
3. Solution I: $0.6 M$ sucrose, 20 mM KH_2PO_4, 0.2% (w/v) BSA, pH 7.5.
4. Solution II: 20 mL solution I, 11.2 mL Percoll, 8.8 mL of 40% (w/v) PVP 40.
5. Solution III: 20 mL solution I, 11.2 mL Percoll, 8.8 mL H_2O.

3. Methods

All procedures are carried out at 4°C, unless otherwise stated. All solutions and apparatus should be prechilled when working at this temperature.

3.1. Isolation of Chloroplasts from Protoplasts

3.1.1. Wheat Protoplast Preparation

1. Take primary leaves from 7–10-d-old plants (*see* Note 1). Finely chop the leaves into thin sections, approx 1–2 mm thick. Use sharp, single-sided razor blades. Spread the leaf sections onto the surface of 30 mL of wheat digestion medium (*see* Note 2) in a shallow dish that provides a large surface area of medium (a crystallizing dish is suitable). Add sufficient leaf material to cover the surface of the medium completely. Incubate for 3 h at 28°C. Do not shake or stir the medium during this period.
2. Following digestion, carefully remove the digestion medium from the dish, leaving the leaf sections in place. Add 20 mL of medium 3 to the leaf sections, and gently swirl to release the protoplasts into the medium. Filter through coarse nylon mesh, e.g., a plastic tea strainer, and retain the filtrate. Return the leaf sections to the dish, add a further 20 mL of medium 3, swirl again, filter, and combine the two filtrates.
3. Centrifuge the filtrates in a swing-out rotor at 150g for 5 min (*see* Note 4). Discard the supernatant, and gently resuspend the pellets in a total volume of 20 mL of medium 1. Divide the suspension equally between four tubes and overlay 2 mL of medium 2, followed by 1 mL of medium 3 onto each suspension to form a discontinuous gradient. Centrifuge at 250g for 5 min in a swing-out rotor.
4. Intact protoplasts collect at the interface between medium 2 and 3. Using a wide-bore Pasteur pipet, carefully remove the layer of protoplasts, transfer to a clean centrifuge tube, and add 2 vol of medium 3. Centrifuge at 150g for 5 min in a swing-out rotor. Resuspend the pellet in medium 4 to give a chlorophyll concentration of approx 1 mg/mL.

3.1.2. Wheat Chloroplast Preparation

1. Resuspend the protoplasts (prepared as in Section 3.1.1.), in 5 mL of medium 4. Pour the suspension into a modified 25-mL disposable plastic syringe (*see* Note 3).

Subcellular Fractionation

2. Break the protoplasts by passing the suspension through 20-μm pore size nylon mesh, and immediately centrifuge at 250g for 1–2 min in a swing-out rotor.
3. Gently resuspend the pellet in medium 4 to give a chlorophyll concentration of approx 1 mg/mL.

3.1.3. Barley Protoplast Preparation

1. Take primary leaves from 7–10-d-old barley plants (*see* Note 1). Carefully remove the lower epidermis, and lay the leaves, lower side down, onto the surface of 50 mL of barley digestion medium (*see* Note 2) in a clear plastic sandwich box. Cover the box with cling film, and incubate at 30°C for 2 h without disturbance.
2. Following digestion, harvest the protoplasts by gently swirling the leaves within the digestion medium. Decant the suspension through nylon mesh, e.g., a plastic tea strainer, into 50-mL centrifuge tubes.
3. Centrifuge at 250g for 5 min in a swing-out rotor, discard the supernatants, and resuspend the pellet in 10 mL of medium 6.
4. Divide the suspension equally between 2 × 20-mL centrifuge tubes. Overlayer each with 2 mL of medium 7 followed by 1 mL of medium 8 to form a discontinuous gradient.
5. Leave the gradients to stand on ice for 1 h, and then centrifuge at 300g for 5 min in a swing-out rotor.
6. Intact protoplasts collect at the interface between media 7 and 8. Carefully remove the protoplasts using a wide-bore Pasteur pipet, transfer to a clean 20-mL centrifuge tube, and add 2 vol of medium 9.
7. Centrifuge at 150g for 5 min in a swing-out rotor, discard the supernatant, and gently resuspend the pellet in medium 9 to give a chlorophyll concentration of approx 1 mg/mL.

3.1.4. Barley Chloroplast Preparation

1. After removing the protoplasts from the gradient (as in step 6), dilute (approx fivefold) in medium 9, and transfer to a modified 25-mL disposable plastic syringe (*see* Note 3).
2. Break the protoplasts by passing the suspension through 20-μm pore size nylon mesh, and immediately centrifuge at 150g for 2 min in a swing-out rotor.
3. Gently resuspend the pellet in medium 9 to give a chlorophyll concentration of approx 1 mg/mL (*see* Note 5).

3.2. Mechanical Isolation of Chloroplasts from Pea Leaves

1. Take 10 g (fresh weight) of pea leaves, add to 40 mL of ice-cold medium 10, and cut into small pieces using razor blades or sharp scissors.

2. Homogenize (5 s full speed.), using an Ultraturrax or Polytron homogenizer. Filter the homogenate through four layers of muslin (prewetted with medium 10).
3. Divide the filtrate equally between two 50-mL centrifuge tubes. Underlay each portion of filtrate with 10 mL of medium 11.
4. Centrifuge at 2500g for 1 min in a swing-out rotor.
5. The chloroplasts form a soft, green pellet at the bottom of the centrifuge tube. Carefully remove the supernatant layers by aspiration without disturbing the pellet. Gently resuspend the pellet (e.g., using a fine paintbrush) in 1.0 mL of medium 10.

3.3. Assay of Chloroplast Intactness

Chloroplast intactness (see Note 6) is assayed according to the method of Lilley et al. (1). The rate of ferricyanide-dependent oxygen evolution is measured using an oxygen electrode (Hansatech, Norfolk, UK).

1. The assay medium, final volume 1.0 mL, is added to an oxygen electrode and consists of medium 12 and 2 mM K$_3$Fe(CN)$_6$ (final concentration, added from a 0.1M stock solution). Intact chloroplasts are added (final concentration of approx 50 µg chlorophyll/mL assay), and the assay is illuminated (PAR of 1000 µmol/m^2/s). Following the addition of 5 mM NH$_4$Cl (final concentration), the rate of O$_2$ evolution is measured, and this gives the "intact rate" of ferricyanide-dependent O$_2$ evolution.
2. To determine the "broken rate," chloroplasts (the same amount as used in the intact assay) are added to 0.5 mL of H$_2$O in the oxygen electrode. After 2 min, 0.5 mL of medium 13 is added, followed by 2 mM K$_3$Fe(CN)$_6$ (as in step 1). Following illumination and the addition of NH$_4$Cl (as in step 1), the rate of O$_2$ evolution is measured.
3. Chloroplast intactness is calculated as:

$$[(\text{Broken rate} - \text{intact rate})/\text{broken rate}] \times 100 = \% \text{ intactness} \qquad (1)$$

3.4. Assay of Photosynthetic Activity (CO_2-Dependent Oxygen Evolution) of Chloroplasts (see Notes 6–8 and 10)

3.4.1. Wheat

1. The assay medium, final volume 1.0 mL, is added to an oxygen electrode (Hansatech, Norfolk, UK), and consists of medium 4, 10 mM KHCO$_3$ (final concentration), 0.15 mM KH$_2$PO$_4$ (final concentration), and chloroplasts (approx 50 µg chlorophyll/mL assay).
2. The rate of oxygen evolution is measured under illumination (minimum PAR of 1000 µmol/m^2/s) at 20°C.

3.4.2. Barley

1. The assay medium, final volume 1.0 mL, is added to an oxygen electrode (Hansatech), and consists of medium 14, 20 mM KHCO$_3$ (final concentration, added from a 1.0M stock), 0.2 mM KH$_2$PO$_4$ (final concentration), and chloroplasts (approx 50 µg chlorophyll/mL assay).
2. Oxygen evolution is measured as for wheat chloroplasts.

3.4.3. Pea (see Note 9)

1. The assay medium, final volume 1.0 mL, is added to an oxygen electrode (Hansatech), and consists of medium 15, 10 mM NaHCO$_3$ (added from a 1.0M stock), and chloroplasts (approx 50 µg/assay).
2. Oxygen evolution is measured as for wheat chloroplasts.

3.5. Assay of Chloroplast Purity

Marker enzymes are assayed in aliquots of the original, crude homogenate, and in the final chloroplast preparation to determine contamination by cytosol, peroxisomes, and mitochondria, the main contaminants of concern.

3.5.1. Cytosolic Contamination

Phosphoenol pyruvate (PEP) carboxylase, exclusive to the cytosol, is assayed as follows: 50 mM HEPES, pH 8.0, 5 mM MgCl$_2$, 0.08 mM NADH, 1 U malate dehydrogenase, and 10–20 µL protoplasts, chloroplasts, or crude extract are added to a cuvet, final volume 1.0 mL. The assay is carried out at 25°C, and the reaction is started with the addition of 2 mM PEP and measured as the decrease in absorbance at 340 nm owing to the oxidation of NADH. The method is modified from that of Foster et al. (2).

3.5.2. Mitochondrial Contamination

Citrate synthase, exclusive to the mitochondria, is assayed as follows: 50 mM Tricine, pH 8.0, 1.0 mM MgCl$_2$, 1.0 mM oxaloacetate (OAA), 0.2 mM acetyl Coenzyme A, 1.5 mM 5,5'-dithio-bis-(2-nitrobenzoic acid; DTNB) and approx 50 µL extract are added to a cuvet, final volume 1.0 mL. The assay is carried out at 25°C, and the reaction is started with the addition of OAA and measured as the increase in absorbance at 412 nm owing to the formation of the 2-nitro-5-thiobenzoate anion. The assay is based on the method of Cooper and Beevers (3).

3.5.3. Peroxisomal Contamination

Glycolate oxidase, exclusive to the peroxisome in leaves (4), is assayed as follows: 50 mM MOPS, pH 7.8, 3 mM EDTA, 0.008% Triton X-100,

0.66 mM reduced glutathione, 0.2 mM flavin mononucleotide (FMN), 0.033% (v/v) phenylhydrazine, and 50 µL extract are added to a cuvet in a final volume of 3.0 mL. The assay is carried out at 25°C, and the reaction is started with the addition of 5 mM sodium glycolate and measured as the increase in absorbance at 324 nm due to of the formation of glyoxylate phenylhydrazine. The assay is based on the method of Behrends et al. *(5)*.

3.6. Isolation of Mitochondria from Pea Leaves (see Note 11)

1. Use sharp razor blades to remove fully expanded leaves by cutting through the petiole. All subsequent steps are carried out at 4°C.
2. Add 600 mL of mitochondrial isolation medium/100 g (fresh weight) of leaves.
3. Cut the leaves into approx 1-cm^2 pieces, using razor blades or sharp scissors.
4. Homogenize (in 150-mL aliquots) using an Ultraturrax homogenizer (Orme Scientific, Manchester, UK) using 2 × 5 s bursts at full speed (mark 10).
5. Filter through four layers of muslin, prewetted with chilled isolation medium.
6. Centrifuge (*see* Note 12) at 2960g for 5 min in a fixed-angle rotor in 250-mL centrifuge tubes, and discard the pellets.
7. Centrifuge the supernatant at 17,700g for 15 min in a fixed-angle rotor in 250-mL centrifuge tubes. Gently, using a fine paint brush, resuspend the pellet in a small volume of wash medium, and dilute to a final volume of 30 mL in the same medium.
8. Centrifuge at 1940g for 10 min in a fixed-angle rotor in 50-mL centrifuge tubes. Discard the pellet. Spin the supernatant at 12,100g for 10 min in a fixed-angle rotor. Gently resuspend the pellet (as before) in a small volume (~1 mL) of wash medium.
9. Load this onto a continuous Percoll/PVP gradient made up as follows: Add 17 mL of solution III to the mixing chamber (i.e., nearest the outlet) of a gradient former and 17 mL of solution II to the left-hand chamber. Form the gradient into a 50-mL centrifuge tube. Centrifuge at 39,200g for 40 min in a fixed-angle rotor. Intact mitochondria form a straw-colored band near to the bottom of the gradient (*see* Note 13).
10. Remove the top half of the gradient, which contains broken chloroplasts (green color), and discard. Using a wide bore Pasteur pipet, carefully remove the mitochondrial fraction without disturbing the rest of the gradient.
11. Add approx 30 mL of wash medium to the mitochondria, and centrifuge at 12,100g for 10 min in a fixed-angle rotor in 50-mL centrifuge tubes.
12. Carefully remove the supernatant by aspiration, taking care not to disturb the soft mitochondrial pellet. Repeat step 11.

13. Carefully remove the supernatant as before, and gently resuspend the purified mitochondria in a minimal volume of wash medium.

3.7. Assay of Mitochondrial Activity

3.7.1. Measurement of Respiratory Control Ratios and P/O Ratios

The assay is carried out in an oxygen electrode (Rank Bros., Cambridge, UK; Hansatech) at 25°C. The reaction mixture consists of $0.3M$ sucrose, 10 mM MOPS, pH 7.2, 10 mM KH_2PO_4, 2 mM $MgCl_2$, 0.1% (w/v) BSA (defatted), mitochondria (0.5–1.0 mg protein/mL assay), and 10 mM substrate (e.g., malate, glycine) in a final volume of 1.0 mL. Oxygen uptake is measured continuously. The state 3 rate of respiration is the rate of oxygen uptake in the presence of ADP. This is determined by adding 100 µM ADP to the above assay and measuring oxygen uptake. The state 4 rate of oxygen uptake is the rate following state 3 when all of the ADP has been converted to ATP (i.e., state 4 is the "ADP-limited" rate). The Respiratory Control Ratio is the ratio of state 3 to state 4 rates (Fig. 1) (*see* Note 14). The ADP/O ratio is calculated from the amount of oxygen consumed during the phosphorylation of a known amount of added ADP. The ADP/O differs for different substrates. For NAD-linked substrates, the ADP/O ratio is 3.0, and for NADH or succinate it is 2.0.

3.7.2. Assay of Mitochondrial Purity and Intactness

To determine mitochondrial purity, the same marker enzymes are assayed as for chloroplasts. Although there are valid assays for mitochondrial intactness, the preferred method of determining mitochondrial quality is to measure the Respiratory Control Ratio (RCR) and ADP/O ratios. For purified leaf mitochondria, a RCR in excess of 2.0 is acceptable, whereas ADP/O ratios should approach theoretical maximum values (2.0 for NADH or succinate, 3.0 for glycine, malate, and so forth).

4. Notes

1. The age of the plant is an important factor in determining the yield of protoplasts; the yield of intact protoplasts decreases rapidly beyond the age ranges presented here. It may be necessary to vary the concentration of digestive enzymes and/or the digestion time if older or younger material is to be used.
2. Different species may well require different digestive enzymes and different periods of digestion, and it is important to optimize these conditions. A simple method is to carry out small-scale digestions with 5 mL of digestion medium. Approximate protoplast yields may be assessed by measur-

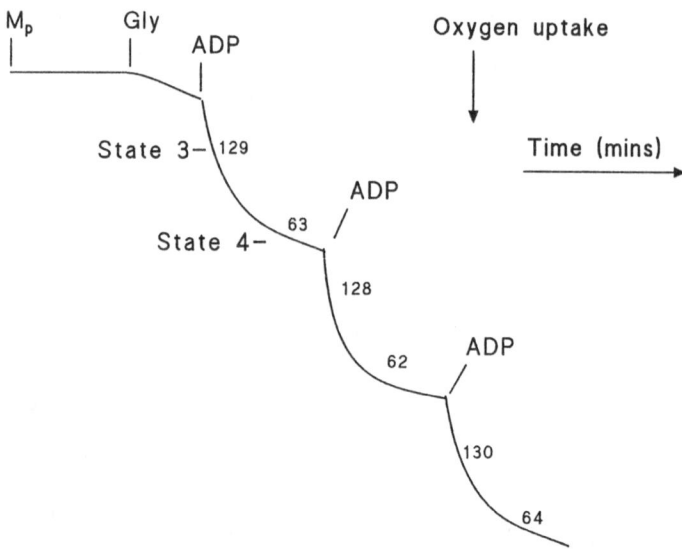

Fig. 1. Oxygen consumption by pea leaf mitochondria. The assay was carried out in an oxygen electrode as described in the text. The following additions were made: purified mitochondria (M_p, 0.4 mg protein), 10 mM glycine (gly), and 100 μmol ADP (where indicated.). State 3 refers to the rate of oxygen uptake in the presence of saturating concentrations of ADP. State 4 occurs when ADP has been depleted. The numbers on the traces refer to the rate of oxygen uptake in nmol/min/mg mitochondrial protein. The RCR is the state 3 rate divided by the state 4 rate. The ADP/O ratio is the amount of oxygen consumed (in natoms) during the phosphorylation of a given amount of ADP (in nmol).

ing chlorophyll content of the medium after the incubation period and by visual inspection of the extract under a microscope. The chlorophyll content will increase with increasing periods of digestion, but eventually the number of intact protoplasts (readily seen as completely circular in outline under the light microscope) will decrease with prolonged digestion. Since the cellulase and cellulysin preparations are not generally very pure, it is advisable to use the minimum amount during digestion.

3. To break the protoplasts, use a 25-mL disposable syringe adapted as follows: cut off the end section to which the syringe needle is usually attached, so that the barrel becomes an open cylinder. Cover the end with a piece of nylon mesh, pore size 20 μm. Remove the syringe plunger, and pour the protoplast suspension into the open end of the syringe barrel. Carefully replace the plunger, and displace it slowly so that the protoplasts are forced through the mesh into a collecting beaker on ice. This is usually sufficient

to completely break the protoplasts. Check by inspecting an aliquot under a light microscope. Broken protoplasts will no longer be circular in outline. If intact protoplasts remain, repeat the passage through the mesh. This procedure has to be carried out with care, since rough handling at this stage will result in chloroplast breakage.
4. All centrifugations for protoplast and chloroplast isolation are carried out with the brakes off.
5. The most serious problem during chloroplast isolation is "clumping," where the organelles aggregate into sticky lumps. Whenever this occurs, the quality of the chloroplasts is poor. One contributory factor is the age of the plant material; the older the leaves, the more likely clumping will occur. I have tried varying the EDTA and Mg ion concentration in the resuspension medium, as suggested by other authors, yet this does not always solve the problem. There also appears to be a difference between cultivars. The barley variety Klaxon is particularly problematic, whereas Maris Mink is more reliable. The peeling of the epidermis prior to digestion was found to reduce the problem of chloroplast clumping in barley significantly, and it also increased the yield of intact protoplasts.
6. Chloroplasts isolated from protoplasts should be at least 90% intact and capable of CO_2-dependent O_2 evolution at rates comparable to those of the isolated protoplasts. For wheat and barley, depending on the growing conditions, rates of at least 40 μmol O_2 evolved/h/mg chlorophyll should be achieved. Rates for pea will be higher than this, at least 50 μmol O_2 evolved/h/mg chlorophyll. The rate of CO_2-dependent O_2 evolution is a better determination of chloroplast quality than is the intactness assay, since it is possible for chloroplasts to break, release their stromal contents, and then reseal. These chloroplasts would appear to be intact, according to the ferricyanide assay, and yet would be incapable of photosynthesis.
7. The conditions required for chloroplasts to carry out CO_2-dependent O_2 evolution are often quite critical, hence the difference in assay media for the different species. The pH, divalent cation concentration, and phosphate concentration should all be optimized for the species being studied. ATP, ADP, and pyrophosphate may also be required to varying extents.
8. The phenomenon of "induction" is frequently observed in isolated chloroplasts, where a significant lag occurs between the start of illumination and the onset of CO_2-dependent O_2 evolution. This lag may last for several min and is owing to the depletion of Calvin cycle intermediates from the chloroplast. Preillumination of the leaves with bright light for 30 min prior to extraction will reduce the induction time if this is thought to be a problem.
9. A high stromal starch concentration may result in poor-quality chloroplasts when using mechanical isolation, since starch grains are thought to disrupt

the chloroplast envelope physically during homogenization. To minimize this effect, leaves should be harvested close to the end of the dark period, when starch content will be at a minimum.
10. Isolated protoplasts are stable for many hours, whereas chloroplasts will only remain active for a short period of time. Barley and wheat chloroplasts should be used as soon as possible after isolation and will rarely remain viable for more than 2 h. Pea chloroplasts will generally last longer, but it is always good practice to carry out experiments on isolated organelles as soon as possible after isolation. If a large number of experiments is planned, it is possible to make one large protoplast preparation in the morning and to break aliquots at intervals during the day to provide freshly prepared chloroplasts. The remaining protoplasts may be kept on ice until required. It is good practice to determine the rate of CO_2-dependent O_2 evolution at the beginning and end of the series of experiments to determine whether deterioration has occurred.
11. The method described for the isolation of mitochondria from pea leaves will also work for spinach.
12. All centrifugations for mitochondrial isolation are carried out with the brake on.
13. A major source of contamination of the mitochondria is from broken chloroplasts. Ideally the mitochondrial band that forms on the density gradient should be straw colored and not green. A high chlorophyll content is the result of thylakoid contamination, often caused by too harsh homogenization or overloading of the gradient with too much extract.
14. Isolated leaf mitochondria will only remain active for approx 2 h. It is good practice to determine the RCR at the start and end of the series of experiments to ascertain whether there has been significant deterioration.

References

1. Lilley, R. McC., Fitzgerald, M. P., Rienits, K. G., and Walker, D. A. (1975) Criteria of intactness and the photosynthetic activity of spinach chloroplast preparations. *New Phytol.* **75,** 1–10.
2. Foster, J. G., Edwards, G. E., and Winter, K. (1982) Changes in levels of Phosphoenol pyruvate carboxylase with induction of Crassulacean acid metabolism in *Mesembryanthemum crystallinum* L. *Plant Cell Physiol.* **23,** 585–594.
3. Cooper, T. G. and Beevers, H. (1969) Mitochondria and glyoxysomes from castor bean endosperm. Enzyme constituents and catalytic capacity. *J. Biol. Chem.* **244,** 3507–3513.
4. Tolbert, N. E. (1981) Microbodies-peroxisomes and glyoxysomes. *Annu. Rev. Plant Physiol.* **26,** 45–73.
5. Behrends, W., Rausch, U., Loffler, H. G., and Kindl, H. (1982) Purification of glycolate oxidase from greening cucumber cotyledons. *Planta* **156,** 566–571.

CHAPTER 8

The Extraction of Enzymes from Plant Tissues Rich in Phenolic Compounds

William S. Pierpoint

1. Introduction
1.1. The Problems

Many enzyme extracts must, of necessity, be made from green leaves, fruits, and other vegetable tissues that contain large amounts of phenols and polyphenols, which hinder or, unless precautions are taken, completely prevent a successful extraction. In spite of this, the processes by which the polyphenols interfere are rarely studied and still very incompletely understood. Some principles are clear, mostly from work done in past decades, and are applied on an *ad hoc* basis to current problems. They can usually be adapted to devise a successful if rather complicated procedure, but there is seldom the time or interest for researchers to establish what the problems really were and if they could have been better overcome. This is unfortunate, but wholly understandable. The reactions involved are often very complex, demand specialist investigations, and may be only relevant to extraction from a particular plant species.

Part of the difficulty is caused by the vast range of phenolic compounds that may be involved. Harborne *(1)* estimated that several thousand structures are known, and summarized the vagaries of their distribution. Simple phenols, such as catechol, are comparatively rare or are present only in traces. Phenolic acids, such as p-OH benzoic and syringic acid are almost ubiquitous, as are the phenyl-propanoids (C_6–C_3–), including caffeic acid and its quinic acid-ester, chlorogenic acid. Thousands of flavonoid compounds are known, differing in hydroxylation pattern, oxidation state of the heterocyclic ring, and degree of glycosylation: some of them are restricted to particular species and particular tissues, whereas others, such as quercetin and cyanidin, are widespread. Perhaps the most notorious phenolic compounds in the present context are the polymeric, astringent tannins, whose protein binding properties have been appreciated in food technology and the leather industry for centuries. Characteristic of some "advanced" orders of cotyledonous plants are the "hydrolyzable" tannins, based on gallic acid residues linked, often as esterified chains, to glucose or some other polyhydric alcohol. More widespread are the "condensed" tannins, oligomers of the flavonoid catechin linked by C_4–C_8 interflavan bonds, which occur in most orders of the vascular plants. These polymers have many of the features of monomeric phenols that predispose them to combine with proteins, but they have more of them and often in a dispostion that facilitates and strengthens this reaction.

The initial binding of phenolic compounds, both monomers and polymers, to enzymes and other proteins is via noncovalent forces, which may, initially, be reversible. It is believed that hydrophobic, ionic, and H-bonds may be involved depending on the specific phenol and protein involved. Fully methoxylated phenolics with a high content of aromatic rings are, of course, more likely to be bound hydrophobically. Free phenolic groups, especially vicinal dihydroxy groups, may form hydrogen bonds with, for instance, the CO and NH of peptide bonds. Such complexes have only been investigated thoroughly in a few simple model cases *(2,3)*, and shown to involve a variety of these linkages and also coordination attachments involving cations. Nevertheless, Haslam and his colleagues have extrapolated from such information to produce a model for the binding of polymeric tannins to proteins. It involves two stages. In the first, the tannin is attached to hydrophobic sites on the surface via its aromatic residues. This bonding is then reinforced by the formation of H-bonds between the phenolic groups and nearby polar functions on the protein. The final product is thus a dissociable complex in which the surface of the protein is rendered less hydrophobic and more susceptible to aggregation and precipitation. If the polyphenol is large enough to interact with more than one protein molecule, the likelihood of aggregation and precipitation will, of course, be much increased. More recent developments of this work *(4)* have emphasized the structural features of proteins, including those of the proline-rich, salivary proteins of herbivores, that predispose them to such coupling with tannins: for most enzymes, this complexing facilitates inactivation.

Phenolic compounds form irreversible covalent linkages with proteins mainly as a consequence of the oxidation of their vicinal dihydroxy groups to quinones or semiquinones. These oxidations may occur nonenzymically in alkaline conditions, especially in the presence of metal ions. They are, however, catalyzed by a variety of enzymes, including o-diphenol oxidases (polyphenoloxidases; PPO), monophenoloxidases (laccases), and by peroxidases in the presence of H_2O_2. The resulting quinone molecules are highly reactive, not only polymerizing with each other, but oxidizing other phenolic compounds and, most relevant in the present context, combining with reactive groups in proteins causing aggregation, crosslinking, and precipitation. These reactions are described as "enzymic-browning," for the products, although complex and poorly defined, are usually brown in color. Leaf extracts that brown rapidly are generally regarded *(5)* as poor sources of active enzymes. Insights on aspects of the browning reactions that affect proteins have been gained from studies of single proteins in simple, model oxidizing systems *(6)*. They emphasize the vulnerability of nucleophilic groups, such as the NH_2– and SH– groups in amino acid side chains to substitution reactions with quinone rings to give protein N- or S-substituted phenols. These may be re-oxidized by excess quinone to the o-quinone state, when they have the potential to react with other nucleophiles producing intra- or interprotein crosslinks, or, in more alkaline conditions, react with quinones giving more complex, greenish, protein N-substituted hydroxyquinone polymers. Other reactions may occur depending on the conditions and especially on the phenols being oxidized and the nature of the oxidizing system. It would take a major analytical effort to characterize satisfactorily the products formed when leaf proteins are exposed to the "natural" enzymic browning reactions of leaf extracts.

1.2. Some Solutions

In most cases, where it is required to extract enzymes from phenol-rich tissue, the specific phenolic compounds and oxidizing systems that are present are unknown or can only be guessed at. An ideal approach would be to establish their nature so that the simplest method of preventing interference could be established. Thus, Gray *(7)* described how the phenolic compounds of bean leaves could be simply extracted and a suitable adsorbent for them chosen. A more usual approach is to follow a general procedure that has worked for other tissues and deals with as many eventualities as possible. Generally, these procedures involve disrupting fresh or deeply frozen (–70°C) tissue as quickly as possible in the presence of polymers that adsorb phenolic compounds, both monomers and polymers, and in conditions that minimize the oxidative reactions which produce quinonoid compounds.

Many polymers, natural and synthetic, have been used to adsorb phenols. They include albumins, hide powders, powdered nylon (ultramid), polyvinylpyrrolidones both soluble (PVP) and insoluble (polyvinylpolypyrrolidone; PVPP), relatively uncharged polystyrene and polyacrylic resins, such as Amberlites XAD-2, -4, and -7, and ion-exchange resins based on polystyrenes, both anion exchangers (Bio-Rad AG1-X8, AG2-X8, and Dowex-1) and cation exchangers (Dowex-50). These polymers are listed in the reviews and papers by Loomis *(5),* Loomis et al. *(8),* Rhodes *(9),* and Smith and Montgomery *(10).* They interact with phenolic compounds in different ways so that PVP, for example, is thought to form stable H-bonds to phenol groups via its –CO–N< linkages, whereas the porous polystyrene resins present large, adsorptive, hydrophobic surfaces. As a consequence, their affinities for different phenols differ. The adsorbents may either be soluble, as are the tannin-binding proteins and PVP, or, more usually, insoluble like PVPP and the resins, so that they can be readily removed from the tissue extracts.

An obvious way of preventing the oxidative reactions is to work in anaerobic conditions, but this is both difficult and cumbersome. It is much easier to make extracts in the presence of low-mol-wt compounds that form unoxidizable complexes with phenols, or that either inhibit oxidases, trap quinones, or reduce quinones back to phenols *(5,6,11).* Borate and germanate have been used to complex phenols, and copper-chelating agents, such as diethyldithiocarbamate (Dieca), used to inhibit copper-dependent PPOs. Quinone trapping agents that have been used include benzene sulfinic acid, and a range of substances, including ascorbate, metabisulfite, and 2-mercaptoethanol used as quinone reductants. However, detailed studies (e.g., *12)* have made it clear that the action of many of these compounds cannot be simply explained, and that they may act in more than one manner. Thus, thioglycollate both inhibits oxidases and reduces quinones, and Dieca inhibits oxidases and reacts with quino-

Phenolic Compounds

nes; even PVP, primarily thought of as a phenol-adsorbent, is also an oxidase inhibitor.

The complexity of leaf extracts and of the reactions that may occur in them thus makes it difficult, if not impossible, to write a procedure suitable for extracting enzymes from all phenol-rich plant tissues. The method described here is the first few stages in the procedure used for extracting the photosynthetic enzyme ribulose bisphosphate carboxylase (Rubisco; EC 4.1.1.39) from green tissue *(13)*. It has been used routinely by Keys and his colleagues for many years and, with a few adaptions, applied successfully to a wide range of plants, including some ferns and mosses. More recently, with the modification mentioned later, it has been used to extract active Rubisco from the "difficult" leaves of Mediterranean tree species *(14)*. The subsequent purification steps specific to Rubisco have been omitted; the extracts, however, should be a suitable starting material for the purification of many other soluble enzymes by appropriate, specific procedures.

2. Materials

1. Use reagents of AR quality wherever possible, and otherwise of the highest standard available.
2. Ammonium sulfate: Use the grade (BDH-Merck, Lutterworth, Leicestershire, UK) especially low in heavy metals that is suitable for enzyme work.
3. PVPP (Polyclar AT; Sigma-Aldrich, Poole, Dorset, UK): free from metal ions and other contaminants by boiling in 10% HCl for 10 min and then washing extensively with glass-distilled water *(5)*. Air-dry for storage and rehydrate for at least 3 h before using: Hydration increases the weight of the polymer about fivefold *(8)*.
4. Extraction buffer: 100 mM HEPES/NaOH, 10 mM NaHCO$_3$, 10 mM MgCl$_2$, 0.1 mM EDTA, 10 mM 2-mercaptoethanol (2-Me), 10 mM sodium diethyldithiocarbamate (Dieca), 0.1% Tween 80. Prepare freshly overnight, adjust its pH to 7.0 (*see* Note 4) before the addition of NaHCO$_3$ and Dieca, and readjust it afterward if necessary. Add 2-Me just before use. NaHCO$_3$ is added specifically to protect Rubisco activity and is unnecessary in extracting other enzymes.
5. Resuspension buffer: Tris-HCl, pH 7.0, containing 1.0 mM dithiothreitol (DTT).

3. Methods

3.1. Extraction of Proteins *(see Note 1)*

1. Cut leaf material (about 100 g), without mincing, into small pieces, and immerse in 500 mL of cold extraction medium contained in the bowl of a chilled Ato-Mix (MSE) or similar blender (*see* Note 2).

2. Add insoluble PVPP (4–8 g), which has been soaked overnight in 250 mL of extraction medium, and homogenize the mixture at high speed for four periods of 15 s with a 30-s interval between the periods.
3. Filter the extract through four layers of muslin into a measuring cylinder, and allow to stand to settle any froth (*see* Note 3).
4. Treat the extract with a further addition of 4 g of PVPP, which has been soaked overnight in the extraction buffer, stir gently for 10 min, and clarify by centrifugation at 10,000*g* for 30 min to remove PVPP and cell residues. The resulting supernatant fluid provides a suitable starting material for purification procedures apppropriate for particular enzymes (*see* Notes 5 and 6).

For enzymes like Rubisco, which can be further purified and concentrated by ammonium sulfate precipitation, this second treatment with PVPP can conveniently be included in the precipitation schedule described in Section 3.2., and the PVPP removed along with a fraction of inactive protein. The ammonium sulfate concentrations are those found to be suitable for Rubisco; other enzymes may require different concentrations.

3.2. Ammonium Sulfate Fractionation of the Extract

1. After the second addition of PVPP, transfer the extract into a beaker and add ammonium sulfate (20 g/100 mL; 35% saturation) in portions with continuous stirring until it has dissolved.
2. Transfer the extract to centrifuge bottles, allow to stand for 30 min, and centrifuge at 25,000*g* for 15 min.
3. Discard the sediment containing PVPP and inactive protein. Filter the supernatant through eight layers of muslin, treat with a further 12 g/100 mL of ammonium sulfate, allow it to stand for 30 min, and centrifuge as before.
4. Resuspend this protein fraction, precipitated between 35 and 55% saturation, in 25 mL of 20 mM Tris-HCl, pH 7.0, containing 10 mM DTT, and treat for a third time with 3 g of PVPP that has been soaked in this resuspension buffer overnight.
5. After allowing it to stand for 10 min with occasional stirring, centrifuge the suspension at 100,000*g* for 30 min, and filter the supernatant liquid, containing active Rubisco, through a 50-μm mesh nylon gauze. This extract can be used for further stages of Rubisco's purification (*see* Note 7).

4. Notes

1. All operations were performed as far as possible at 5°C either in a cold room or on a cold bench. Apparatus and solutions were cooled overnight. Operations were performed as quickly as possible to minimize exposure of enzymes to phenolic compounds.

2. The use of relatively large amounts of extraction liquid, usually five to seven times the weight of the leaf material, ensures that cells are disrupted while they are submerged, that the concentration of liberated phenols and oxidative enzymes is immediately diluted, and that there is an adequate supply of phenol adsorbent and oxidase inhibitors. The volume of the medium should be such that the speed of blender blades, which is adequate to disrupt the leaves, does not produce undue vortexing and the sparging of air into the homogenate.
3. The original extraction medium contained 10 times as much Tween 80, but this often gave rise to excessive frothing.
4. The pH of the extraction medium will affect some interactions of phenols with proteins and adsorbents. Hydrophobic interactions will presumably be unaffected. Ionization of phenolic groups in alkaline conditions will prevent the formation of H-bonds with proteins, but it will also prevent the formation of H-bonds with PVPP. Alkaline conditions (pH >7.5) will increase the auto-oxidation of dihydroxy phenols to quinonoids, and also the auto-oxidation of protective thiol reagents. In the instances cited by Loomis *(5)*, enzyme extraction was generally optimal at pH values between 7.0 and 6.5. The optimal pH to extract an uncharacterized enzyme from an unfamiliar, phenol-containing plant, is a matter for experimentation.
5. Choice of polyphenol-adsorbents:
    a. Insoluble adsorbents: PVPP was chosen as it is an excellent adsorbent of preformed tannins as well as many monomeric phenols. Its affinity for different classes of compounds is illustrated in Fig. 1 taken from the experiments of Gray *(7)*. It is less effective in absorbing hydroxycinnamic acids, such as chlorogenic acid, that are common, widespread substrates for the browning reaction. In this respect, ion-exchange resins, such as the anion-exchanger Dowex-1 (Fig. 1) or the uncharged polystyrene Amberlites XAD-2 and XAD-4 (Rohm and Haas Co., Philadelphia, PA) are more effective. Dowex-1,X10 (Bio-Rad, Hercules, CA) was the preferred adsorbent used for the extraction of mevalonate kinase from the leaves of French beans *(Phaseolus vulgaris; 7)*; it was preequilibrated with the buffered (sodium phosphate; 50 mM, pH 7.5) extraction medium, and used at the rate of 20 g/100 mL of extraction medium. The Amberlite resins, especially XAD-4, appear to have been used more widely, presumably because of their great porosity and hydrophobic surface area and relative freedom from charged groups. They require extensive washing with water and methanol following manufacturer's instructions *(4)*, but this can be done in bulk so that they may be stored, ready for use, in a moist condition. They have been used in specific extraction procedures in quantities ranging from 0.1 g *(15)* to 1–2 g *(8)* hydrated weight/g of fresh tissue. It must be remembered

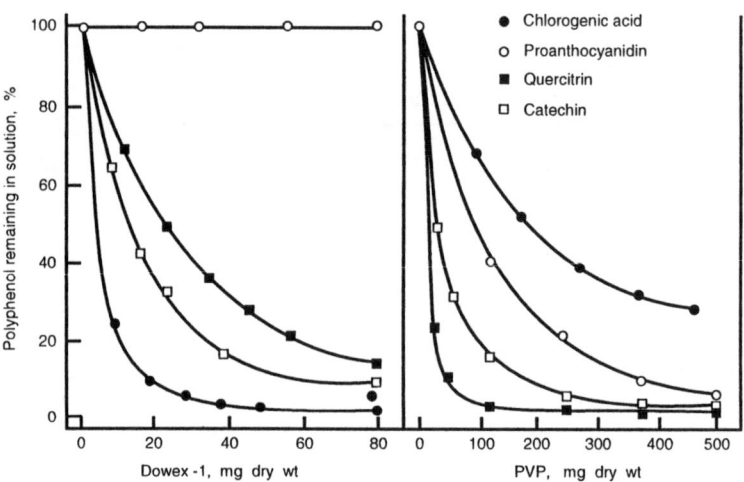

Fig. 1. Adsorption of different types of polyphenols by Dowex-1 (left) and PVPP (labeled PVP, right). Two milligrams of polyphenol in 3 mL phosphate buffer (50 mM; pH 7.5) were blended for 30 s with different amounts of hydrated adsorbents. The mixture was centrifuged and unadsorbed polyphenol in the supernatant estimated from $A_{280\ nm}$. Reprinted from Gray (7) by kind permission of the author and the editor on behalf of Pergamon Press.

that these resins may absorb protein as well as polyphenols, although tests using bovine serum albumin (BSA) and plant peroxidases suggest that this is negligible with the Amberlite resins at broadly neutral pH values (8). The ability of these resins to adsorb leaf components other than polyphenols, which interfere with enzyme extraction, such as the monoterpenes from peppermint leaves and the isothiocyanates of horseradish (8), are additional advantages of their use.

A possible disadvantage of the presence of either PVPP or polystyrene polymers during leaf disruption, especially when blenders are used, is a small degree of breakdown. Thus, PVPP is reported to produce soluble PVPP–protein complexes in the extract, and the resins release UV-absorbing compounds, such as benzoic acid, following blending. The adsorbents could, of course, be added after leaf disintegration or used as columns through which the extracts could be filtered. Any advantages of these procedures would be offset by the disadvantage of not having adsorbents present during leaf disruption when cell proteins are first exposed to liberated polyphenols.

Because the adsorptive abilities of PVPP and polystyrene resins are different and to some extent complementary (e.g., Fig. 1), they have often been used together when extracting refractory plant tissues. Thus, PVPP and XAD-4 (5 and 1 g hydrated weight, respectively, per g tissue) have been used successfully in extracting active enzymes from such unpromising material as apple fruits, peppermint leaves, and even the hulls of walnuts *(8)*. More recently, a mixture of PVPP and XAD-4 was shown to be most effective in extracting polyphenoloxidases from strawberry fruits *(16)*.

  b. Soluble polyphenol absorbents: A recent procedure for extracting Rubisco from the leaves of Mediterranean trees uses casein (2% [w/v]) as well as PVPP (1% [w/v]) in the extraction medium *(14)*. A more usual protein additive is BSA used at about 1%. These proteins are known to react with tannins as well as with enzyme-generated quinones *(17)*. BSA is especially helpful in the extraction of organelles, such as mitochondria (*see* ref. *5*), when the protein, along with adsorbed and reacted polyphenols, is effectively removed during centrifugal sedimentation of the organelles. The separation of added casein from Rubisco is, apparently, satisfactorily accomplished in the subsequent chromatographic stages of Rubisco purification. Rubisco, however, is the major protein component of leaves. Researchers attempting to purify enzymes present in very small amounts are often reluctant to add large amounts of protein to their extracts.

  Soluble forms of PVP have been used, and indeed were the first forms of this type of polymer to be used as polyphenol adsorbents. They adsorb phenolic compounds, but are unlikely to react with generated quinones. They are commercially available in three forms, with molecular weights about 10,000, 40,000, and 360,000, respectively, so that a size suitable for separation by gel filtration from a specific enzyme can be conveniently chosen. Earlier publications *(5,8)* emphasized the necessity of using only pharmaceutical grades of these chemicals: the grades supplied by a major supplier in the UK (Sigma) are presumed to be satisfactory, as they have been tested for suitability in such demanding processes as plant tissue culture and nucleic acid manipulation.

6. Choice of oxidase inhibitor/antioxidant: The procedure described uses both 2-ME (10 mM) and Dieca (10 mM) to prevent the browning reactions. They both inhibit PPO, and this may be the principal cause of their effectivness, although they also combine with quinones, either reducing them, complexing them, or both. A more recent modification of the procedure *(14)* omits the Dieca and increases the concentration of 2-Me to 100 mM. This modification also includes casein (2% [w/v]) in the extraction medium and this, as mentioned, also reacts with quinones.

A brief survey of extraction procedures described in the then current volumes of *Phytochemistry (6)*, suggested that 2-Me was a common choice of antioxidant, as were dithiothreitol (DTT) and dithioerythritol (DTE), the two sterioisomers of Cleland's reagent. Ascorbate was used less commonly. DTT, like 2-Me, appears to act as an enzyme inhibitor and also as a quinone scavenger, but the action of neither DTT nor DTE has been studied in detail. Ascorbate probably acts principally by reducing quinones back to re-oxidizable phenols, and so is readily removed by low concentrations of enzyme-produced quinones *(12)*. It should therefore be used in relatively high (>50 mM) concentrations. The thiol reagents, while they may be very effective in many, if not most cases, may be much less so in others. 2-Me activates unwanted proteases in extracts of tobacco leaves *(6)*, and the technical literature on DTT and DTE lists a number of enzymes that are inhibited by these reagents.

Many other compounds have proven useful as inhibitors/antioxidants, but have been less generally used. Mercaptobenzathiazole, a copper chelator that powerfully inhibits PPO, has been advocated by Palmer (cited in ref. *5*), and at low concentrations (0.1 mM) aids in extracting enzymes from tobacco leaves and mitochondria from potato tubers. Inorganic sulfites, especially metabisulfite, are effective in the extraction of enzymes from many leaves *(9,18)*, although their potential to cleave bonds in proteins demands a careful choice of concentration (approx 4 mM). Benzene sulfinic acid is a milder reagent that is thought principally to be a quinone scavenger, and is effective in extracting acylases from tobacco leaves *(19)*. KCN powerfully inhibits some oxidases, but is, understandably, now seldom used for this purpose. At 15 mM, it is reported to be a less effective inhibitor of the PPO from apple peel than Dieca, potassium metabisulfite, 2-Me, or ascorbic acid. It was used (1 mM), in conjunction with sodium metabisulfite (10 mM), in successfully extracting mevalonic acid kinase from the phenol-rich leaves of French beans *(20)*.

7. All the extracts, including the final one are likely to contain active proteinases, which may modify or inactivate sensitive enzymes. Rubisco in extracts from young green tissues is apparently insensitive to them. Another photosynthetic enzyme, the phosphatase that hydrolyzes 2'-carboxy-D arabinitol 1-phosphate, an inhibitor of Rubisco that is induced in leaves during darkness, is very sensitive. During its purification from the leaves of French beans, using a method similar to that described in Section 3. *(21)*, it is necessary to include the proteinase inhibitors phenylmethylsulfonyl fluoride (PMSF; 1 mM) and either benzamidine hydrochloride (1 mM) or p-aminobenzamidine dihydrochloride (1 mM) in the extraction buffer, in the resuspension buffer, and also in the liquid (1 L) against which the resuspended

ammonium sulfate precipitate is dialyzed overnight before chromatography on an affinity column. These inhibitors are added to buffer solutions and dialysis water along with 2-Me (50 mM) just before use. PMSF is usually dissolved in the minimal amount of ethanol before addition to the liquids.

Acknowledgment

The author acknowledges the help and encouragement of A. J. Keys and M. A. J. Parry.

References

1. Harborne, J. B. (1980) Plant phenolics, in *Encyclopedia of Plant Physiology* vol. 8 (Bell, E. A. and Charlwood, B. V., eds.), Springer-Verlag, Berlin, Germany, pp. 329–402.
2. Spencer, C. M., Cai, Y., Martin, R., Gaffney, S. H., Goulding, P. N., Magnolato, D., Lilley, T. H., and Haslam, E. (1988) Polyphenol complexation—some thoughts and observations. *Phytochemistry* **27**, 2397–2409.
3. Haslam, E. (1989) *Plant Polyphenols: Vegetable Tannins Revisited.* Cambridge University Press, Cambridge, UK.
4. Luck, G., Liao, H., Murray, N. J., Grimmer, H. R., Warminski, E. E., Williamson, M. P., Lilley, T. H., and Haslam, E. (1994) Polyphenols, astringency and proline-rich proteins. *Phytochemistry* **37**, 357–371.
5. Loomis, W. D. (1974) Overcoming problems of phenolics and quinones in the isolation of plant enzymes and organelles. *Methods Enzymol.* **31**, 528–544.
6. Jervis, L. and Pierpoint, W. S. (1989) Purification technologies for plant proteins. *J. Biotechnol.* **11**, 161–198.
7. Gray, J. C. (1978) Absorption of polyphenols by polyvinylpyrrolidone and polystyrene resins. *Phytochemistry* **17**, 495–497.
8. Loomis, W. D., Lile, J. D., Sandstrom, R. P., and Burbott, A. J. (1979) Adsorbent polystyrene as an aid in plant enzyme isolation. *Phytochemistry* **18**, 1049–1054.
9. Rhodes, M. J. C. (1977) The extraction and purification of enzymes from plant tissue, in *Regulation of Enzyme Synthesis and Activity in Higher Plants* (Smith, H., ed.), Academic, London, UK, *Proc. Phytochem. Soc. Eur.* **14**, 245–269.
10. Smith, D. M. and Montgomery, M. W. (1985) Improved methods for the extraction of polyphenol oxidase from d'Anjou pears. *Phytochemistry* **24**, 901–904.
11. Anderson, J. W. (1968) Extraction of enzymes and subcellular organelles from plant tissues. *Phytochemistry* **7**, 1973–1988.
12. Pierpoint, W S. (1966) The enzymic oxidation of chlorogenic acid and some reactions of the quinone produced. *Biochem. J.* **98**, 567–580.
13. Bird, I. F., Cornelius, M. J., and Keys, A. J. (1982) Affinity of RuBP carboxylases for carbon dioxide and inhibition of the enzymes by oxygen. *J. Exp. Bot.* **33**, 1004–1013.
14. Delgado, E., Medrano, H., Keys, A. J., and Parry, M. A. J. (1995) Species variation in Rubisco specificity factor. *J. Exp. Bot.* **46**, 1775–1777.
15. Hallahan, D. L., Dawson, G. W., West, J. M., and Wallsgrove, R. M. (1992) Cytochrome P-450 catalysed monoterpene hydroxylation in *Nepeta mussinii. Plant Physiol. Biochem.* **30**, 435–443.

16. Wesche-Ebeling, P. and Montgomery, M. W. (1990) Strawberry polyphenoloxidase: extraction and partial characterization. *J. Food Sci.* **55**, 1320–1324.
17. Pierpoint, W. S. (1969) *o*-Quinones formed in plant extracts: their reaction with bovine serum albumin. *Biochem. J.* **112**, 619–629.
18. Anderson, J. W. and Rowan, K. S. (1967) Extraction of soluble leaf enzymes with thiols and other reducing agents. *Phytochemistry* **6**, 1047–1056.
19. Pierpoint, W. S. (1973) An *N*-acylamino acid acylase from *Nicotiana tabacum* leaves. *Phytochemistry* **12**, 2359–2364.
20. Gray, J. C. and Kekwick, R. G. O. (1973) Mevalonate kinase in green leaves and etiolated cotyledons of the French bean *Phaseolus vulgaris. Biochem. J.* **133**, 335–347.
21. Kingston-Smith, A. H., Major, I., Parry, M. A. J., and Keys, A. J. (1992) Purification and properties of a phosphatase in French bean *(Phaseolus vulgaris)* leaves that hydrolyzes 2'-carboxy-D-arabinitol 1-phosphate. *Biochem. J.* **287**, 821–825.

CHAPTER 9

Avoidance of Proteolysis in Extracts

Robert J. Beynon and Simon Oliver

1. Introduction

Proteolytic enzymes (or proteases, peptidases, or proteinases) hydrolyze the peptide bond in proteins and peptides. The nomenclature is imprecise, but there is a broad acceptance that endopeptidases break bonds that are "internal" in the primary sequences, whereas exopeptidases trim one, two, or perhaps three amino acids from the amino or carboxy terminus of the substrate. Every cell and subcellular compartment has its own complement of proteolytic enzymes, and in normal circumstances, the activities of the proteolytic enzymes are well regulated. When a tissue is disrupted, however, this control is lost, and the proteinases may then attack proteins at a rate that leads to a loss of those proteins within the time scale of the study. Adventitious proteolysis is a technical problem that may require modification to methodology to minimize the assault on the protein of interest *(1–4)*.

The isolation of almost every protein will incur losses, and in some circumstances, those losses become unacceptable. Exactly what constitutes "unacceptable" will be dictated by individual needs, and will depend on downstream requirements and time scale. Few purification protocols generate pure protein, but rather, reduce contaminant proteins to the level where they cannot be detected by routine methods—consider the difference between a Coomassie-blue-stained gel and a silver-stained gel—the additional sensitivity of the latter will usually identify many contaminants in a preparation that looked homogenous by the former. Even apparently pure proteins may contain low levels of contaminating peptidases, which may explain the tendency to assume that any loss of protein, or of biological activity associated with that protein is owing to proteolytic destruction. Loss of material or activity can, of course, be owing to other, nonproteolytic processes—a strategy based on suppression of proteolysis will only be effective if proteolysis is really the culprit. In this chapter, therefore, we advocate the use of methods to prove that proteolysis is a problem and to measure proteolytic activity directly in the samples, as a prerequisite to prevention.

As mentioned previously, loss of material, as protein, or as a biological activity has many potential causes. Eliminate the possibility of, for example, thermal denaturation, oxidative damage, adsorption onto surfaces, or persistent binding to the column matrix. More subtle reasons for losses of activity might be complexation with an inhibitor in a time-dependent process, or loss of an essential cofactor or accessory protein as a consequence of a purification step. In crude extracts, and particularly when the protein has yet to be purified, discriminating between proteolysis and other events is difficult, but it is as well to be aware of the range of traumas that a protein can undergo after it has been removed from the normal cellular milieu, substantially diluted and removed from the protective effects of other proteins and low-mol wt ligands, such as natural reductants. Even if recoveries are initially good, proteolysis can become a problem during storage. Slow proteolytic degradation might destroy the material, or worse, allow it to retain some biological functions, but change others. Awareness of this possibility is the first step in its control.

Proteolysis of a protein may also occur in vivo before the cells are disrupted. This is sometimes evident in heterologous expression systems, where a protein is exposed to different proteolytic systems in the host cell. Strategies for prevention of this type of intracellular degradation are very different from those for prevention of postextraction degradation and are not discussed here.

1.1. Recognition of Proteolytic Attack

1.1.1. Why are Proteins Susceptible to Digestion?

The classical studies of Linderström-Lang and colleagues, over 40 years ago, opened the way to an understanding of the attack by proteinases on native structures *(5,6)*. Two types of proteolysis were defined: a "zipper" mechanism wherein the proteinase "nicks" one substrate molecule, and then attacks a further molecule, often at the same site, or, alternatively, "all or none" proteolysis in which the initial proteolytic attack renders the products much more likely to be further digested, such that they are quickly and completely digested to limit peptides. In the first mode, the products are resistant to further digestion, and accumulate as partially degraded proteins that may retain some biological function—the Klenow fragment of DNA polymerase, for example. In the second mode, the products of the first proteolysis are destabilized and become more susceptible to further digestion, such that transient intermediate products are unlikely to be observed.

Compared to unfolded polypeptides, most proteins are in fact relatively resistant to attack by endo- and exopeptidases. A denatured protein is usually much easier to digest than the native counterpart—denaturation and enhanced digestibility are in part the function of the low pH of the stomach and the lysosome. Folding of a protein chain into a compact tertiary or quaternary structure protects many peptide bonds because they are internalized in the protein. Moreover, there is accumulating evidence that proteolytic susceptibility can, in part, be attributed to the ability of the whole or a segment of a protein chain to be flexible *(7,8)*. The more flexible a loop, the more likely it is to be accommodated in the active site of the proteinases in a productive interaction. Thus, proteolytic attack can be dependent on the ability to cut a specific site on the protein that is flexible and susceptible. After the first nick, the behavior of the protein can adopt either of the two extreme models described by Linderström-Lang. Treatments that diminish this flexibility will protect the protein, and thus achieve the same goal as inhibition or removal of the proteinases.

1.1.2. Experimental Evidence for Proteolytic Degradation

Evidence for proteolytic degradation of a protein in a crude extract or during the course of a protein purification comes from several sources. A few guidelines may be provided herein, but we stress that each case must be considered unique, and that there is no general solution that will apply in all cases.

1. Loss of biological activity: If the only parameter that can be measured is biological activity, proving that proteolysis is responsible may be difficult. As discussed in Section 1., there are many reasons why the activity of a protein may be diminished. It may be necessary to conduct some analyses of the time-course of inactivation. Is the inactivation first order? Is the rate of inactivation reduced by addition of stabilizers, such as glycerol or non-

ionic detergents? Can inactivation be enhanced or prevented by addition of proteinase activators or inhibitors (*see* Section 1.3.)?
2. Loss of protein: Loss of the protein of interest is harder to assess, particularly if the protein is in low abundance in a crude mixture, such as a cell extract. If it is naturally abundant, or is expressed heterologously at high levels, it may be possible to identify the protein on 1-D gel electrophoresis. Under circumstances in which, for example, 50% of the biological activity is lost, it ought to be possible to *see* a similar loss of material on the gel. Of course, this strategy cannot discriminate proteolytic destruction of the protein from, for example, absorption to glass or plastic surfaces. Experiments should be conducted with siliconized glassware, in the presence of a protective protein of different size, such as albumin, or in the presence of a nonionic detergent, such as Tween-80 at 0.1% (w/v).

    Monitoring of proteolysis is simpler if an antibody to the protein can be used for Western blotting. The protein can be identified at low concentrations, and degradation will be manifest as loss of the band of the correct molecular weight, possibly with the appearance of degradation products. Reducing and nonreducing gels may give different results, depending on the relative disposition of nicksites and disulfide-linked cysteine residues, and both types of gels should be used if the protein is suspected or known to contain disulfide bonds.
3. Microheterogeneity: Proteolysis can manifest itself as a degree of heterogeneity in a finished product. Multiply sized bands on SDS-PAGE, or on isoelectric focusing may indicate limited proteolytic attack. A blotting antibody can help to confirm that the heterogeneity is owing to the protein of interest.

    If sufficient material is available for electrospray or MALDI-TOF mass spectrometry, then the precise mass of the isolated protein and the extent of heterogeneity can be checked directly. It will be important to know the extent of posttranslational modification that the protein might have undergone.

1.2. Strategies for Prevention of Proteolysis

Although individual cases will require unique solutions, there are some general principles that can be addressed. First, cells differ in the intracellular concentrations of proteinases, and artifactual proteolysis can be diminished by the use of a cell/tissue in which proteinase levels are low. For example, liver and kidney contain much higher amounts of proteolytic activity than skeletal or cardiac muscle. In many single-cell systems, mutant strains, defective in the expression of one or several proteinases, have been developed.

Many proteinases coexist with naturally occurring inhibitors. Complexation between proteinase and inhibitor may reduce the proteolytic activity in crude broken cell preparations, and this can be advantageous. However, in a subsequent purification step, the proteinase and inhibitor may be separated, and the proteolytic degradation may manifest itself late in the purification protocol.

Proteins offer protection to each other by competing for the active site of the proteinase. The early stages of a protein purification, in which the goal is to eliminate much of the bulk, unwanted protein, may therefore have the effect of rendering a protein more readily digested by a contaminating proteinase. Proteolytic losses can therefore, paradoxically, become more of a problem as the protein is purified.

Unlike many procedures for purification of nucleic acids, which use potent denaturants to destroy nucleases, the need to preserve the native structure of proteins means that such denaturants cannot be used. Also, many proteinases are more stable to a range of denaturants than their substrates. A denaturing step, which has the aim of inactivating the proteinase to diminish proteolysis, may in fact have the opposite effect, and render the protein more susceptible to digestion. This behavior is often seen in sample preparation for SDS-PAGE—when the sample is heated in SDS containing buffer, the protein is unfolded and undergoes rapid destruction, with apparent loss of material from the expected position on the gel. We routinely precipitate samples with trichloroacetic acid (TCA) before SDS-PAGE. This concentrates the samples, eliminates a great deal of buffer variability, and denatures proteinases and substrates in a rapid step that does not allow time for extensive artifactual proteolysis. Typically, a final concentration of TCA of 5% (w/v) at 4°C for 10 min will be adequate. The precipitated proteins are recovered by brief centrifugation (10,000g for 1 min, in a benchtop microcentrifuge), and the TCA is poured away. Residual TCA, which would otherwise change the pH of the sample buffer, is removed by repeated gentle washing with acetone or diethyl ether, and after three washes, the protein pellet is redissolved in SDS-PAGE sample buffer.

Routine additions to buffers can modify the activity of proteinases in the preparation. For example, reducing agents, such as 2-mercaptoethanol, activate cysteine proteinases, but inhibit metalloproteinases. Chelators also inhibit metalloproteinases. Some proteinases are activated by the presence of calcium ions. Thus, the choice of buffer may influence proteinase activity.

Thus, judicious choice of appropriate methodology may help to diminish the risk of proteolysis. However, there are many circumstances in which it is not possible to separate proteins and potentially damaging proteinases. Under these circumstances, the only real option is to inhibit the proteinases *in situ*.

1.3. Proteinases and Their Inhibition

A strategy of prevention of proteinase activity is usually based on a combination of two approaches: inhibition of the proteinases *in situ* and separation of the proteinases from the protein of interest. Proteinases bring about the hydrolysis of peptide bonds by several catalytic mechanisms, and inhibition of their action will be different for each mechanistic class *(9)*. Mechanism-based inhibitors will comprise a reactive group that is susceptible to attack by the active site residues, and high-affinity interaction of the reactive group with the enzyme can be achieved by additional functional groups that bind, for example, to the specificity pocket or the extended subsites of the proteinases. Most of these inhibitors bring about covalent modification of the active site residues, and are thus irreversible. Once the proteinase has been inhibited, it should not be necessary to add further inhibitor.

Another class of inhibitors capitalizes on the affinity of the proteinases for a substrate analog, without covalent interaction between them. These noncovalent inhibitors can be proteins or low-mol wt peptide analogs. Such inhibitors may also be usefully immobilized on columns for removal of the proteinases from the protein solution, although this is rarely used as an option. The complex formed between the proteinase and the inhibitor is reversible, and depends on the concentration of inhibitor in solution. If that inhibitor concentration is reduced by, for example, inactivation or dialysis, then the proteinase will recover activity. Strategies that add inhibitor to a crude preparation, which then undergoes precipitation (by ammonium sulfate or polyethylene glycol), column chromatography, or dialysis, must ensure that fresh inhibitors are added when appropriate.

In a few instances, the noncovalent binding is so tight that to all practical intents and purposes, the inhibitor can be considered as practically irreversible, and it will only be necessary to sustain low concentrations of inhibitor in buffers.

Serine proteinases use a nucleophilic serine residue to attack the carbonyl carbon of the scissile bond. The serine residue is a much stronger nucleophile than other serine residues in proteins, and this extreme property is facilitated by input of electrons from an aspartate residue via a histidine residue—the charge relay system. The serine and the histidine residues are both targets for inhibition. The inhibitors used most commonly are sulfonyl fluorides of which the most common is phenylmethyl sulfonyl fluoride (PMSF). All sulfonyl fluorides are unstable in solution, and a common mistake is to prepare stock solutions of PMSF in buffer or to assume that the inhibitor continues to work for extended times. At neutral pH values and 25°C, PMSF has a half-life of <1 h. Provided all of the serine proteinases are inhibited within this time scale, PMSF is effective, but exposure of cryptic proteinase activity later in a purification would need PMSF to be added afresh. Stock solutions are normally stored at −20°C in dry organic solvents, in which state PMSF is stable for weeks. Sulfonyl fluorides are also potent inhibitors of acetylcholinesterase and are thus toxic. A second class of proteinase inhibitors are derived from coumarins, of which 3,4 dichloroisocoumarin (3,4-DCI) is best known. Unlike PMSF, 3,4-DCI does not react quickly with acetylcholinesterase and is relatively nontoxic. Although 3,4-DCI is inactivated quite quickly in aqueous buffers (half-life about 20 min at neutral pH values), stock solutions in organic solvents are relatively stable.

Cysteine proteinases use a cysteine residue as the nucleophile to attack the scissile bond, and most inhibitors target this cysteine residue. The reactivity of the cysteine residue means that it reacts well with general-purpose —SH reactive reagents, such as iodoacetamide or iodoacetic acid, but these also modify free —SH groups in other proteins. If a cysteine proteinase is suspected, it may be preferable to use the epoxide inhibitor E-64 (L-trans-epoxysuccinyl-leucylamide-[4-guanidino]-butane, N-[N-L-3-transcarboxyrane-2-carbonyl]-L-leucyl-agmatine). This is a potent irreversible inhibitor of cysteine proteases that does not react with other —SH groups in proteins and that will not react with, and be consumed by, other low-mol wt thiols. Stock solutions are stable for days in aqueous buffers.

Some peptide aldehydes, such as leupeptin, chymostatin, and elastatinal, have been used to inhibit serine or cysteine proteases. These are reversible inhibitors and will need to be maintained at a reasonable working concentration in the buffers. Also, these peptides are prone to inactivation, and it is preferable to develop strategies based on irreversible inhibition.

Aspartic proteinases use a pair of aspartic residues to polarize the scissile bond. In general, aspartic proteinases are less of a problem than other proteinases, and they often have acidic pH optima. The nearest to a general-purpose inhibitor of aspartic proteinases is pepstatin, a reversible, but tight-binding inhibitor.

Metalloproteinases use a bound zinc ion as an electrophile to polarize a water molecule, which then attacks the scissile bond. The most general inhibitors of metalloproteinases are chelators, such as EDTA or 1,10 phenanthroline. If chelators cannot be included in the sample buffers, more complex strategies, based on some awareness of the type of metalloproteinase, may dictate an inhibition strategy. For example, some metalloproteinases are inhibited by phosphoramidon.

Method 1 provides a recipe for an inhibitor cocktail that ought to prevent proteolysis under many circumstances. However, it should be stressed that such cocktails are not guaranteed to work under all conditions. Detailed literature, the mechanism of action, and manipulation of a range of commercially available proteinase inhibitors can be found elsewhere (*see*, for example, *3,4,9*).

1.4. Assay of Endopeptidase Activity

There is some virtue in adopting a simple, yet sensitive assay for proteolytic activity, for tracking proteolytic activity during sample work-up, or for monitoring the efficiency of inhibition strategies. In a previous publication *(3),* the use of a radioiodinated peptide was advocated, but it is appreciated that few laboratories would be willing to take on this method unless they were already equipped for this type of radiochemical work. Accordingly, we present here a different assay, using a protein labeled with fluorescein isothiocyanate (FITC) based on a previously published protocol *(10,11)* (Methods 2 and 3). The fluorescent labeled protein is digested with a proteinase containing sample, and the undigested and large FITC-peptides are separated from small FITC products by precipitation with TCA. The FITC peptides in the supernatant fraction are monitored in a fluorimeter.

A second method (Method 4), which does not work for all endopeptidases, is zymography. In this method, a proteinase sample is electrophoresed in an SDS-PAGE into which has been copolymerized a protein, such as gelatin or casein. After the gel has been run, it is incubated in a nonionic detergent to remove the SDS, and then in a "refolding" buffer that may allow the proteinases to recover their structure and activity. If the proteinase is active, the gelatin is digested and converted to low-mol wt peptides that are washed out of the gel. Subsequent staining with

Coomassie blue indicates the zone of lysis as a clear region on a uniform blue background.

Zymography is quite sensitive, and a visible zone of lysis can be seen when nanogram amounts of enzyme are loaded. If excess proteinase is loaded, the zone of lysis can become large and diffuse. Also, if the proteinase is small, excess loading can show up as an unstructured zone of staining because the proteinase can diffuse out of the gel and digest the gelatin over all or part of the surface. In general, zymography does not work well if the gel or sample is reduced, possibly because of the opportunities for incorrect disulfide pairing, and the role of preformed disulfide bonds in directing folding. Not all proteinases will refold correctly and, thus, will not be detectable by this method.

2. Materials

2.1. Method 1: Stock Inhibitor Solutions

1. PMSF stock solution: $0.2M$ in dry methanol or propanol. Dissolve 38 mg ($M = 174.2$) of PMSF in 1.0 mL of solvent. PMSF is toxic! Weigh this compound in a fume hood, and wear disposable gloves and a mask. Store at $-20°C$.
2. 3,4-DCI stock solution: 10 mM in DMSO. Dissolve 2.2 mg ($M = 215$) of 3,4-DCI in 1.0 mL of DMSO. Store at $-20°C$.
3. Iodoacetic acid stock solution: 200 mM in water. Dissolve 42 mg ($M = 207.9$) of sodium iodoacetate in 1.0 mL of water. Use immediately.
4. E64-c stock solution: 5 mM in water. Dissolve 1.8 mg of E64-c ($M = 357.4$) in 1.0 mL water. Store at $-20°C$.
5. 1,10 phenanthroline stock solution: 100 mM in methanol. Dissolve 19.8 mg 1,10 phenanthroline ($M = 198.2$) in 1.0 mL of methanol. Store tightly capped at room temperature or 4°C.
6. EDTA stock solution: $0.5M$ in water. Dissolve 18.6 g EDTA (disodium salt, dihydrate, $M = 372.2$) into 70 mL water, titrate to pH 7.0 or 8.5, and make up to 100 mL. Stable at room temperature or 4°C.
7. Pepstatin stock solution: 10 mM in DMSO. Dissolve 6.9 mg pepstatin ($M = 685.9$) in 1.0 mL of DMSO. Store at $-20°C$.

2.2. Method 2: Preparation of FITC-Casein Substrate

1. Casein (purified powder, Sigma [St. Louis, MO] cat. no. C-5890).
2. Fluorescein isothiocyanate (Sigma cat. no. F-7250).
3. $0.05M$ Sodium carbonate, $0.15M$ NaCl, pH 9.5.
4. $0.05M$ Tris-HCl, pH 8.6.
5. Sephadex G-25 column, equilibrated with assay buffer.

2.3. Method 3: Assay of Enzyme Activity with FITC-Casein

1. Stock assay buffer (5X final concentration), e.g., $0.1M$ HEPES, pH 7.5.
2. FITC-casein: 1 mg/mL, in water or buffer, or casein fluorescein isothyocyanate (Sigma cat. no. C-0528).
3. TCA: 5% (w/v) in water.
4. Neutralizing buffer: $0.5M$ Tris-HCl, pH 8.6.

2.4. Method 4: Zymography

1. Stock gelatin solution: 1.2% (w/v) in water. Store at 4°C and microwave gently on the defrost setting to melt before use. Use gelatin electrophoresis-grade, Type A, Sigma cat. no. G-8150.
2. Triton-X100: 2.5% (w/v) in water.
3. 10X Refolding buffer: $0.5M$ Tris-HCl, $2M$ NaCl, 5.5% $CaCl_2$, 0.67% (w/v) Brij35, pH 7.6. Adjust pH before adding Brij35.

3. Methods

3.1. Method 1: Working Inhibitor Cocktails

1. From the stock solutions described in Section 2.1., make a working inhibitor cocktail in water (not buffer, since some buffer compounds accelerate decomposition of the inhibitors). For 1.0 mL of working solution, and for each class of proteinases, proceed as follows:
   a. Serine: 200 µL PMSF (20 mM final) or 200 µL 3,4-DCI (2 mM final);
   b. Cysteine: 200 µL iodoacetate (40 mM final) or 200 µL E64c (1 mM final);
   c. Metallo: 100 µL 1,10 phenanthroline (10 mM final) or 100 µL EDTA (50 mM final); and optionally
   d. Aspartic: 100 µL pepstatin (1 mM final).
   Make up the final volume to 1.0 mL with water.
2. Use the working cocktail within 1 h of preparation. Dilute this by 20-fold into the sample (*see* Notes 1 and 2).

3.2. Method 2: Preparation of FITC-Casein Substrate

1. Dissolve 2 g of casein in 100 mL of sodium carbonate buffer, pH 9.5. Add 100 mg of FITC. Mix gently for 1 h at room temperature.
2. Dialyze the FITC-casein against several changes of 2 L $0.05M$ Tris-HCl, pH 8.5, followed by the buffer that is preferred for routine assay. Alternatively, dialyze against Tris-HCl, and then distilled water to exchange the substrate into any buffer in the future. The Tris buffer is used to consume excess reagent.
3. Determine the protein concentration by standard procedures. If needed, calculate the number of FITC residues/casein molecule from the A_{490nm} of

the conjugate at pH 8.6. The extinction coefficient is 61,000/M/cm. The conjugate can now be frozen at −20°C.
4. To change the buffer in which the substrate is dissolved and to remove residual unbound FITC, apply the conjugate to a Sephadex G-25 column, equilibrated and eluted with a suitable assay buffer. Typically, the column volume should be about 10 times greater than the volume of FITC-casein for good buffer exchange. Monitor the elution profile at 280 nm, and collect the protein peak.
5. This substrate, approx 5 mg/mL, can be diluted to 1 mg/mL, and stored in aliquots at −20°C.

3.3. Method 3: Assay of Enzyme Activity with FITC-Casein

1. Combine 10–50 µL of enzyme sample with 10 µL FITC-casein in a microfuge tube. Add further assay buffer to make a reaction volume of 100 µL. Include a blank consisting of the buffer only.
2. Incubate at the desired temperature for 1–24 h.
3. Stop the reaction by adding 200 µL of 5% (w/v) TCA with mixing. Incubate the tubes at 4°C for 1 h to allow proteins to flocculate. Sediment the precipitated proteins by centrifugation at approx 10,000g for 10 min in a benchtop centrifuge at room temperature.
4. Add 100 µL of the supernatant to 2.9 mL of 0.5M Tris-HCl, pH 8.6. This strong buffer neutralizes the TCA, because the fluorescence of FITC is pH dependent.
5. Mix thoroughly, and measure the fluorescence with the excitation wavelength set at 490 nm and the emission wavelength set at 525 nm. Slit widths of 10 nm or less should be used.
6. The fluorescence of a sample, incubated in the absence of a proteinase and substituting water for the TCA, but otherwise processed identically, gives the total fluorescence of the substrate and allows quantitation of the casein digestion as weight solubilized/time (*see* Notes 3–5).

3.4. Method 4: Zymography

1. If necessary, desalt samples into a low salt buffer (e.g., 0.02M HEPES, pH 7.5) using spun columns. This will improve the sharpness of the bands.
2. Add 5 µL of nonreducing SDS-PAGE sample buffer to 20 µL of the enzyme sample, and incubate at 37°C for 1 h. Do not heat the sample in a boiling water bath.
3. While the sample is incubating, cast the gel in a standard kit, including gelatin (*see* Note 6) at a final concentration of 0.6% (w/v).
4. Typically, load between 1 and 20 µL of each sample. In early experiments, two different loadings (e.g., 1 and 15 µL) could usefully be included.

5. Run the electrophoresis as usual. There may need to be some adjustment of running time. At the end of the run, remove the gel without touching the surface.
6. Soak the gel in 2.5% Triton-X 100 for 1 h at room temperature to wash out SDS.
7. Rinse the gels in deionized water until foaming ceases (at least three times).
8. Incubate the gels in 1X refolding buffer, overnight at 37°C (*see* Notes 7 and 8).
9. Rinse the gels three times in deionized water.
10. Stain the gels with Coomassie brilliant blue for 1 h.
11. Destain the gels (several hours of destaining may be needed before the bands can be seen) (*see* Note 9).

4. Notes

1. The inhibitor cocktail can introduce salt (NaEDTA) and organic solvents to the sample. The 20-fold dilution is designed to minimize the effects of these constituents, but there may be circumstances in which this could still be problematic.
2. Although the serine and cysteine proteinase inhibitors are irreversible, the aspartic and metalloproteinase inhibitors are reversible. Even if the inhibitors are added at an early stage, they may be lost during purification. Alternatively, cryptic proteinases may be exposed at a later step by, for example, zymogen activation or dissociation of an endogenous inhibitor. Be prepared to add further inhibitor cocktail throughout the purification scheme.
3. The FITC-casein assay is very sensitive (ng to sub-ng) amounts of proteinase. The substrate should be stored at $-20°C$, because even a low level of bacterial contamination will give higher blank values. Always include a blank sample in FITC assays. If a standard is needed, a protease solution of 20 ng/mL trypsin will give a strong fluorescence in 2–3 h.
4. FITC-casein, like casein, is not very soluble at pH values below 4.0. This assay is not suitable for proteinases that have pH optima at low pH values.
5. This assay uses small amounts of substrate, which are usually far below saturating concentrations for the enzyme. As such, linearity over time will be lost if more than 10–20% of the substrate is digested. If digestion is limited, then fluorescence signal is proportional to the amount of enzyme added.
6. For zymography, other proteins, such as fibronectin or casein, can be copolymerized into the gel. The presence of the protein in the gel may alter the mobility of the proteinase, and a mol-wt estimate obtained by zymography should be carefully checked. It is possible to include a lane of mol-wt markers at sufficiently high a concentration that they can be seen as even darker bands on the uniform blue background.

7. The "refolding" step seems to be important, and a temperature of 37°C gives much improved recovery of activity over an incubation at room temperature. For metalloproteinases, there is apparently no necessity to add zinc ions to the refolding buffer, and this should be discouraged because many metalloproteinases are inhibited by excess zinc.
8. Zymography is not routinely used for analysis of cysteine proteases because of the need to add reducing agents to activate the enzyme. It might be worthwhile adding a reducing agent (10 mM dithiothreitol) for the last few hours of incubation in refolding buffer.
9. Because zymography has the potential to separate noncovalently bound inhibitors from proteinases, it may indicate proteolytic activity when none is apparent in soluble extracts.

References

1. Pringle, J. R. (1975) Methods for avoiding proteolytic artifacts in studies of enzymes and other proteins from yeast, in *Methods in Cell Biology,* vol 12 (Prescott, D. M., ed.), Academic, New York, pp. 149–184.
2. Pringle, J. R. (1978) Proteolytic artifacts in biochemistry, in *Limited Proteolysis in Microorganisms?* (Cohen, G. N. and Holzer, H., eds.), US Department of Health, Washington, DC, pp. 191–196.
3. Beynon, R. J. (1988) Prevention of unwanted proteolysis, in *Methods in Molecular Biology, Vol 3: New Protein Techniques* (Walker, J., ed.), Humana, Clifton, NJ, pp. 1–23.
4. North, M. J. (1989) Prevention of unwanted proteolysis, in *Proteolytic Enzymes—A Practical Approach* (Beynon, R. J. and Bond, J. S., eds.), IRL, Oxford, pp. 105–124.
5. Neurath, H. (1978) Limited proteolysis—an overview, in *Limited Proteolysis in Microorganisms* (Cohen, G. N. and Holzer, H. eds.), US Department of Health, Washington, DC, pp. 191–196.
6. Price, N. C. and Johnson, C. M. (1989) Proteinases as probes of conformation of soluble proteins, in *Proteolytic Enzymes—A Practical Approach* (Beynon, R. J. and Bond, J. S., eds.), IRL, Oxford, pp. 163–191.
7. Hubbard, S. J., Campbell, S. F., and Thornton J. M. (1991) Molecular recognition: conformational analysis of limited proteolytic sites and serine proteinase inhibitors. *J. Mol. Biol.* **220,** 507–530.
8. Hubbard, S. J., Eisenmenger, F., and Thornton, J. M. (1994) Modeling studies of the change in conformation required for cleavage of limited proteolytic sites. *Protein Sci.* **3,** 757–768.
9. Salvesen, G. and Nagase, H. (1989) Inhibition of proteolytic enzymes, in *Proteolytic Enzymes—A Practical Approach* (Beynon, R. J. and Bond, J. S., eds.), IRL, Oxford, pp. 25–55.
10. Sarath, G., De La Motte, R. S., and Wagner, F. W. (1989) Protease assay methods, in *Proteolytic Enzymes—A Practical Approach* (Beynon, R. J. and Bond, J. S., eds.), IRL, Oxford, pp. 83–104.
11. Twining, S. S. (1984) Fluorescein isothiocyanate-labeled casein assay for proteolytic enzymes. *Anal. Biochem.* **143,** 30–34.

CHAPTER 10

Concentration of Extracts

Shawn Doonan

1. Introduction

In a typical purification starting from 1 kg of tissue, the volume of the initial homogenate might be 2 L, and the final product of the procedure might be 2 mL of pure protein solution; that is, a volume decrease of 1000-fold is required. This emphasizes the fact that water is a major contaminant during protein purification, and at several steps during the procedure the need will arise for concentration of the active extract or fraction. Concentration may occur as a concomitant of a step in purification, for example, fractional precipitation (*see* Chapter 13) or ion-exchange chromatography (*see* Chapter 14); in the latter case, protein from a dilute solution may be absorbed onto the resin and subsequently eluted in a smaller volume by application of a salt gradient. In general, however, one or more steps specifically aimed at concentration of the protein solution will need to be done.

For concentration of initial cell extracts, particularly when working on a large scale (1–5 L of homogenate), the method of choice is precipitation by ammonium sulfate. Virtually all proteins are precipitated from solution at sufficiently high salt concentrations. This arises from the fact that protein surfaces tend to have hydrophobic patches, which, in solution, are surrounded by ordered water molecules. When salt is added to the protein solution, water is recruited to solvate the ions of the dissociated salt, thus progressively exposing the hydrophobic regions of the protein surface. At some point, these patches will start to interact, leading to aggregation and ultimately to precipitation of the proteins from solution (for a more detailed discussion, *see* ref. *1*). The requirements for the salt are that it be highly soluble in water, that its component ions be innocuous to proteins, and that it has a low heat of solution; ammonium sulfate satisfies these criteria most completely. The technique is, then, to precipitate the proteins from solution by addition of ammonium sulfate, recover the proteins by centrifugation, resuspend in the minimum amount of water or buffer, and remove residual ammonium sulfate by dialysis.

A second technique for protein concentration that is also applicable to relatively large volumes, but that is usually used at later stages in a purification protocol (e.g., after column chromatography) is forced dialysis or ultrafiltration. In this use is made of semipermeable membranes, which allow passage of water and other small molecules, but not of proteins. If a protein solution is placed in a bag of such material and pressure is applied, then small molecules will be forced out and the protein molecules retained. The rate of passage of small molecules will, of course, be severely reduced by precipitation of protein onto the walls of the dialysis bag; this is quite likely to occur with crude protein solutions, and it is for this reason that the method is not generally used at early stages in a purification procedure.

The particular version of membrane filtration described in the present chapter requires a minimal amount of specialized equipment and can be used to concentrate relatively large volumes of solution in an overnight experiment. A variant of the method that requires the use of commercial pressure cells, but has the advantage that it can be used for fractionation as well as for concentration, is described in Chapter 12. For small protein samples, drying by lyophilization is an alternative means of removal of water, but has the disadvantage that buffer ions (unless volatile) will remain in the sample; this technique is described in Chapter 31.

2. Materials

2.1. Precipitation with Ammonium Sulfate

1. Ammonium sulfate (preferably Analar grade).
2. Electric paddle stirrer.
3. Refrigerated centrifuge and rotor, e.g., a 6 × 250 mL angle rotor.
4. Screw-top plastic tubes for the rotor to be used.
5. Visking dialysis tubing (14- or 19-mm inflated diameter; 26- or 31-mm flat width). This is available from most suppliers of laboratory materials (e.g., Fisons, Crawley, UK).

Concentration of Extracts

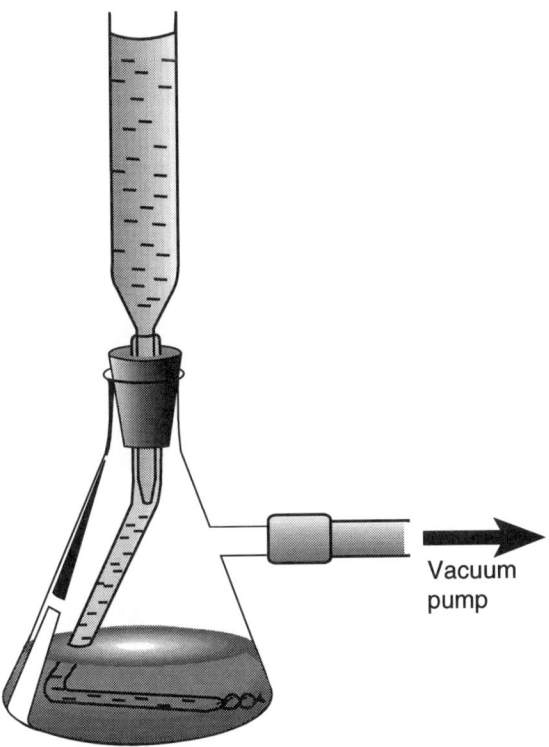

Fig. 1. Experimental arrangement for concentration by forced dialysis. The dialysis tubing should extend up the stem of the funnel so that it is above the top of the bung. In this way, it will be gripped tightly and will not slip off when the vacuum is applied.

2.2. Forced Dialysis

1. Heavy-walled Buchner flask (in the range 500-mL to 5-L capacity depending on the volume of solution to be concentrated—*see* Note 6).
2. Rubber bung to fit the flask and bored to accommodate the stem of a glass funnel.
3. Glass funnel (*see* Fig. 1 and Note 8).
4. Visking dialysis tubing (6-mm inflated diameter; 10-mm flat width).
5. Efficient water vacuum pump.

3. Methods

3.1. Precipitation with Ammonium Sulfate

1. Measure the volume of the protein solution to be concentrated, and pour it into a glass beaker of capacity about twice the measured volume of solution.

2. Weigh out 0.5 g of solid ammonium sulfate for every 1 mL of protein solution. If the ammonium sulfate contains lumps, then these should be broken up using a pestle and mortar (*see* Note 1).
3. Place the beaker containing the protein on ice, and stir either manually or with a paddle stirrer. A slow rate of stirring should be used to avoid foaming.
4. Add the ammonium sulfate to the protein solution in small batches over a period of several minutes ensuring that one lot of ammonium sulfate has dissolved before adding the next.
5. After addition is complete, leave to stand for 10 min to ensure complete precipitation and then recover precipitated protein by centrifugation at about 5000g for 30 min using screw-top plastic tubes (*see* Note 2).
6. Decant off the supernatant solution from each tube, and suspend the protein pellets in the minimum volume of water or of an appropriate buffer (*see* Note 3).
7. Transfer the protein suspension to a dialysis sack and dialyze against at least two changes of buffer using 100 times the volume of the sample and allowing 3–4 h for equilibration (*see* Note 4).
8. After dialysis, remove any precipitated material by centrifugation.

3.2. Forced Dialysis

1. Cut a length of the 6-mm diameter dialysis tubing (*see* Note 5) about twice as long as the height of the flask (*see* Note 6).
2. Soak the dialysis tubing in the buffer to be used for a few minutes (*see* Note 7). Tie two firm knots in one end. Open the other end by rubbing gently between the fingers, and then push the open end onto the stem of the funnel (*see* Note 8) so that the tubing extends 3-4 cm up the stem. The stem of the funnel should be moistened with buffer to facilitate this and a towel used to grip the dialysis tubing. Care must be taken not to tear the dialysis tubing.
3. Pour buffer (*see* Note 9) into the Buchner flask to a depth of about 5 cm. Thread the knotted end of the dialysis tubing through the hole in the rubber bung, pull through, and then push the stem of the funnel through the hole until the end has projected through. Place the bung in the flask (*see* Fig. 1).
4. Evacuate the flask using a water pump (*see* Note 10), seal the outlet, and test for leaks in the system by rotating and inverting the flask so that all parts of the dialysis tubing are sequentially immersed in the buffer. Pinhole leaks will be apparent from a stream of air bubbles emerging from the tubing. If any are present, reject the piece of tubing and start again.
5. Release the pressure, and pour protein solution into the funnel. Air will usually be trapped in the tubing. To remove this, take the bung out of the flask, and gently squeeze the tubing to expel air bubbles.
6. Replace the tubing and bung in the flask, evacuate, and then seal off the outlet. Place the flask in a cold room overnight or until the volume of solution has decreased to the desired extent (*see* Note 11).

Concentration of Extracts

7. Release the vacuum, cut off the dialysis tubing from the funnel, and transfer the concentrated solution to an appropriate container (*see* Note 12).

4. Notes

1. This amount of ammonium sulfate gives a concentration of about 75% saturation at 0°C, which should be high enough to precipitate most proteins. An initial experiment should be carried out to confirm that the protein of interest is indeed precipitated at this concentration. Higher values can be used (up to a maximum of 0.7 g/mL, which corresponds to 100% saturation), but this increases the density of the solution and makes it more difficult to sediment the protein.

    Note that strong solutions of ammonium sulfate are acidic, so it is important that the protein solution be buffered at a neutral pH and a buffer concentration of 0.05–0.1M. Note also that ammonium sulfate (even Analar grade) contains traces of metal ions; hence, if the protein of interest is metal-sensitive, EDTA (10 mM) should be added to the protein solution.

2. Care must be taken with this step to protect both the centrifuge and the operator! The tubes must be balanced to within 0.1–0.2 g across the rotor axis (*see* Note 3; Chapter 2), and it is particularly important not to counterbalance a tube full of 75% ammonium sulfate with a tube full of water because of the large difference in density. Either the ammonium sulfate/ protein suspension should be divided between two tubes or the suspension should be balanced using a solution of ammonium sulfate of the same concentration. It is also most important to avoid spillage of ammonium sulfate solutions into the centrifuge head. Such solutions are extremely corrosive to the materials of which rotors are constructed and can lead to irreversible damage rendering the rotors unsafe to use. After centrifugation of ammonium sulfate (or other salt) solutions, rotors should always be removed and washed in warm water.

    Note that proteins precipitated in 70% ammonium sulfate are usually stable, and hence it is often convenient to store the suspension overnight at 4°C in this form before proceeding with the purification schedule. Indeed, pure proteins are often stored at –20°C in 70% ammonium sulfate for long periods.

3. This step is crucial to achieving a satisfactiory degree of concentration. Add a very small volume of water or buffer (about 10 mL if using a 250-mL tube), and resuspend the protein pellet using a glass rod. If several centrifuge tubes have been used, then transfer the protein suspension from tube to tube, resuspending the pellet each time; only add more water or buffer if the suspension becomes too thick to transfer readily. After transfer of the final suspension to the dialysis bag, a further small aliquot of water or buffer can be used to rinse out the tubes and the rinsings combined with the

suspension. A common error is to attempt to redissolve the protein pellets after centrifugation; since the pellets contain considerable quantities of ammonium sulfate, this takes a large volume of buffer, and it is easy to end up with a volume comparable to that at the outset!

4. Visking tubing comes in several sizes. For a given volume of solution, the larger the diameter of the dialysis tubing used, the shorter the piece that will be required; attainment of equilibrium, however, will be slower with short, fat bags than with long, thin ones. The sizes recommended are a compromise between these two factors.

For most applications, it is only necessary to soak the dry tubing in water or in the buffer and it is ready to be used. The tubing does, however, contain significant quantities of sulfur compounds and of heavy metal ions; the latter may be a problem if the protein of interest is metal-sensitive. They may be removed by boiling the dialysis tubing in 2% (w/v) sodium bicarbonate 0.05% (w/v) EDTA for about 15 min, washing with distilled water, and then boiling in distilled water twice for 15-min periods. Prepared tubing can be stored indefinitely at 4°C in water or buffer containing 0.1% (w/v) sodium azide.

For most applications, Visking tubing is perfectly adequate, but problems will arise if the protein of interest is small. The pores in this type of tubing are of such a size that the nominal mol-wt cutoff (NMWC) is about 15,000, although larger proteins may still pass through the pores if they have an elongated shape. As a rule of thumb, Visking tubing can be used with confidence if M_r >20,000. For dialysis of smaller proteins, Spectropor tubings with a range of NMWC values starting at 1000 are available (e.g., from Pierce and Warriner, Chester, UK), and the appropriate tubing should be selected. These tubings are much more expensive than Visking tubing, and the rate of dialysis decreases with pore size; hence, they should only be used if strictly necessary.

To remove ammonium sulfate from the protein suspension, a piece of dialysis tubing with a volume about twice that of the suspension is taken and securely closed with a double knot at one end. The suspension is then poured into the bag using a funnel, air is removed from the top part of the bag by running it between the fingers, and the top secured with a double knot. It is very important to have this space in the bag to allow for expansion, since water will flow in while the internal salt concentration is high. If insufficient space is left, the bag can become very tight owing to this inflow. Bags rarely burst since the membranes are quite strong, but they are tricky to open in this state; the best way is to insert one end into a measuring cylinder and then prick the bag with a scalpel.

Equilibrium will be reached in about 3–4 h, but only if the system is stirred; otherwise about 6 h should be allowed. Care should be taken when stirring to ensure that the magnetic pellet or stirring paddle does not tear the dialysis bag.

The volume of dialysis solution to be used depends, of course, on the sample volume and the final concentration of ammonium sulfate desired. If a ratio of 1:100 (sample:dialysis fluid) is used, then at equilibrium the ammonium sulfate concentration will have been reduced 100-fold (this is not strictly true because it ignores the Donnan effect, but will do as an approximation). A second dialysis will then result in a total decrease of 10,000-fold and so on.

The buffer to be used for dialysis is usually dictated by the requirements of the next step in the purification schedule. For example, if this is to be ion-exchange chromatography, then column equilibration buffer is the logical choice.

5. The question of NMWC is also important here, and if the protein of interest has an $M_r < 20,000$, then one of the Spectropor tubings must be used. This will decrease the rate of concentration, but reduce losses of material.

Passage of low-mol wt material through the dialysis tubing can sometimes be used to advantage if it is required to carry out a crude separation of large and small components of a mixture. For example, an effective separation of the peptide components from the protein components of bee venom was achieved using forced dialysis through Visking tubing (2).

6. The optimum size of flask and length of tubing to be used depend on the scale on the experiment to be carried out. For volumes of the order of 100–200 mL, a 500-mL flask is ideal and the total length of tubing in this case would be about 25 cm. A final volume of concentrate of 2–3 mL is easy to achieve. For very large volumes a 5-L flask should be used with correspondingly long lengths of dialysis tubing (otherwise, concentration will take a very long time). With large flasks, it is perfectly feasible to mount up to three funnels in the bung, and thus increase the capacity and speed of the system.

7. If desired, the tubing can be pretreated as described in Note 4 above to remove metal ions.

8. Funnels of the type shown in Fig. 1 can be made from heavy glass tubing of about 25-mm diameter drawn down to a stem of about 7-mm diameter so that the dialysis tubing is a tight push fit; to do this requires an experienced glassblower. An alternative is to purchase cylindrical dropping funnels of the appropriate size from a laboratory supplier. These will have a tap between the stem and the cylindrical part, but this is not a problem. Whatever the type of funnel used, it is crucial that the end of the stem be fire-glazed; otherwise, the sharp glass edge will inevitably cut the dialysis tubing.

9. The buffer chosen will usually be that for the next step in the purification schedule. It should be noted, however, that the composition of the buffer will change during ultrafiltration owing to efflux of buffer ions originally in the sample so that it cannot be asssumed that the concentrated sample is properly equilibrated for further purification.
10. Do not use an oil pump; otherwise, it is likely that the pressure will be reduced too much and the dialysis bag may break (or the flask may implode, but this is not likely if it is made of heavy glass). As a useful guide, evacuate the flask until bubbles start to form in the buffer in the flask; this will occur at about 15 mmHg depending on the ambient temperature.

    Care should always be taken when using glass vessels under low pressure. Ideally, a cage should be used, but as a minimum safety glasses should be worn. Do not drop evacuated flasks!
11. Using a single flask of 2-L capacity with a single funnel and about 40 cm of tubing, it should be possible to reduce 500 mL of protein solution to about 5 mL overnight. If the capacity of the funnels available is too small to take the amount of protein solution to be concentrated, then it is simple to set up a syphon between solution in the funnel and extra solution in an external reservoir. Using a single 5-L flask with three funnels each one being replenished from a reservoir by means of a syphon, it is possible to concentrate 2 L of dilute protein solution down to about 20 mL overnight.
12. If left too long, the retentate may reduce in volume to such an extent that it is necessary to wash the protein out of the tubing with a small amount of buffer. On occasion, pure proteins may even crystalize in the tubing (*see* ref. *3* for an example); in this case, care should be exercised in handling the tubing so that the crystals do not puncture it.

References

1. Englard, S. and Seifter, S. (1990) Precipitation techniques. *Methods Enzymol.* **182,** 285–300.
2. Shipolini, R. A., Callewaert, G. L., Cottrell, R. C., Doonan, S., Vernon, C. A., and Banks, B. E. C. (1971) Phospholipase A from bee venom. *Eur. J. Biochem.* **20,** 459–468.
3. Barra, D., Bossa, F., Doonan, S., Fahmy, H. M. A., Martini, F., and Hughes, G. J. (1976) Large-scale purification and some properties of the mitochondrial aspartate aminotransferase from pig heart. *Eur. J. Biochem.* **64,** 519–526.

Chapter 11

Making and Changing Buffers

Shawn Doonan

1. Introduction

Control of pH is a central consideration in handling proteins, and this requires the use of buffer solutions. Furthermore, a typical protein purification schedule will contain several steps generally needed to be carried out in buffers of different pH values and ionic strengths (*see* Note 1). Hence, it is necessary to have available methods for changing buffers. The present chapter reviews the properties of buffers and preparation of buffer solutions, and then deals with ways in which a protein solution can be changed from one buffer to another.

Buffer solutions consist of a weak acid and a salt of that acid, or of a weak base and a salt of that base. For example, in a solution of a weak acid HA and its sodium salt, the following occur:

$$HA \rightleftharpoons H^+ + A^- \text{ (incomplete)} \tag{1}$$

$$NaA \rightarrow Na^+ + A^- \text{ (complete)} \tag{2}$$

The species A^- is referred to as the conjugate base of the acid HA.

Addition of H^+ to the buffer moves equilibrium (1) to the left using A^- supplied by (2), whereas added OH^- (or other base) combines with H^+ provided by equilibrium (1) moving to the right; in either case, change of H^+ concentration, and hence, of pH, is resisted. The Henderson-Hasselbalch equation describes the relationship between the pH, the pK_a of the buffer and the relative concentrations of the free acid and of the salt as follows (*see* Note 2):

$$pH = pK_a - \log[(\text{acid})/(\text{salt})] \tag{3}$$

From: *Methods in Molecular Biology, Vol. 59: Protein Purification Protocols*
Edited by: S. Doonan Humana Press Inc., Totowa, NJ

Hence, if, for example, a solution of a weak acid of concentration (a) M is partially neutralized by addition of a strong base to a concentration of (b) M, then the result is a buffer solution where the pH is given by:

$$\text{pH} = pK_a - \log\{[(a) - (b)]/(b)\} \quad (4)$$

It is immediately apparent from Eq. (4) that when (b) = 0.5(a), then pH = pK_a, i.e., the pH is numerically equal to the pK_a value. The equation also suggests that the pH of a buffer is unaffected by dilution; this is true to a first approximation for some buffers, but with others, the pH changes quite markedly with dilution (*see* Note 3).

What is not so obvious is that the ability of the buffer to resist changes of pH is maximum at the pK_a and falls off on either side such that the buffering power is small to negligible outside the range $pK_a \pm 1$. This can be seen by considering the parameter β (the buffer capacity or buffer value), which is defined as:

$$\beta = db/d\text{pH} \quad (5)$$

and is given by:

$$\beta = 2.303\{(a)K_a(H^+)/[K_a + (H^+)]^2\} + 2.303[(H^+) + (OH^-)] \quad (6)$$

where (a) is the total concentration of the free acid plus the salt. In the pH range of approx 3–11, the value of β is determined entirely by the first term in the equation, and under those circumstances a more convenient form is:

$$\beta = 2.303\{(b)[(a) - (b)]/(a)\} \quad (7)$$

This is obtained by taking the first derivative of the Henderson-Hasselbalch equation and inverting it. The buffer value is a measure of the amount of base needed to produce a unit change in the pH of the buffer, and the larger its value, the greater the the resistance to pH change. Table 1 shows the buffer values for a solution of a weak acid to which various quantities of base have been added and emphasizes the restricted range of pH values over which the solution has good buffering properties. Obviously, the stronger the buffer, the greater the buffer value, but for practical reasons, buffers of strength >$0.1M$ are not usually used.

There are, then, a variety of considerations in selecting a buffer for a particular application. These include:

1. The desired pH: The pK_a of the buffer must be as close to this as possible and certainly not outside the range pH ± 1.

Table 1
pH and Buffer Values of a 0.1M Solution of a Weak Acid (pK_a = 7) as a Function of Concentration of an Added Base b

(b), M	pH	Buffer value
0.005	5.72	0.0109
0.010	6.05	0.0207
0.020	6.40	0.0368
0.030	6.63	0.0484
0.040	6.82	0.0553
0.050	7.00	0.0576
0.060	7.18	0.0553
0.070	7.37	0.0484
0.080	7.60	0.0368
0.090	7.95	0.0207
0.095	8.28	0.0109

2. The pH of the buffer should change as little as possible with temperature, with dilution, and with added neutral salt (see Note 4).
3. The buffer should be chemically unreactive.
4. The buffer should not absorb light at 280 nm, particularly if it is to be used for chromatographic procedures where column monitoring will usually be carried out by absorbance measurements.
5. For cation ion-exchange chromatography, an anionic buffer should be used and vice versa.
6. Buffers that might interact with components of the protein mixture (e.g., borate with glycoproteins) should be avoided.

Table 2 gives a list of commonly used buffer compounds with their pK_a values; values for other susbstances are given in ref. *1*. Included in the table are a variety of "Good" buffers (MES, ADA, PIPES, ACES, BES, MOPS, HEPES, TAPS, CHES, CAPS). These zwitterionic buffers are chemically unreactive, do not absorb at 280 nm, and their pH values vary only slightly with temperature and dilution *(2)*. They are, therefore, ideal buffers in many respects, but are very expensive (see Note 5) and their use is often not feasible when large volumes of buffer are required for procedures such as ion-exchange chromatography and dialysis.

As mentioned, there will frequently be a need during protein purification for changing the buffer in which a protein mixture is dissolved; this may be to change the pH, the buffer ionic strength, or the concentration of a neutral salt. There are two major ways in which this can be done.

Table 2
Buffer Compounds and Their pK_a Values at 25°C

Compound	Trivial name	pK_a
Phosphoric acid (pK_1)	–	2.15
Glycine (pK_1)	–	2.35
Fumaric acid (pK_1)	–	3.02
Citric acid (pK_1)	–	3.13
Formic acid	–	3.75
Succinic acid (pK_1)	–	4.21
Fumaric acid (pK_2)	–	4.38
Citric acid (pK_2)	–	4.76
Acetic acid	–	4.76
Succinic acid (pK_2)	–	5.64
2-(N-morpholino)ethanesulfonic acid	MES	6.10
Carbonic acid (pK_1)	–	6.35
[Bis(2-hydroxyethyl)imino]-tris(hydroxymethyl)methane	Bis-Tris	6.46
N-2-acetamidoiminodiacetic acid	ADA	6.59
Piperazine-N,N'-bis(2-ethanesulfonic acid)	PIPES	6.76
N-2-acetamido-2-hydroxyethanesulfonic acid	ACES	6.78
Imidazole	–	6.95
N,N-bis-(2-hydroxyethyl)-2-aminoethanesulfonic acid	BES	7.09
3-(N-morpholino)propanesulfonic acid	MOPS	7.20
Phosphoric acid (pK_2)	–	7.20
N-2-hydroxyethylpiperazine-N'-2-ethanesulfonic acid	HEPES	7.48
N-ethylmorpholine	–	7.67
Triethanolamine	–	7.76
Tris(hydroxymethyl)aminomethane	Tris	8.06
N-[Tris(hydroxymethyl)methyl]glycine	Tricine	8.05
N,N-bis(2-hydroxyethyl)glycine	Bicine	8.26
3-{[Tris(hydroxymethyl)methyl]-amino}propanesulfonic acid	TAPS	8.40
2-Amino-2-methylpropan-1,3-diol	–	8.79
2-Aminoethylsulfonic acid	Taurine	9.06
Boric acid	–	9.23
Ammonia	–	9.25
Ethanolamine	–	9.50
Cyclohexylaminoethanesulfonic acid	CHES	9.55
3-Aminopropanesulfonic acid	–	9.89
β-alanine	–	10.24
Carbonic acid (pK_2)	–	10.33
3-(Cyclohexylamino)propanesulFonic acid	CAPS	10.40
γ-Aminobutyric acid	–	10.56
Piperidine	–	11.12
Phosphoric acid (pK_3)	–	12.33

Table 3
Standard Buffer Solutions

Buffer	Composition, g/L	Concentration, M	pH 5°C	15°C	25°C
Phthalate	10.12 g of $KHC_8H_4O_4$	0.05	4.00	4.00	4.01
Phosphate	3.39 g of KH_2PO_4 + 3.53 g of Na_2HPO_4	0.025	6.95	6.90	6.87
Borate	3.80 g of $Na_2B_4O_7 \cdot 10H_2O$	0.01	9.40	9.28	9.18

The first of these is dialysis in which the protein solution is enclosed in a bag of a semipermeable membrane, i.e., one that allows the passage of small molecules but not of large ones, and is then equilibrated in two or more changes of a large excess of the target buffer solution. Ultimately, equilibrium will be reached where the internal and external buffers may approximate to the same pH and concentration (see Note 6). A related technique uses membrane UF; this is described in Chapter 12.

The second method is gel filtration. This uses a column of porous beads designed such that water and low-mol wt solutes have access to the interior of the beads whereas larger molecules do not. In gel filtration the solution of protein is applied to such a column equilibrated in the target buffer and the proteins are then eluted with target buffer. The protein molecules will move ahead of the buffer in which they were originally dissolved and will emerge in the target buffer. This technique can also be used as a method for fractionation of protein mixtures on the basis of size as described in Chapter 25.

2. Materials
2.1. Preparation of Buffers

1. pH meter capable of accuracy at two decimal places and with temperature compensation (see Note 7).
2. Magnetic stirrer and pellet.
3. Standard buffer solutions with pH values bracketing that of the buffer to be made. These can either be prepared using the information in Table 3 or purchased from most suppliers of chemicals.
4. Analar HCl or NaOH pellets depending on the buffer to be made.

2.2. Dialysis

1. Visking dialysis tubing (either 14- or 19-mm inflated diameter; 26- or 31-mm flat width). This is available from most suppliers of laboratory materials (e.g., Fisons, Crawley, UK).

2. Target buffer (at least 200 times the volume of protein solution is required).

2.3. Gel Filtration

1. Sephadex G-25 (medium grade) (*see* Note 8). About 1 g will be required for every 5 mL column volume. This material is produced by Pharmacia Biotech (Milton Keynes, UK), but is available from general suppliers.
2. Appropriate size chromatography column. The packed volume will need to be about five times greater than the volume of protein solution to be treated (*see* Note 9).
3. Peristaltic pump, fraction collector and absorbance detector (optional—*see* Chapter 36).
4. Target buffer (about 10 times the volume of the column bed).

3. Methods

3.1. Preparation of Buffers

It is usually convenient to make stock solutions of buffers that can then be diluted for use (*see* Note 3). Recipes are available for many of the more common buffers (*see* ref. *1*), or compositions can be calculated using the Henderson-Hasselbalch equation. More usually, however, buffers are made by weighing out the required amount of the buffering substance, dissolving in water, and then adjusting the pH to the desired value by adding HCl or NaOH as appropriate; other acids or bases may, of course, be used. For example, to make 1 L of $2M$ sodium acetate buffer, pH 5.0, proceed as follows.

1. Standardize the pH meter using standard buffers of pH 4.01 and 6.87, assuming that the buffer is to be made at room temperature.
2. Weigh out 2 mol (120 g) of glacial acetic acid, and transfer to a 1-L beaker.
3. Add about 800 mL of distilled water, place on a magnetic stirrer, and insert the pH meter electrodes.
4. Slowly add solid NaOH pellets (preferably from a freshly opened bottle to ensure that they are not contaminated with sodium carbonate), making sure that one lot dissolves before adding the next. When the pH has reached about 4.8, allow the buffer to cool to room temperature (heat will have been generated by reaction of acetic acid with the NaOH), and then finally adjust the pH to 5.0 using a solution (about $4M$) of NaOH.
5. Transfer the solution quantitatively to a graduated flask or measuring cylinder (the latter is sufficiently accurate), and make up the volume to 1 L.
6. Store in a glass container at 4°C. Preservatives are not necessary because the high concentration inhibits bacterial growth.

3.2. Changing Buffers by Dialysis

It should be borne in mind that proteins are polyelectrolytes whose state of ionization varies with pH because of their content of ionizable side chains. Hence, particularly if the protein concentration is high and the solution is at a pH far removed from that of the target buffer, it is not possible to equalize the pH and ionic strength values of the internal and external buffers by dialysis because of the Donnan effect (*see* Note 10). In the following, it is assumed that dialysis will be carried out in a cold room or refrigerator at about 5°C using a diluted stock buffer.

1. Calibrate the pH meter with standard buffers (Table 3) cooled to 5°C.
2. Dilute stock buffer solution to the desired concentration using cold distilled water, and check that the pH has not been altered by dilution or the decrease in temperature. If it has, adjust it with the acidic or basic component of the buffer as appropriate. The volume of buffer should be about 100 times that of the sample to be dialyzed.
3. If the target buffer is to contain a neutral salt, then add this before checking and adjusting the pH.
4. Take a length of Visking dialysis tubing able to hold about 1.5 times the volume of the protein solution, and soak it in the buffer for a few min. For some applications, it may be desirable to pretreat the dialysis tubing and/or use tubing with a lower nominal mol wt cutoff (*see* Note 4, Chapter 10).
5. Tie two tight knots in one end of the tubing, and then pour in the protein solution with the aid of a funnel. Squeeze out air above the protein solution to allow room for expansion, and seal the bag with two more knots.
6. Place the dialysis bag in the buffer solution, and leave for at least 4 h in the cold to reach equilibrium. If a magnetic stirrer is used, be sure that the pellet does not tear the dialysis bag.
7. Replace the buffer with a fresh lot, and repeat the dialysis for another 4 h.
8. Check the pH of the dialyzed solution and, if necessary, adjust it to the target value by very careful addition of dilute acid or base (*see* Note 10). Use constant stirring to ensure that local high concentrations of acid or base do not occur.
9. Remove any insoluble material from the dialyzed sample by centrifugation at about 5000g for 15 min.

3.3. Changing Buffers by Gel Filtration

This is a more rapid procedure (if a column is already available) and, hence, to be preferred particularly if time is important, e.g., if the protein of interest is unstable. It should not be used, however, with crude protein mixtures, since there is a strong possibility that protein will precipitate during the procedure and ruin the column.

1. Take 1 g of Sephadex G-25 (medium grade) for every 5 mL of desired column volume, and stir carefully into 10 vol of the target buffer. Leave at 5°C overnight for the gel to swell to its maximum extent. Resuspend by stirring, allow to settle, and remove any fines by aspiration.
2. Resuspend the gel in about 2 vol of buffer and pour into the chromatography column (*see* Note 11) with the outflow blocked off. Allow a few centimeters of settled bed to form, and then start flow through the column at the rate to be used for gel filtration (*see* Note 12). As clear liquid forms above the gel suspension, remove it by aspiration and replace it with fresh suspension. After the gel bed has reached the desired height, attach a buffer reservoir and pass two column volumes of buffer through the column to ensure equilibration of the bed.
3. Apply the sample to the column, and then continue elution with about one column volume of the target buffer. Collect appropriate sized fractions (about 1/20th of the column volume).
4. If using an automatic absorbance monitor, the fractions of interest will be obvious. Otherwise, measure the absorbance of individual fractions at 280 nm, and combine fractions containing protein (*see* Note 13).
5. Check the pH of the combined fractions, and if necessary, adjust as described in Section 3.2., step 8 (*see* Note 14).
6. Re-equilibrate the column by passage of two column volumes of buffer.

4. Notes

1. Sometimes, by judicious choice of the sequence of steps, it is possible to avoid changing the buffer between one step and the next. This is desirable, because it saves time and can improve the overall yield of the purification. An example of where this approach has been used is given in ref. *3*.
2. This equation is strictly only valid in the pH range 3–11. Outside this range the self-ionization of water becomes significant.
3. The reason for this is that the concentration terms in the Henderson-Hasselbalch equation should properly be thermodynamic activities; these are concentration-dependent to varying extents.
4. K_a values are temperature-dependent, and hence, so are the pH values of buffer solutions. The magnitude of the effect depends on the particular buffer substance chosen. Similarly, added salt can affect the pH of a buffer because of differential effects on the thermodynamic activities of the component buffer ions. If possible, buffers should be selected where these effects are minimal (*see* ref. *1*).
5. Prices of "Good" buffers are in the range $300–750/kg. Common organic acids and bases used as buffers cost about one-tenth of this, and inorganic buffers are even cheaper.

6. This is not strictly true because of the so-called Donnan equilibrium. The protein, which is a nonpermeant polyelectrolyte, will affect the distribution of the permeant ionic species. Consider a protein solution volume V_1 at a pH such that it is negatively charged, and suppose that the concentration of charges on the protein is $C_1 M$; these negative charges will be neutralized by positive ions (e.g., Na^+) also at a concentration of $C_1 M$. If this solution is enclosed in a dialysis bag and placed in a volume V_2 of NaCl solution concentration $C_2 M$, then NaCl will cross the membrane until equilibrium is attained, i.e., until the activities of the NaCl in the two compartments are equal. It can be shown that, to a first approximation, the equilibrium condition is:

$$(Na^+)_{in} \times (Cl^-)_{in} = (Na^+)_{out} \times (Cl^-)_{out} \qquad (8)$$

where in and out refer to inside and outside of the dialysis bag. If the change of concentration of NaCl inside the dialysis bag is $+x\ M$, then it follows that the change outside will be $-x\ (V_1/V_2)\ M$ and the equilibrium condition is:

$$x(C_1 + x) = \{C_2 - [x\ (V_1/V_2)]\}^2 \qquad (9)$$

Hence if, for example, $C_1 = 20$ mM, $V_1 = 10$ mL, $C_2 = 100$ mM, and $V_2 = 1000$ mL, then $x = 89.6$ mM, i.e., the NaCl concentration inside the dialysis bag at equilibrium will be 89.6 mM and that outside will be 99.1 mM. Repeating dialysis with a fresh NaCl solution will result in only a marginal increase of internal NaCl concentration to 90.5 mM. This example is, perhaps, a little extreme because it would require a 50 mg/mL solution of protein, average $M_r = 50,000$, each molecule of which carried a net 20 negative charges. It does, however, serve to illustrate the point that the distribution of salts, including buffer ions, in dialysis will be affected by the presence of protein and that even very extensive dialysis will not achieve equality of internal and external concentrations. One consequence of this is that dialyzed solutions of high and low concentrations of the same protein mixture will not behave identically on, for example, ion-exchange chromatography because their ionic strengths and pH values will differ.

7. Note that the temperature compensator corrrects for the conversion of measured emf to pH and not for the variation of buffer pH with temperature. If Tris buffers are to be used and very accurate pH values are required, then special glass electrodes are needed that do not respond to Tris itself; these are available from several suppliers. Care must be taken of glass electrodes if good pH measurements are to be made. They must never be allowed to dry out (store in saturated KCl solution), and protein must not be allowed to accumulate on them (wash with detergent solution if this occurs).

8. This exclusion limit is appropriate for proteins of $M_r > 20,000$. For smaller proteins, Sephadex G-10 should be used. This latter material produces a smaller bed volume/g dry wt and hence is more expensive to use.

9. For desalting or changing buffers with small volumes of protein solution (up to about 2 mL), Pharmacia Biotech markets prepacked Sephadex G-25 columns in either disposable or reuseable form. These are very convenient, but moderately expensive. Information can be obtained from the company's technical publications.
10. The situation is complex and it is very difficult to analyze precisely because of the change in net charge of a protein with changing pH. However, it should be clear from the discussion in Note 6 that, since proteins will carry a net charge at all pH values except the isoelectric point, the presence of protein inside the dialysis bag will affect the distribution of ions, including those of the buffer. Therefore, for example, dialysis of a negatively charged protein mixture with a buffer formed from a weak acid will lead to a situation where the concentration of conjugate base is lower inside than outside, and hence, the pH inside will be lower than in the bulk buffer. The higher the protein concentration and the greater the difference between the starting pH of the protein solution compared with the target pH, the greater the effect will be. Hence, it is not possible to obtain complete equality of pH and buffer concentration by dialysis even for relatively dilute protein solutions. Given that for most applications the pH rather than the ionic strength is the key factor, it is best to dialyze to equilibrium, and then to adjust the pH to the target value if necessary by addition of acid or base.
11. The ratio of the length to the diameter of the column should be in the range 10:1–20:1. Longer but narrower columns may give slow flow rates, whereas short, fat columns may give poor separations and greater dilutions of the emergent protein solution because of imperfections in the packing. Pharmacia Biotech markets columns of these dimensions, as do other companies, but homemade columns provide a cheaper satisfactory alternative (*see* Chapter 36).
12. Flow rates should be approx 0.1–0.2 mL/min/cm^2 of cross-sectional area. Faster flow rates will lead to incomplete exchange between the bulk solvent and the beads, whereas slower values will result in more band spreading and dilution.
13. If a column is to be used several times for desalting or for exchange of buffers, then it may be convenient to calibrate it by passing through a mixture of a standard protein (e.g., bovine serum albumin) and a salt (e.g., NaCl) and determining the volume ranges over which the protein and the salt emerge. For detecting the protein absorbance values at 280 nm are used, whereas detection of the presence of NaCl could be done by addition of silver nitrate to aliquots of the fractions and observing the precipitation of silver chloride; if the buffer contains chloride or phosphate ions (silver phosphate is insoluble also), then some alternative low-mol wt substance

such as acetone (which can be determined spectroscopically) would have to be used. Thereafter, provided the same equilibration buffer is used, the protein fraction can be collected simply on the basis of elution volume.
14. The point here is that protein solutions are themselves buffers by virtue of the ionizable side chains that they contain. In gel filtration, the protein may be transferred from a buffer at one pH to an only slightly larger (about 1.5 times) volume of a buffer at a different pH. Unless the second buffer is very strong it is unlikely to have sufficient buffering capacity to effect the required change in pH particularly if the starting and target pH values are widely different. For example, a change from pH 8.0 to 6.0 will require protonation of all or most of the histidine residues in a protein, thus withdrawing protons from the target buffer with a consequent rise in pH. Hence, again the pH should be checked after gel filtration and adjusted if necessary.

References

1. Dawson, R. M. C., Elliot, D. C., Elliot, W. H., and Jones, K. M. (1986) *Data for Biochemical Research,* 3rd ed., Oxford University Press, Oxford, UK, pp. 417–448.
2. Good, N. E., Winget, G. D., Winter, W., Connolly, T. N., Izawa, S., and Singh R. M. M. (1966) Hydrogen ion buffers for biological research. *Biochemistry* **5,** 467–477.
3. Cronin, V. B., Maras, B., Barra, D., and Doonan, S. (1991) The amino acid sequence of the aspartate aminotransferase from baker's yeast *(Saccharomyces cerevisiae). Biochem. J.* **277,** 335–340.

CHAPTER 12

Purification and Concentration by Ultrafiltration

Paul Schratter

1. Introduction

Membrane ultrafiltration (UF) is a pressure-modified, convective process that uses semipermeable membranes to separate species in aqueous solutions by molecular size, shape, and/or charge. It separates solvents from solutes (i.e., the dissolved species) of various sizes. The result of removing solvent from a solution is solute concentration or enrichment. Repeated or continuous dilution and reconcentration are used to remove salts or exchange solvent (in such applications as buffer exchange). Definitions of some terms used in UF are given in Table 1.

UF is a low-pressure procedure, generally more gentle for the solutes than nonmembrane processes. It is more efficient than such processes, and can simultaneously concentrate and desalt solutions. It does not require a phase change, which often denatures labile species and can be performed at cold-room temperatures.

UF should be viewed as an excellent tool for efficient separation of biological substances into groups, according to molecular weight and size. For finer separation, it must be followed by a more selective process, such as chromatography or electrophoresis.

1.1. Membrane Processes

UF is one of a spectrum of membrane separation techniques that include reverse osmosis, dialysis, and microfiltration. Reverse osmosis separates solvents from low-mol-wt solutes (typically <100

Table 1
Some Definitions of Terms Used in UF[a]

Concentration
    Enrichment of a solution by solvent removal.
    The relative amount of a molecular species in a solution, expressed in percent.

Concentration polarization
    Accumulation of rejected solute on the membrane surface; depends on interactions of pressure, viscosity, crossflow (tangential) velocity, fluid flow conditions, and temperature.

Cutoff (MWCO)
    The molecular weight at which at least 90% of a globular protein is retained by the membrane.

Fouling
    Irreversible decline in membrane flux owing to deposition and accumulation of submicron particles and solutes on the membrane surface; also, crystallization and precipitation of small solutes on the surface and in the pores of the membrane, not to be confused with concentration polarization.

Permeate
    The solution passing through the membrane, containing solvent and solutes not retained by the membrane.

Rejection
    The fraction of solute held back by the membrane.

Retentate
    The solution containing the retained (rejected) species.

Yield
    Amount of species recovered at the end of the process as a percentage of the amount present in the sample.

---

[a]Table courtesy of Amicon, Inc.

Dalton) at relatively high pressures. Reverse osmosis membranes are normally rated by their retention of sodium chloride, whereas UF membranes are characterized according to the molecular weight of retained solutes.

Dialysis is a diffusive process employing a second liquid (dialysate) on the opposite side of the membrane from the sample. The permeation rate of molecules from sample to dialysate is in direct ratio to their concentration and inversely proportional to their molecular weights. Therefore, the rate of transport of a salt through the dialysis membrane diminishes as salt concentration in the sample declines during the process, so that desalting by dialysis tends to be quite slow. In UF, all completely

Ultrafiltration

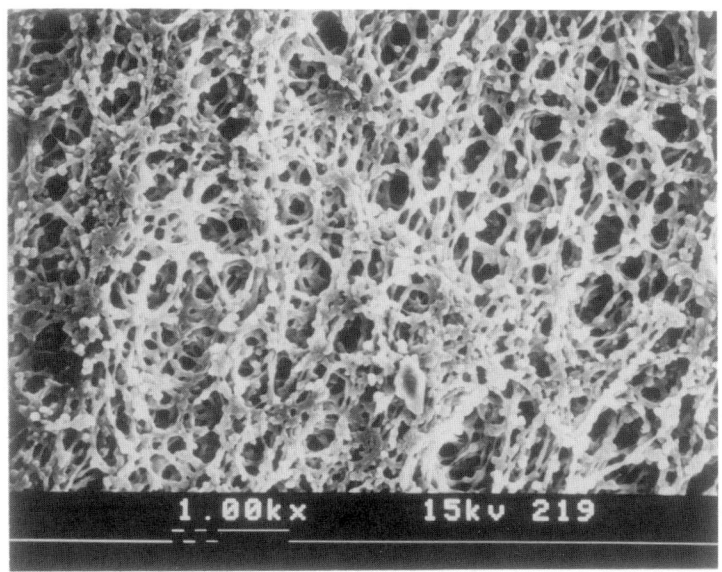

Fig. 1. Cross-section of microporous membrane. Particles are trapped on its surface or within pores. Electron micrograph.

membrane-permeating species pass equally with the solvent, independent of their concentration.

Microporous membranes (microfilters) are generally rigid, continuous meshes of polymeric material with pore diameters that are two or three orders of magnitude larger than those of ultrafilters. Species are either retained on the membrane surface or trapped in its substructure (Fig. 1). Microfilters retain bacteria, colloids, and particles upward of 0.025 µm in diameter, depending on rated pore size. Pore sizes typically used for microfiltration are 0.22 and 0.45 µm.

1.2. UF Materials and Devices

UF is fundamentally a very simple process. It marries the selective permeability of a membrane structure to a device or system that applies the required pressure, minimizes buildup of retained material on the filter, and provides for access and egress of the fluid. The surface of the UF membrane contains pores with diameters small enough to distinguish between the sizes and shapes of dissolved molecules. Those above a predetermined size range are rejected, whereas those below that range pass through the membrane with the solvent flow.

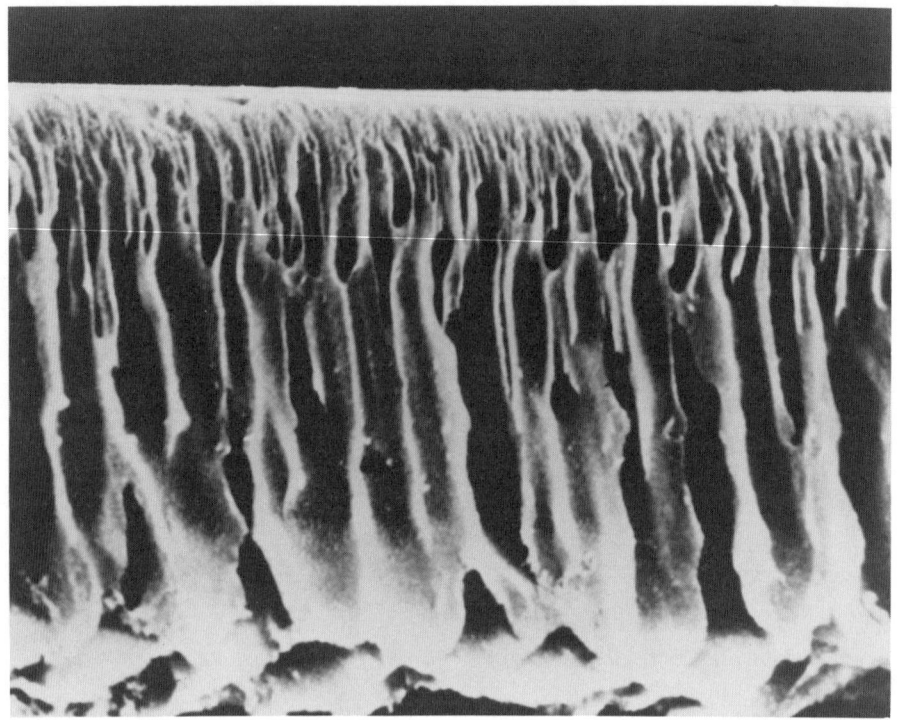

Fig. 2. Cross-section of anisotropic UF membrane. UF takes place in the top layer. Open-celled structure is highly permeable. Electron micrograph.

1.2.1. Membrane Ultrafilters

UF membranes are made of various polymers. They generally have two distinct layers. On the side in contact with the sample or fluid stream is a very thin (0.1–1.5 µm) dense "skin" with extremely fine pores whose diameters are in the range of 10–400 Å (1×10^{-6} to 4×10^{-5} mm). Below this is a much thicker (50–250 µm) open-celled substructure of progressively larger voids, largely open to the filtrate side of the ultrafilter (Fig. 2). Any species capable of passing through the pores of the skin (whose sizes are precisely controlled in manufacture) can therefore freely pass the membrane. That arrangement provides a unique combination of selectivity and exceptional throughput at modest pressure. It resists clogging because retained substances are rejected at the smooth membrane surface.

Most membranes are cast on tough, porous substrates for improved handling qualities and repeated use. They offer dependably controlled retention, water permeability, and solute transport. The best membranes are inert, noncytotoxic, and do not denature biological materials.

Some membranes with high flow characteristics are made of inert, nonionic polymer. They do not adsorb ionic or inorganic solutes, but may adsorb steroids and hydrophobic macromolecules. Advanced hydrophilic membranes have exceptionally low nonspecific protein binding properties. They should be used where maximum solute recovery is of special importance.

1.2.2. Devices for UF

1.2.2.1. STIRRED CELLS

Pressurized cells are a convenient means of UF for volumes in the range of 3 mL to 2 L. They are capable of final concentrate volumes of 50 µL to 60 mL, with concentration factors typically in the range of 60–80-fold.

A stirred cell (Fig. 3) is generally a vessel containing the solution to be filtered with a means of installing the membrane at the bottom, supported by a polymer grid. An access port permits pressurization, normally with nitrogen, in the range of 2.7–3.4 atm (40–50 psi). The pressure on the surface of the liquid forces the sample through the ultrafilter where separation between retained and passing solutes takes place. Formation of a layer of retained material on the membrane surface is minimized by means of a magnetic stirrer that is propelled by mounting the cell on a magnetic stirring table. The gentle stirring action assures minimal exposure of labile solutes to degradation by shear effects. The cell can also be connected to a pressurized reservoir that continuously refills the cell as filtrate flows from it, for desalting or buffer exchange.

Stirred cells accommodate membranes of various diameters, normally from 25–90 mm. Choice of membrane diameter involves two conflicting aspects. The larger the membrane (and therefore the cell), the faster the run will be accomplished. However, if the molecules to be concentrated are dilute and their maximum recovery is very important, a large membrane and cell-wall surface area may expose them to nonspecific adsorption and possibly significant loss. Therefore, if speed is important, a large cell should be used; if high recovery from a dilute solution is vital, the smallest possible cell (and membrane area) should be employed. Since large membranes and cells are more costly, that may also be a factor in size selection. Normally, the choice is a trade-off between the two extremes.

Fig. 3. Stirred UF cell on magnetic stirrer. The reservoir and selector valve are used for desalting or buffer exchange.

Stirred cells are excellent for solutions with up to 10% solute concentration. They are used widely for concentration or desalting of dilute proteins, enzymes, polypeptides, viruses, yeasts, bacteria, and so forth.

1.2.2.2. CENTRIFUGAL DEVICES

A selection of devices is offered by UF equipment manufacturers that employs centrifugal force to exert the needed pressure on the sample to obtain UF. The smallest of these devices—for initial volumes in the microliter range and up to 2 mL—consist of small, capped sample reservoir tubes with the UF membrane sealed across the bottom, supported by a polymer grid. The tube fits into a vial to capture the filtrate (Fig. 4). These units only require ordinary laboratory centrifuges, with centrifugal force of up to 14,000g. Use of a fixed-angle centrifuge rotor, rather than a swinging-bucket type, can reduce the time required for a separation run. Since

Ultrafiltration

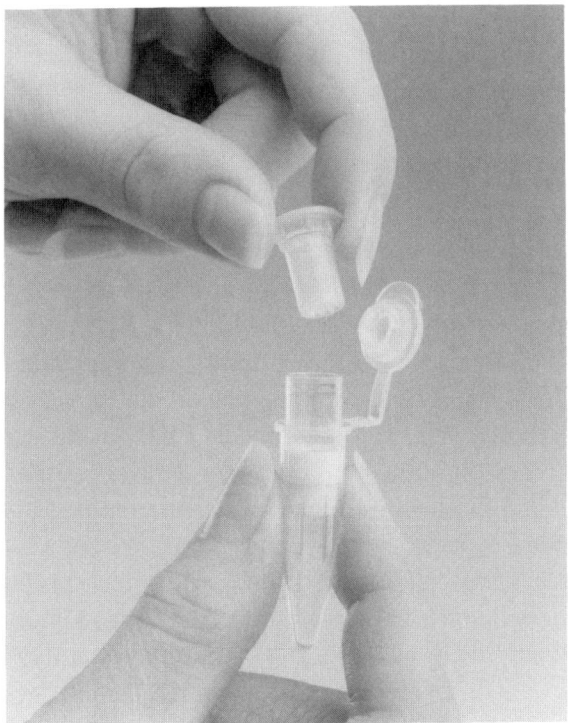

Fig. 4. Centrifugal microconcentrator with microporous insert (Micropure, top) for simultaneous separation of protein from electrophoresis gel and protein concentration or desalting.

centrifuge rotors can hold multiple units, multiple samples can be processed quickly at the same time. Some devices include a brief extra centrifugation step with the UF element reversed so that the side containing the retained material faces the filtrate vial. This drives every bit of the retentate into the vial for maximal recovery.

Centrifugal ultrafilters presently range in volume capacity from 0.5–15 mL. The smallest allow concentration from 0.5 mL initial volume to as little as 5 µL (100-fold concentration). Normal spin time for concentration at room temperature is 6–60 min, depending on the selected molecular weight cutoff (MWCO) as well as solute viscosity. Use of low-adsorption membranes and plastics makes these devices very efficient, typically delivering solute recoveries of over 90%. They concentrate the product without change in ionic environment or denaturation.

Some larger centrifugal devices include more elaborate mechanisms for separation. One, for example, is so constructed that centrifugal force drives dense material in the sample away from the ultrafilter surface. It is, therefore, effective in concentrating particle-laden samples with high recovery. This is useful, for example, for the concentration of cell supernatants, lysates and extracts.

1.2.2.3. STATIC DEVICES

For concentration of samples without any centrifuge, pressurization, or other accessory equipment, some devices use capillary action as the driving force to transport the liquid through the membrane. By letting the membrane form the wall of one or several chambers and backing it with an absorbent material, the solvent (water) in the samples in the chambers is pulled through the membrane by capillary action. This causes the individual samples to be reduced in volume and the retained molecules to be concentrated (Fig. 5). This type of device is widely used in clinical laboratories for the concentration of urine and cerebrospinal fluid in order to make it easier to detect very dilute disease-indicating species in those samples. Samples are loaded into individual chambers. The unit is left unattended for 1 or 2 h when the concentrated samples can be withdrawn with a pipet. A treatment near the bottom of the membrane prevents accidental drying of the sample. Graduation lines indicate the degree of concentration, which may be up to 200-fold. These devices can also be used to desalt the sample by repeated dilution and concentration.

1.3. Operating Parameters

1.3.1. Molecular Weight Cutoff

The MWCO of a membrane is defined as the molecular weight of hypothetical globular solutes (proteins) that will be 90% rejected by the membrane (Table 2). For example, a 10,000 MWCO ultrafilter will nominally reject 90% of molecules with a mol wt of 10,000 Dalton. Since rejection is actually a function of physical size, shape, and electrical characteristics of the molecule, the MWCO is only a convenient indicator, based on model solutes. Linear molecules, such as polysaccharides, will tend to slip through a membrane that would reject globular molecules of the same molecular weight.

Although two membranes can be claimed to have the same cutoff, they can exhibit quite different rejection behavior owing to distribution of pore diameters. Solute retention is not absolute. A "sharp" cutoff mem-

Ultrafiltration 123

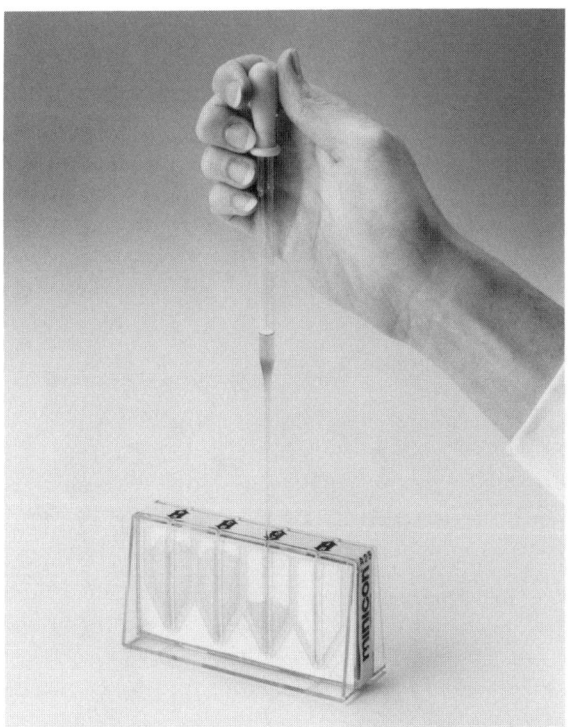

Fig. 5. Static UF device for use in clinical sample concentration.

brane will have minimal retention for species below its nominal MWCO rating. A "diffuse" cutoff membrane can significantly retain species of a size below the nominal MWCO or allow passage of some species above its cutoff. For concentration of retained species, either sharp or diffuse cutoff membranes will generally work equally well, but where the permeate is of interest, the final product may be markedly different. To determine cutoff sharpness, solute rejection tables for different membranes should be compared (see Note 1).

1.3.2. Solute Retention

Retention or rejection of a solute by an UF membrane defines the degree to which given molecules will be held back from the passing solution by the membrane and, hence, remain in the "retentate." For each membrane of a given MWCO, there is a specific degree of rejection of biomolecules. For example, cytochrome c (M_r = 12,400) may be rejected >98% by a 3000 MWCO membrane, >95% by a 10,000 MWCO mem-

Table 2
Typical Solute Rejection with UF Membranes[a]

Solute	M_r	YC05	YM1	YM3	YM10	YM30
NaCl	58	<20	0	0	—	
Dextrose	180	>70	0	0	—	—
Sucrose	342	>85	45	20	—	—
Raffinose	504	95	65	25	—	—
Bacitracin	1400	98	92	>80	20	—
Inulin	5000	>98	95	—	25	—
Cytochrome c	12,400	>98	>98	>98	>95	<15
Myoglobin	17,000	>98	>98	>98	>98	—
α-Chymotripsinogen	24,500	>98	>98	>98	>98	>80
Albumin	67,000	>98	>98	>98	>98	>98
IgG	160,000	>98	>98	>98	>98	>98
Apoferritin	480,000	>98	>98	>98	>98	>98
IgM	960,000	>98	>98	>98	>98	>98

[a]Table courtesy of Amicon, Inc.

brane, and <15% by a 30,000 MWCO membrane. Concentration proceds in direct proportion to volume reduction, i.e., solute concentration doubles at 50% volume reduction. For guidance, manufacturers provide rejection tables or curves in their literature.

For freely membrane-permeating species, such as sodium chloride, the concentration of solute in the retentate and the permeate will be equal. If, for example, a solution containing 1% protein and 2% salt is processed with a membrane that is totally retentive to the protein, doubling the protein concentration by reducing the starting solution by 50% will result in a retentate containing 2% protein and 2% salt.

Many biological macromolecules tend to aggregate, or change conformation, under varying conditions of pH and ionic strength, so that their effective size may be much larger than that of the "native" molecule. This will cause increased retention at the membrane. The degree of hydration, counterions, and steric effects can also cause molecules with similar molecular weights to exhibit very different retention behavior.

Solute/solvent and solute/solute interactions in the sample can also change effective molecular size. For example, some proteins will polymerize under certain concentration and buffer conditions, whereas others (such as heme proteins) may dissociate into corresponding subunits.

Ultrafiltration

Ionic interactions or p-p stacking can cause small molecules to behave similarly to molecules of greater M_r. When this occurs, as in the case of phosphate ions with a 500 MWCO membrane, the small molecules may not effectively permeate the membrane.

1.3.3. Concentration Polarization

As solute concentration increases during the process of enrichment, solute at the membrane surface forms a gel that is permeable to solvent under pressure. This effect is called concentration polarization. At moderate to high concentration of retained solutes, the flow resistance of the gel layer will reach a level where it significantly exceeds that of the membrane, in effect forming a secondary membrane. As solute continues to accumulate at the membrane/liquid interface, resistance grows. When net transport of solute by convection equals the back diffusion of solute toward the bulk solution, owing to the concentration gradient, further increase in transmembrane pressure will not increase flow through the membrane and may cause it to decrease.

All efficient UF devices or systems must provide the means to minimize the effect of concentration polarization. The most important of these are magnetic stirring, pumped tangential flow across the membrane surface in narrow channels, and positioning the membrane surface at an acute angle with respect to the force vector acting on the fluid. The latter is achieved in centrifugal UF devices by using an angle-head rotor. Centrifugal force causes the gel film to slide across the angled surface, keeping the rest of the membrane surface relatively clear for permeation by solvent and membrane-passing molecules at relatively high rates. Cone-shaped membranes also employ the force vector at an angle during centrifugal separation.

Flow rate decrease owing to concentration polarization should not be confused with the effect of membrane fouling. Fouling is the deposition and accumulation of submicron particles and solute on the membrane surface, or crystallization and precipitation of smaller solutes on or within the pores of the membrane. There may, in addition, be chemical interaction with the membrane.

1.3.4. Maximizing Solute Recovery

Although UF membranes are normally inert, adsorptive losses may occur. Additional losses, caused by formation of concentrated solute gel or cake on the membrane, can be counteracted by polarization control, as indicated in Section 1.3.3., by operating at modest pressures, and by a final agitation cycle at zero transmembrane pressure.

Effects of adsorption are more noticeable with dilute solutions where adsorption may severely diminish the amount of the desired product. Since adsorption is largely a function of membrane and device surface area, the relation of sample concentration and volume to surface area should be considered before choosing a system. Small, dilute samples should be concentrated by using membranes or devices with minimum surface area, while maintaining reasonable flow characteristics.

Buffer can affect membrane adsorption. Phosphate buffers can cause increased losses. Tris or succinate buffers allow better recovery. This may relate to lyotropic effects on hydrophobic bonding to the membrane or device.

Although rejection is used to characterize membrane performance, it does not always directly correlate with solute recovery from a sample or volume. Actual solute recovery—the amount of original material recovered after UF—is generally based on mass balance calculations (*see* Note 2).

1.3.5. Temperature and pH

Raising the operating temperature normally increases UF rates. Higher temperature increases solute diffusivity (typically 3–3.5%/°C for proteins) and decreases solution viscosity. One normally operates at the highest temperature tolerated by the solutes and the equipment. However, UF equipment is often used in cold rooms.

Changing solution pH often changes molecular structure. This is especially true for proteins. At the isoelectric point, the protein begins to precipitate in some cases, causing a decrease of filtrate flow.

1.4. Applications

1.4.1. Concentration

In macromolecular concentration, the membrane enriches the content of a desired biological species or provides filtrate cleared of retained substances. Microsolute (e.g., salt) is removed convectively with the solvent. Pressure, created by external means, forces liquid through the ultrafilter. Solutes larger than the nominal MWCO of the membrane are retained. The required pressure can be generated by use of compressed gas, pumping, centrifugation, or capillary action. With dilute solutions (1 mg/mL or less), flow rates are directly proportional to applied pressure. At higher concentrations, increased viscosity and polarization con-

Table 3
Desalting by Repetitive Concentration and Redialysis[a]

Spin number[b]	Protein recovery	NaCl concentration
Start point	—	500 mM
1	95.1%	140 mM
2	94.2%	25 mM
3	94.0%	5 mM

[a]Table courtesy of Amicon, Inc.
[b]Start: 15 mL of γ-globulin. Spin 1: reduced to 3 mL and rediluted to 15 mL. Spins 2 and 3: repetition of reduction to 3 mL and dilution to 15 mL. After third spin, salt concentration reduced 100-fold.

centration act to reduce the flow, requiring steps to reduce concentration polarization (see Note 3).

1.4.2. Salt Removal or Buffer Exchange (Diafiltration)

Removal of small molecules from a solution by alternating UF and redilution, or by continuous UF and dilution to maintain constant volume, is called diafiltration.

Ultrafilters are ideal for removal or exchange of salts, sugars, nonaqueous solvents, separation of free from protein-bound species, removing materials of low molecular weight, or rapid change of ionic and pH environment. In contrast with dialysis, the rate of microsolute removal or "wash-out" by diafiltration is a function of the UF rate, and independent of microspecies (e.g., salt) concentration. This greatly reduces desalting times by using convective salt removal or exchange at flow rates equal to those of solvent passage. Diafiltration is also used for efficient microsolute exchange, or "wash-in."

Membrane-permeating solutes (i.e., those significantly smaller than the cutoff, especially salts) pass through the membrane pores at the same rate as water. Transport through the membrane is by convection, not by diffusion, so that the rate of permeation is independent of molecular size.

Diafiltration washes microspecies from the solution, purifying the retained species (see Notes 4 and 5). In the discontinuous method, the sample is diluted before concentration or it is diluted after concentration and reconcentrated; this can be repeated one or more times, each time obtaining further desalting or solvent exchange (Table 3). Small volumes may be easily desalted in one step by sample dilution before concentration. The continuous method of diafiltration is to connect the UF device (such as a stirred cell) to a pressurized reservoir containing solvent, normally buffer or water. As filtration proceds, solvent automatically flows from the reservoir into the device, at the same rate as the rate of filtration.

UF does not change salt concentration or buffer composition. A solution volume with 100 mM salt still contains 100 mM salt after concentration. Discontinuous diafiltration (rediluting the retentate with water and concentrating again) effectively decreases the salt concentration of the sample by the concentration factor of the UF. To achieve more complete salt removal, multiple concentration and redilution are required. For example, if a 1-mL sample containing 100 mM salt is diluted to 2 mL before concentration in a centrifugal device, the salt concentration in the 2-mL sample will be 50 mM. When reduced to 25 µL (80 times), the concentrate will still contain 50 mM salt. If more complete salt removal is desired, the sample can be rediluted with water or buffer to 2 mL before reconcentration. At this point, the salt concentration will have been reduced to 0.625 mM (50 mM/80), which will remain after the second concentration to 25 µL. Each further such dilution and reconcentration step would reduce salt concentration by 1/80, in the present example. For most samples, three concentration/reconstitution/reconcentration cycles will remove about 99% of the initial salt content. With very small sample volumes, dilution of the sample before the initial concentration step can often decrease salt concentration to an acceptable level.

1.4.3. Separation of Free from Protein-Bound Microsolute

In the past, free drug levels in serum or plasma samples were not widely measured, partly for want of a convenient means of separating free from protein-bound drug. The chosen technique was generally equilibrium dialysis, a time-consuming procedure, subject to effects of dilution and buffers. However, this method does not directly indicate the free drug concentration in the sample. Such other techniques as ultracentrifugation and gel filtration are no less time consuming, and there is inadequate standardization of results.

Today's better alternative for free/bound separation is UF with a centrifugal UF device specially designed for filtrate recovery. Spun in a standard laboratory centrifuge (preferably angle-head), free drugs readily pass the membrane for collection and analysis. The sample is not diluted in the process. Multiple samples are conveniently handled, typically in 10-min runs/set.

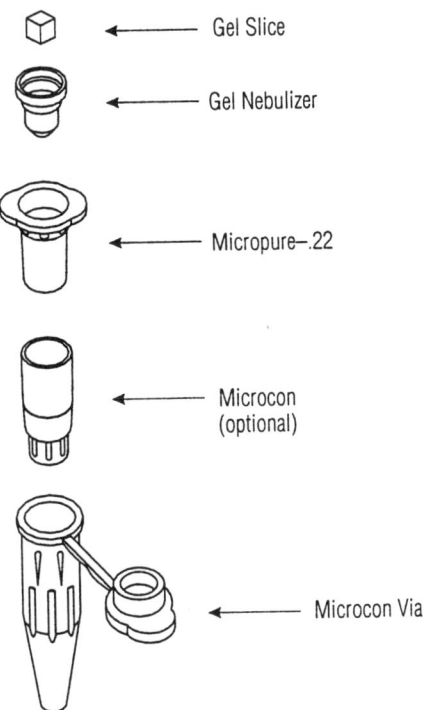

Fig. 6. Stacked elements of device for extraction of DNA, RNA, or proteins from gel.

1.4.4. Recovery from Electrophoresis Gels

The availability of microporous inserts for centrifugal UF devices offers an easy method of recovering proteins (or other macromolecules) from electrophoretic gels. The gel piece containing the protein of interest is crushed or macerated to increase its surface area, then placed into the insert, and mixed with an elution buffer. During subsequent incubation, protein diffuses out of the gel. Centrifugation of the combined devices causes the buffer containing the protein to flow through the microporous membrane of the insert, which retains the gel particles. At the membrane surface of the UF device, the protein is retained for recovery, while the buffer and any salts pass into the filtrate vial. An added new device, placed into the insert described (Gel Nebulizer), makes the process even easier by converting pieces of gel into a spray of fine particles during centrifugation (Fig. 6). This makes extraction of the proteins more efficient.

1.4.5. Purification and Fractionation

Macromolecular mixtures may be separated into size-graded classes by UF, provided the species to be separated have at least a 10-fold difference in molecular weight (*see* Note 6). This can be accomplished either by direct UF, where the permeating solute is obtained in its initial concentration in the ultrafiltrate, or by diafiltration, where the retained solute is obtained in its initial concentration in the retentate and the diluted permeating solute in the ultrafiltrate. Normally, polarization effects require predilution or multiple dilution and reconcentration of the sample.

1.4.6. Detergent Removal

UF membranes efficiently remove detergents from protein solutions. The chemical nature of most detergents causes micelle formation above a critical concentration limit (Critical Micelle Concentration), causing aggregation of the detergent and leading to gross changes in molecular structure. This affects the amount of the detergent that can be removed from solution with membranes of specific cutoff (*see* Note 7).

2. Materials

All UF equipment and membranes may be obtained from Amicon (Beverly, MA). In addition, the following may be required depending on the application:

1. Oxygen-free nitrogen with pressure regulator (for stirred cells) (*see* Note 8).
2. Microcentrifuge (variable speed) and microcentrifuge tubes.
3. Homogenizer (e.g., Eppendorf fitting pestle, VWR cat. no. KT749515–0000 or KT749520–000 and pestle mixer motor, VWR cat. no. KT749540–0000).

The following solutions are required for staining and recovering proteins from polyacrylamide gels:

1. Gel-staining solution: 0.1% Coomassie brilliant blue R250, 50% methanol, 10% acetic acid.
2. Destaining solution: 7% acetic acid, 12% methanol.
3. Wash solution: 50% methanol.
4. Extraction buffer: 100 mM NaHCO$_3$, 8M urea, 3% SDS, 0.5% Triton X-100 (reduced), 25 mM DTT.

3. Methods
3.1. Using a Stirred Cell for Concentration

1. Prefilter or centrifuge any solution containing particulate matter, such as cell debris or precipitates (*see* Note 9).
2. Fill a cell of appropriate size (*see* Section 1.2.2.) with the sample, and then secure it on a magnetic stirring table. Connect the inlet line to a regulated

Ultrafiltration 131

gas pressure source providing 2.7–3.4 atm (40–50 psi). Nitrogen is recommended (*see* Note 8).
3. Pressurize the cell according to instructions. Keep within cell pressure limits. When operating with either hazardous or specially valuable materials, always pressure-check the cell first to assure that all components are properly assembled.
4. Turn on the stirring table, and adjust the stirring rate until the vortex created is approx one-third the depth of the liquid volume.
5. When the run is completed, continue stirring for a few min after depressurization to maximize recovery of retained substances. This will resuspend the polarized layer at the membrane surface.

3.2. Protein and Peptide Recovery from Polyacrylamide Gels

The following protocol is recommended by Amicon for its Micropure™ microporous inserts and Microcon® microconcentrators:*

1. Stain the gel with staining solution. Remove excess dye by soaking the gel in destaining solution with gentle agitation for approx 2 h or until a clear background is obtained.
2. Cut out the gel piece containing the band of interest with a clean scalpel or razor blade.
3. Place the gel piece in a microcentrifuge tube. Remove dye from the protein by adding 1 mL of wash buffer and sonicating for 5–15 min at 50–60°C or until the gel clears.
4. Remove the wash buffer. Add 50–100 µL of extraction buffer. Incubate for 20–30 min at 65°C.
5. Homogenize the gel using a motor-driven pestle homogenizer.
6. Incubate the tube with homogenized gel in a bath at 50–60°C for 2–3 h or overnight.
7. Place the Micropure-0.22 insert into the Microcon sample reservoir (use Microcon-100 or Microcon-50 for large proteins and Microcon-10 or Microcon-3 for polypeptides). Add 100 µL wash solution. Transfer the gel slurry into Micropure with a pipet. Rinse the tube with 100 µL of wash solution to remove remaining slurry. Transfer to Micropure.
8. Spin the assembly until all liquid is removed from the Micropure insert (13,000g for 20–30 min in Microcon-10; for others, *see* the Centrifugation Guidelines in Table 4). Discard Micropure. Proteins or peptides are retained above the Microcon membrane, depending on selected membrane MWCO.

*All Amicon tradenames mentioned are trademarks of which Minicon, Centricon, and Microcon are registered.

Table 4
Centrifugation Guidelines[a]

Membrane MWCO[b]	Maximum g-force	Spin times[c] 4°C	25°C
3000	14,000	185	95
10,000	14,000	50	35
30,000	14,000	20	12
50,000	14,000	10	6
100,000	3000[d]	35	15

[a]Table courtesy of Amicon, Inc.
[b]MWCO in Daltons.
[c]Spin time in minutes; 500-µL samples concentrated to 10 µL.
[d]500g for DNA/RNA samples.

9. To remove the extraction buffer from the protein, add 400 µL of wash solution to the Microcon units. Spin as indicated in the Centrifugation Guidelines (Table 4). To complete buffer exchange, add 400 µL of the desired buffer and spin at 13,000g for 30 min (for Microcon-10; for others, see Centrifugation Guidelines in Table 4).
10. To recover concentrated sample, invert Microcon into a new vial. Spin at 1000g for 3 min.

3.3. Concentration of Antibodies from a Hybridoma

By inserting a microporous filter into a centrifugal UF unit, the sample can be run through two stages of separation at the same time. The insert, with a porosity of 0.22 or 0.45 µm, retains all particulate matters (such as bacteria, cell fragments, or electrophoretic gel particles), but lets the solution containing the solutes of interest—such as proteins—flow into the UF element where the protein is concentrated. The small microporous insert can also be used by itself, with a standard microcentrifuge vial, to free samples of particles. The following is a protocol offered by Amicon for its Micropure microporous inserts and Microcon microconcentrators.

1. Place the Micropure-0.45 insert into the Microcon sample reservoir (use Microcon-50 for IgG or Microcon-100 for IgM).
2. Pipet up to 350 µL of sample into the Micropure insert.
3. Spin at 500g for 5 min or until there is no more liquid in Micropure insert.
4. Discard Micropure. Antibody is retained above the membrane in the Microcon.
5. If the antibody is to be desalted or if buffer is to be exchanged, fill the Microcon sample reservoir with 450 µL of desired buffer. Spin Microcon-50 at 12,000g for 8 min. For Microcon-100, spin at 3000g for 15 min.

6. To recover concentrated antibody, invert Microcon into a new vial. Spin at 1000g for 3 min.

4. Notes

1. The 90% rejection standard and the significance of molecular shape rather than weight make it very important that the selected membrane cutoff be well below the molecular weight of the solute to be retained. As a rule, it is best to choose a membrane or device with cutoff at about half of the molecular weight of the protein to be concentrated. This provides a balance between high protein recovery and minimal filtration time. For example, if bovine serum albumin (M_r = 67,000) were the protein of interest, a 30,000 MWCO (rather than 50,000 MWCO) membrane or device would result in the most efficient concentration and recovery of the protein in the retentate. For highest possible retention, a membrane cutoff one-tenth of the M_r of the solute should be used (10,000 in the example above). Of course, the tighter the membrane, the slower the filtration rate. When species of low molecular weight are to be exchanged, the membrane cutoff should be substantially above that of the passing solute.
2. Recovery of protein can sometimes be improved by passivation of the membrane. For very dilute protein solutions (in the range of 1 µg/mL), concentrate recovery is often not quantitative (Fig. 5). This loss of protein can be caused by nonspecific binding of the protein to exposed binding sites on the plastic of the concentration device. The extent of nonspecific binding varies with the relative hydrophobicity of individual protein conformations. Pretreating (passivating) the plastic by blocking the available binding sites before concentration can often improve the recovery yield from dilute protein solutions. For passivation procedures, manufacturers' recommendations should be followed. The procedure is simple, requiring the filling of devices with a prescribed solution, leaving them overnight, and then rinsing thoroughly. Among recommended passivation solutions are 1% IgG in PBS, Tween-20, or Triton-X in distilled water, bovine serum albumin, and even powdered milk in distilled water.
3. Highly viscous solutions filter slowly, as do solutions containing particulate matter, such as colloids. Where a viscous agent (sucrose, glycerin, etc.) is to be removed, flow can often be increased by predilution.
4. For rapid desalting (diafiltration), the UF cell can be connected to an auxiliary reservoir containing a diafiltrate solution with the desired microsolute concentration, if any. The auxiliary reservoir is then gas pressurized to 2.7–3.4 atm (40–50 psi). The cell's fluid volume, as well as the macrosolute concentration, remain constant as filtrate is automatically replaced by diafiltrate solution. This technique provides a simple means for rapid microsolute exchange. It is typically used as substitute for dialysis. A Concentration/Dialysis Selector Switch (Amicon) permits simple switching between concentration and diafiltration (Fig. 3).

5. Fully membrane-permeating molecules pass at the same rate as salt because their transport through the membrane is independent of their individual molecular sizes. Proteins of $M_r > 3000$ may be separated from each other and from smaller molecules if they differ by a factor of about 10 in molecular size and are not associated in solution. For example, one cannot diafilter bovine serum albumin ($M_r = 67,000$) from IgG ($M_r = 160,000$), but can easily diafilter biotin ($M_r = 244$) from cytochrome c ($M_r = 12,400$). Small volumes should be prediluted to assist in diafiltration.
6. UF is not primarily a fractionation technique. It can only separate molecules that differ by at least an order of magnitude in size and is best done in a diafiltration arrangement, with multiple washes to separate the mixture of macromolecules. Molecules of similar size cannot be separated by UF. During concentration polarization, the gel layer on the membrane surface superimposes its own rejection characteristics on those of the membrane. Usually, concentration polarization increases rejection of lower M_r species. A membrane with a nominal 100,000 MWCO may reject 10–20% of albumin ($M_r = 67,000$) in a 0.1% solution of pure albumin. However, in the presence of larger solutes, such as IgG, it may reject 90% of the albumin.
7. For example, the monomer of Triton X-100 ($M_r = 500$–650) should pass readily through a 3000 MWCO membrane. However, at concentrations above its Critical Micelle Concentration of 0.2 mM, Triton X-100 forms micelles composed of approx 140 monomeric units. During UF, the micelles behave like globular proteins of $M_r = 70,000$–$90,000$. Therefore, above the Critical Micelle Concentration of Triton X-100, a 100,000 MWCO membrane would be required to removed the detergent effectively.
8. Use of compressed air can cause large pH shifts owing to dissolution of carbon dioxide. With sensitive solutions, oxidation can occur, leading to other potential problems.
9. Because of their unique design, which uses gravitational force to counteract deposition of suspended particles on the membrane, certain concentrators (Centriprep®, Amicon) are especially useful for UF of solutions with high solid content. The feature counteracts membrane fouling and eliminates the need for prefiltration of samples. It can also be obtained with a microporous filter for separation of antibodies from cell-culture supernatant or for desalting of bacteria. Another centrifugal UF device uses a low-adsorptive, polypropylene sample reservoir, which minimizes nonspecific adsorption of solutes to the walls of the devices. It also contains a low-adsorptive, hydrophilic membrane. No passivation is required before processing very dilute protein solutions. By employing a final inverted-spin feature, high solute recovery is possible, even from dilute solutions in the 1 µg/mL range (Centriplus™).

CHAPTER 13

Bulk Purification by Fractional Precipitation

Shawn Doonan

1. Introduction

The solubility of a particular protein in aqueous solution depends on the solvent composition and on the pH; hence, variation in these parameters provides a way of purifying proteins by fractional precipitation. The factors that dictate solubility are complex, since the surface of a protein is itself complex containing ionized residues, polar regions, and hydrophobic patches, all of which will interact with the solvent in ways that are not completely understood. Hence, it is not possible to elaborate a theoretical approach to fractional precipitation *(1)*; rather, the methods are used in an essentially empirical fashion. The most widely used procedures are precipitation by addition of salt (salting out) and by addition of organic solvents; these are the focus of this chapter (*see* Note 1).

Addition of high concentrations of salt to a protein solution causes precipitation largely by removing water of solvation from hydrophobic patches on the protein's surface, thus allowing these patches to interact with resulting aggregation. For a pure protein, the relationship between solubility S (in g/kg of water) and the ionic strength I (in mol/kg water) is given by:

$$\log S = \beta - K_s[(I/2)] \qquad (1)$$

where β and K_s are constants for a particular protein at a particular pH and temperature. The point here is that a protein will precipitate over a range of ionic strength values (determined by the value of K_s) and that different proteins will precipitate over different, but frequently overlapping, ranges. This highlights the fact that it is rarely possible to purify a particular protein from a complex mixture by using fractional precipitation alone (although heroic attempts were made to do this in the early days of protein pruification). Rather, the value of the method is that it provides a simple procedure for enrichment of the protein of interest from large volumes of extracts and, at the same time, can be used to concentrate the fraction (*see* Chapter 10); with large-scale purifications in particular, this is important in reducing the problem to a manageable scale. Hence, fractional precipitation by salt is almost invariably used at an early stage of a purification procedure, often on the clarified extract, to obtain an initial purification and concentration (*see* Note 2). Further purification is then achieved by chromatographic methods, which have higher resolving power, but generally lower capacity.

A variety of salts has been used in the past for this purpose, including NaCl, Na_2SO_4, KCl, $CaCl_2$, and $MgSO_4$, and these are still sometimes used for particular applications. By far the most frequently used salt is, however, ammonium sulfate ($[NH_4]_2SO_4$). The reasons for this include its high solubility in water (about $4M$ at saturation), its low heat of solution, the fact that the density of saturated solutions (1.235 g/mL) is less than that of proteins, hence allowing for their collection by centrifugation, and the essentially innocuous nature to proteins of its constituent ions. Concentrations of the salt are traditionally expressed in terms of percentage saturation at a particular temperature; Table 1 gives the amounts of solid ammonium sulfate required to obtain the required concentration at 0°C *(2)*, the temperature at or near which fractional precipitation is usually carried out (*see* Note 3).

The precipitation of proteins by addition of organic solvents is a more complex process. Factors involved probably include the decrease in dielectric constant, which promotes aggregation by charge interaction, as well as sequestration of water of solvation of the protein. The same caveats about the usefulness of the method apply as discussed for salt fractionation, although solvent fractionation has achieved some noteworthy successes, such as the classical Cohn fractionation of plasma proteins (summarized in ref. *3*). There is with this method, however, the added problem that organic solvents can cause protein denaturation by interaction with hydrophobic residues in the protein's interior. Hence, it is usually essential to carry out solvent precipitation at a low temperature to minimize denaturation. The solvents used need

Table 1
Amounts of Solid Ammonium Sulfate Required to Change the Concentration
of a Solution from a Given Starting Value to a Desired Target Value at 0°C

Initial percentage saturation at 0°C	Target percentage saturation at 0°C[a]																
	20	25	30	35	40	45	50	55	60	65	70	75	80	85	90	95	100
0	106	134	164	194	226	258	291	326	361	398	436	476	516	559	603	650	697
5	79	108	137	166	197	229	262	296	331	368	405	444	484	526	570	615	662
10	53	81	109	139	169	200	233	266	301	337	374	412	452	493	536	581	627
15	26	54	82	111	141	172	204	237	271	306	343	381	420	460	503	547	592
20		27	55	83	113	143	175	207	241	276	312	349	387	427	469	512	557
25			27	56	84	115	146	179	211	245	280	317	355	395	436	478	522
30				28	56	86	117	148	181	214	249	285	323	362	402	445	488
35					28	57	87	118	151	184	218	254	291	329	369	410	453
40						29	58	89	120	153	187	222	258	296	335	376	418
45							29	59	90	123	156	190	226	263	302	342	383
50								30	60	92	125	159	194	230	268	308	348
55									30	61	93	127	161	197	235	273	313
60										31	62	95	129	164	201	239	279
65											31	63	97	132	168	205	244
70												32	65	99	134	171	209
75													32	66	101	137	174
80														33	67	103	139
85															34	68	105
90																34	70
95																	35

[a]g of solid ammonium sulfate/L of solution.

to be miscible with water in all proportions and nontoxic; acetone and ethyl alcohol best meet these requirements (*see* Note 4).

Whichever of the two methods is used, there will necessarily be a trade-off between degree of purification achieved and yield. An ammonium sulfate or solvent "cut" over a 10% concentration range might give a yield of 70% of the desired protein with a purification factor of fivefold. Increasing the concentration range may increase the yield, but with a corresponding decrease in purification factor. For an example of the use of both methods in purification of the same protein, *see* ref. *4*.

2. Materials
2.1. Fractional Precipitation with Ammonium Sulfate
1. Solid ammonium sulfate (*see* Note 5).
2. Magnetic stirrer or electrical paddle stirrer.

3. Ice bath.
4. Refrigerated centrifuge and rotor, e.g., a 6 × 250-mL angle rotor.
5. Screw-top plastic bottles for the rotor to be used.
6. Visking dialysis tubing (14- or 19-mm inflated diameter; 26- or 31-mm flat width).

2.2. Fractional Precipitation with Acetone

1. Analar acetone precooled to −20°C (*see* Note 6).
2. Electrical paddle stirrer.
3. Ice-salt cooling bath.
4. Centrifuge, rotor, bottles, and dialysis tubing as in Section 2.1.

3. Methods

3.1. Fractional Precipitation with Ammonium Sulfate

It is assumed that a trial experiment has been carried out in order to determine the optimal concentration range of ammonium sulfate for the particular protein sample to be fractionated (*see* Note 7). Suppose that the trial showed the best results to be obtained with the fraction precipitated between 35 and 50% saturation. Proceed as follows.

1. Measure the volume of the protein solution to be fractionated, and pour it into a glass beaker of capacity about twice the measured volume of solution.
2. Weigh out 0.194 g of solid ammonium sulfate for every 1 mL of protein solution (Table 1). If the ammonium sulfate contains lumps, then these should be broken up using a mortar and pestle.
3. Place the beaker containing the protein on ice, and stir either with a magnetic stirrer or with a paddle stirrer. A slow rate of stirring should be used to avoid foaming.
4. Add the ammonium sulfate to the protein solution in small batches over a period of several minutes ensuring that one lot of ammonium sulfate has dissolved before adding the next (*see* Note 8).
5. After addition is complete, leave to stand for 10 min to ensure equilibrium, and then remove precipitated protein by centrifugation at about 5000*g* and 4°C for 30 min using screw-top plastic tubes (*see* Note 9).
6. Decant off the supernatant solution from each tube into a measuring cylinder, and determine the total volume (*see* Note 10).
7. Pour the combined supernatants into a beaker in the ice bath and add 0.087 g of ammonium sulfate/mL of protein solution (the amount required to take the concentration from 35 to 50% saturation; *see* Table 1) using the same precautions as in steps 2–4.
8. Recover the precipitated protein by centrifugation as in step 5, decant the supernatant solution into a beaker (*see* Note 10), and suspend the protein

pellets in the minimum volume of water or of an appropriate buffer (*see* Note 11).
9. Transfer the protein suspension to a sack of Visking dialysis tubing, and dialyze against at least two changes of buffer using 100 times the volume of the sample and allowing 3–4 h for equilibration (*see* Note 12).
10. After dialysis, remove any precipitated material by centrifugation.

3.2. Fractional Precipitation with Acetone

Again, a trial experiment must be carried out to determine the optimum precipitation range for the particular protein fraction (*see* Note 7). Assuming this to be between 37.5 and 50% (v/v), proceed as follows.

1. Measure the volume of the protein solution, pour it into a beaker (*see* Note 13) immersed in an ice-salt bath, and stir until the temperature reaches 0°C.
2. For each 1 mL of protein solution, add 0.60 mL of acetone (precooled to –20°C) dropwise with constant stirring and at such a rate that the temperature does not rise above 0°C (*see* Notes 8 and 14). After addition of acetone is complete, continue sitrring for 10 min with constant control of temperature.
3. Remove precipitated protein by centrifugation at 0°C for 10 min at 3000g using precooled centrifuge tubes (*see* Note 15).
4. Measure the volume, and return the combined supernatant solutions to the beaker in the ice-salt bath. Add a further 0.25 mL of acetone/mL of protein solution using the precautions described in steps 1 and 2.
5. Recover the precipitated protein by centrifugation as in step 3. Pour off the supernatant solutions, and invert the centrifuge tubes over filter paper to drain; blot off any drops of solution adhering to the walls of the tubes.
6. Resuspend the pellets in water or an appropriate bufffer, and remove residual acetone by dialysis, membrane filtration, or gel filtration (*see* Note 16).

4. Notes

1. An alternative that is sometimes used is precipitation by alteration of pH. Proteins will generally have minimum solubility at their isoelectric points, where the net charge is zero and there is no electrostatic repulsion. The approach is to incubate the protein fraction at various pH values to see if the species of interest precipitates; obviously, if the isoelectric point of the protein is known, then that pH should be used. A problem with the method is that the protein may not be stable at its isoelectric point if this is far removed from neutrality. In addition, the protein may not precipitate, particularly if its concentration is low, in the absence of added salt; this makes establishing the conditions for pH fractionation quite difficult.
2. There is a further advantage in using ammonium sulfate fractionation immediately after homogenization and clarification (*see* Chapter 2). Some

tissues give homogenates that are very difficult to clarify because of the presence of membrane fragments and nucleoprotein complexes, which resist sedimentation. This particulate matter usually aggregates at low concentrations of ammonium sulfate and sediments readily in the first fraction. Hence, unless the protein of interest is contained in this fraction, the protein solution obtained from this procedure will be devoid of suspended matter as required for use of such techniques as column chromatography. Fractionation with organic solvents confers the same advantage. In this case, it is because of the low density and viscosity of water/solvent mixtures, which facilitate sedimentation of particulate matter.

3. Reference 3 gives a corresponding table for 25°C, which may be used if it is preferable to carry out fractionation at room temperature. Alternatively, the following formula can be used:

$$g = [533 (P_2 - P_1)/(100 - 0.3 P_2)] \qquad (2)$$

where g is the number of grams of ammonium sulfate required to change the concentration of 1 L of solution from P_1 to P_2% saturation at 20°C.

4. An alternative to using organic solvents is provided by fractionation with water-soluble polymers of which polyethylene glycol (PEG) is the most commonly employed. The mechanism of precipitation seems to be by steric exclusion, i.e., the protein is concentrated in the extra-polymer space of the solution until its solubility limit is exceeded and precipitation occurs *(5)*. Consistently, larger proteins tend to precipitate earlier than smaller ones, and precipitation is relatively insensitive to pH and ionic strength. The commonly used precipitants are PEG 4000 or PEG 6000 (i.e., PEGs with mol-wt averages of 4000 or 6000). The great advantage of these precipitants is that they have little tendency to denature proteins. For further details, *see* ref. 5.

5. Analar ammonium sulfate is more than twice as expensive as the general-purpose grade (around $100 for 5 kg as compared to about $45 for GPR). It is doubtful whether the extra cost is worth it when dealing with large volumes of crude protein mixtures; Aristar grade is certainly not worth using at about $225/kg. The major problem with cheaper grades is the presence of low amounts of heavy metal contaminants; if the protein of interest is metal-sensitive, then EDTA (10 m*M*) can be added to the solution to remove these contaminants.

6. Use Analar grade, which is only marginally more expensive than the general-purpose grade. If carrying out fractionation with ethanol, then either absolute ethanol (>99%) or the more commonly available 96% variety may be used; in the latter case, the volume added to achieve the desired concentration would need to be adjusted to allow for the water content.

Fractional Precipitation

7. Optimum in this context means the range of ammonium sulfate concentrations that gives the desired balance between yield and purification. If the protein of interest has been purified previously, then the required range may be available in the literature, but it should be borne in mind that precipitation will depend on pH, on buffer composition, and on the protein concentration and composition of the fraction. Hence, unless these are identical with those in published procedures, it is unwise to assume that the protein will behave identically in your purification.

    Ideally, the trial ammonium sulfate fractionation should be carried out exactly as described in Section 3.1., but using a small volume (~20 mL) of fraction. That is, ammonium sulfate should be added to concentrations of 0–20, 20–30%, and so on, in 10% steps at each stage removing precipitated protein by centrifugation before increasing the salt concentration. The recovered precipitates should be dissolved in buffer and assayed for the protein of interest and for total protein; the latter can be done most simply by measuring the absorbance of a suitably diluted sample at 280 nm and using the approximate relationship that an absorbance of one corresponds to a protein concentration of 1 mg/mL (accurate enough for present purposes). If the test for activity of the protein is sensitve to NH_4^+ or SO_4^{2-} ions, then the dissolved pellets will have to be dialyzed before assay. This trial should give the necessary information to proced to large-scale work, but if, for example, the protein of interest precipitates equally in the ranges 20–30% and 30–40%, it may be worthwhile checking to see if the range 25–35% gives better results than simply taking a 20–40% cut. If the protein precipitates over a very broad range so that only a low degree of purification can be achieved, then it is probably better to abandon the idea of fractionation and use precipitation only as a means of concentration if large volume is a problem (*see* Chapter 10).

    A quicker way of carrying out a trial is to take several samples of the protein fraction and add 20% ammonium sulfate to the first, 30% to the second, and so on. Then, after equilibration, recover the precipitated protein by centrifugation, dissolve, and assay as above. The problem with this is that the precipitation behavior of a particular protein will depend on the precise protein composition of the solution, so that the conditions in a trial of this sort will not properly reflect the conditions in the large-scale procedure. The use of this method is not, therefore, recommended. The above considerations apply equally to trial solvent fractionation experiments.

8. Important: If the salt or organic solvent is added too quickly, then high local concentrations will develop and proteins will be precipitated that would remain soluble at the target ammonium sulfate or solvent concentration. Such proteins may not readily redissolve, and the result will be

decreased purity of the active fraction and possible loss of the protein of interest in lower fractions. With solvents, there is also the increased risk of denaturation. Some authors recommend the use of saturated solutions of ammonium sulfate for fractionation, but this has the disadvantage of leading to large volume increases and should not be necessary if due care is taken when adding the solid salt.

9. Care must be taken with this step to protect both the centrifuge and the operator! The tubes must be balanced to within 0.1–0.2 g across the rotor axis (*see* Note 3, Chapter 2), and it is particularly important not to counterbalance a tube full of ammonium sulfate solution with a tube full of water because of the large difference in density between the two. Either the ammonium sulfate/protein suspension should be divided between two tubes, or the suspension should be balanced using a solution of ammonium sulfate of the same concentration. It is also most important to avoid spillage of ammonium sulfate solutions into the centrifuge head. Such solutions are extremely corrosive to the materials of which rotors are constructed and can lead to irreversible damage, rendering the rotors unsafe to use. After centrifugation of ammonium sulfate (or other salt) solutions, rotors should always be removed and washed in warm water.

Note that proteins precipitated in ammonium sulfate are usually stable, and hence, it is often convenient to store the suspension overnight at 4°C in this form before proceding with the purification schedule.

10. The pellets of precipitated protein from this step may be discarded, but to be on the safe side, it is worth keeping them until it has been established that the protein of interest is indeed obtained in the next fraction. If something has gone wrong, then it is easier to reprocess the 0–35% precipitate than to go back to the beginning of the preparation and start again! Similarly, do not throw the supernatant from the 35–50% cut away until you are sure that your protein has been precipitated.

11. This step is crucial to obtaining a concentrated fraction. Add a very small volume of water or buffer (about 10 mL if using a 250-mL tube), and resuspend the protein pellet using a glass rod. If several centrifuge tubes have been used, then transfer the protein suspension from tube to tube, resuspending the pellet each time; only add more water or buffer if the suspension becomes too thick to transfer readily. After transfer of the final suspension to the dialysis bag, a further small aliquot of water or buffer can be used to rinse out the tubes and the rinsings combined with the suspension. A common error is to attempt to redissolve the protein pellets after centrifugation; since the pellets contain considerable quantities of ammonium sulfate, this takes a large volume of buffer and it is easy to end up with a volume comparable to that of the original fraction. When doing a

trial fractionation (*see* Note 7), it is acceptable to redissolve the precipitates, since the volume will not generally be important.
12. Visking tubing comes in several sizes. For a given volume of solution, the larger the diameter of the dialysis tubing used, the shorter the piece that will be required; attainment of equilibrium, however, will be slower with short, fat bags than with long, thin ones. The sizes recommended are a compromise between these two factors. For most applications, it is only necessary to soak the dry tubing in water or in the buffer to be used for dialysis, and it is ready to be used. The tubing does, however, contain significant quantities of sulfur compounds and of heavy metal ions; the latter may be a problem if the protein of interest is metal sensitive. They may be removed by boiling the dialysis tubing in 2% (w/v) sodium bicarbonate, 0.05% (w/v) EDTA for about 15 min, washing with distilled water, and then boiling in distilled water twice for 15-min periods. Prepared tubing can be stored indefinitely at 4°C in water or buffer containing 0.1% (w/v) sodium azide. For most applications, Visking tubing is perfectly adequate, but problems will arise if the protein of interest is small. The pores in this type of tubing are of such a size that the nominal mol-wt cutoff (NMWC) is about 15,000, although larger proteins may still pass through the pores if they have an elongated shape. As a rule of thumb, Visking tubing can be used with confidence if M_r >20,000. For dialysis of smaller proteins, Spectropor tubings with a range of NMWC values starting at 1000 are available (e.g., from Pierce and Warriner, Chester, UK), and the appropriate tubing should be selected. These tubings are much more expensive than Visking tubing, and the rate of dialysis decreases with pore size; hence, they should only be used if strictly necessary.

To remove ammonium sulfate from the protein suspension, a piece of dialysis tubing with a volume about twice that of the suspension is taken and securely closed with a double knot at one end. The suspension is then poured into the bag using a funnel, air is removed from the top part of the bag by running it between the fingers, and the top secured with a double knot. It is very important to have this space in the bag to allow for expansion, since water will flow in while the internal salt concentration is high. If insufficient space is left, the bag can become very tight owing to this inflow. Bags rarely burst, since the membranes are quite strong, but they are tricky to open in this state; the best way is to insert one end into a measuring cylinder, and then prick the bag with a scalpel. Equilibrium will be reached in about 3–4 h, but only if the system is stirred; otherwise, about 6 h should be allowed. Care should be taken when stirring to ensure that the magnetic pellet or stirring paddle does not tear the dialysis bag. The volume of dialysis solution to be used depends, of course, on the sample volume and the final concentration of ammonium sulfate desired. If a ratio of 1:100 (sample:dialysis fluid) is used, then at equilibrium, the ammonium sulfate concentration will have been reduced 100-fold (this is not strictly true, because it ignores the Donnan effect, but will do as an approximation). A second dialysis will then result in a total decrease of 10,000-fold, and so on. The buffer to be used for dialysis is usually dictated by the requirements of the next step in the purification schedule. For example, if this is to be ion-exchange chromatography, then column equilibration buffer is the logical choice.

13. Glass is acceptable, but stainless steel is to be preferred because of the more rapid heat transfer. Plastic will not do.
14. Addition of ethanol or acetone to water leads to a volume reduction of about 5%, but this is usually ignored in calculating percentage concentrations. To calculate the volume (v) in mL of solvent required to change the concentration of 1 L of solution at P_1 to P_2%, use the formula:

$$v = [1000 \, (P_2 - P_1)/(100 - P_2)] \qquad (3)$$

15. Only short centrifugation times are required because of the low density and viscosity of water/solvent mixtures.
16. Do not attempt to resuspend the precipitates in too small a volume (in distinction to the practice with salt fractionation), since the result would be a high solvent concentration with attendant risk of denaturation. Similarly, if the protein is sensitive to solvents, it is necessary to remove residual solvent quickly, so membrane filtration or gel filtration may be preferable to dialysis.

References

1. Englard, S. and Seifter, S. (1990) Precipitation techniques. *Methods Enzymol.* **182,** 285–300.
2. Dawson, R. M. C., Elliot, D. C., Elliot, W. H., and Jones, K. M. (1986) *Data for Biochemical Research,* 3rd ed. Oxford University Press, Oxford, UK, pp. 537–539.
3. Green, A. A. and Hughes, W. L. (1955) Protein fractionation on the basis of solubility in aqueous solutions of salts and organic solvents. *Methods Enzymol.* **1,** 67–90.
4. Banks, B. E. C., Doonan, S., Lawrence, A. J., and Vernon, C. A. (1968) The molecular weight and other properties of aspartate aminotransferase from pig heart muscle. *Eur. J. Biochem.* **5,** 528–539.
5. Ingham, K. C. (1990) Precipitation of proteins with polyethylene glycol. *Methods Enzymol.* **182,** 301–306.

CHAPTER 14

Ion-Exchange Chromatography

David Sheehan and Richard FitzGerald

1. Introduction

Proteins contain charged groups on their surfaces, which enhance their interactions with solvent water and hence their solubility. At physiological pH, some of these charged groups are cationic (positively charged, e.g., Lysine), whereas others are anionic (negatively charged, e.g., Aspartate). Since proteins differ from each other in their amino acid sequence, the net charge possessed by a protein at physiological pH is determined ultimately by the balance between these charges (i.e., negatively charged proteins possess more negatively charged groups than positively charged groups). This also underlies the different pIs of proteins (*see* Chapter 23). Ion-exchange chromatography *(1)* separates proteins first on the basis of their charge type (cationic or anionic) and, second, on the basis of relative charge strength (e.g., strongly anionic from weakly anionic).

The basis of ion-exchange chromatography (Fig. 1) is that charged ions can freely exchange with ions of the same type. In this context, the mass of the ion is irrelevant. Therefore, it is possible for a bulky anion like a negatively charged protein to exchange with chloride ions. This process can later be reversed by washing with chloride ions in the form of a NaCl or KCl solution. Such washing removes weakly bound proteins first, followed by more strongly bound proteins with a greater net negative charge.

Like most column chromatography techniques, ion-exchange chromatography requires a stationary phase which is usually composed of insoluble, hydrated polymers, such as cellulose, dextran, and Sephadex *(2)*.

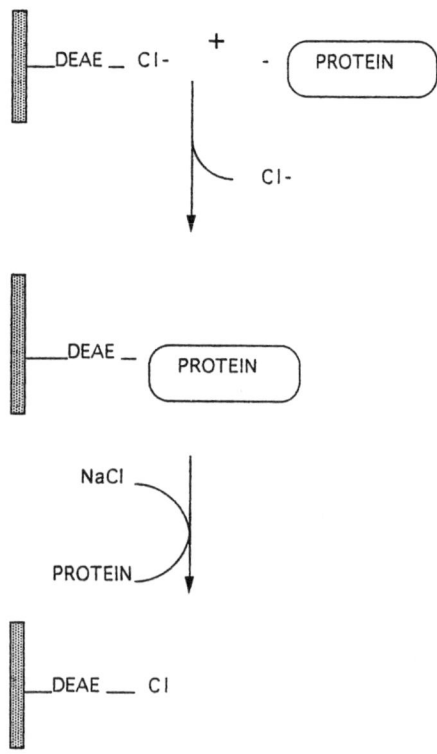

Fig. 1. Ion-exchange chromatography of anionic protein. The protein exchanges with the Cl⁻ ion and binds to the resin. This process is reversed when an NaCl gradient is applied to the column. Each protein has slightly different charge characteristics and, therefore, elutes at a slightly different Cl⁻ ion concentration. This, therefore, results in separation of proteins.

The ion-exchange group is immobilized on this stationary phase, and some of the chemical structures of commonly used groups are shown in Table 1.

In this chapter, the use of microgranular diethylaminoethyl (DEAE)-cellulose manufactured by Whatman (Maidstone, UK) is described. Variations on this method suitable for other resins are summarized in Section 4. A novel format for ion exchange is the immobilization of DEAE or carboxymethyl (CM) groups on filters packed in a cartridge (MemSep,™ PerSeptive Biosystems, Farmingham, MA). These cartridges may be used in a low-pressure format or with an FPLC system (*see* Chapter 26), and offer the advantages of high resolution and high flow rates with very low back pressure. This is owing to the use of thin

Ion-Exchange Chromatography

Table 1
Some Commonly Used Ion Exchange Groups[a]

Group	pH range	Structure
Anion exchangers		
Q (quaternary ammonium)	2–12	$-CH_2-\overset{+}{N}(CH_3)_3$
DEAE (diethylaminoethyl)	2–9	$-O-CH_2-CH_2-\overset{+}{N}H(C_2H_5)_2$
QAE (quaternary aminoethyl)	2–12	$-O-(CH_2)_2-\overset{+}{N}(C_2H_5)_2-CH_2-CHOH-CH_3$
Cation exchangers		
SP (sulphopropyl)	2–12	$-(CH_2)_2-CH_2-SO_3^-$
S (methyl sulphonate)	2–12	$-CH_2-SO_3^-$
CM (carboxymethyl)	6–11	$-O-CH_2-COO^-$

[a]These groups may be immobilized on various stationary phases, such as cellulose, dextran, agarose, and glass beads.

filters rather than beads as the stationary phase and can result in separations in as little as 3–5 min.

2. Materials

1. The buffers used will depend on the characteristics of the protein of interest (*see* Note 1). Buffers are prepared fresh before use, and all reagents are of Analar grade. Deionized/distilled water is used. Typically, the following buffers may be employed for a column that requires operation at pH 8.0.
   Buffer A: 200 mM Tris-HCl, pH 8.0.
   Buffer B: 10 mM Tris-HCl, pH 8.0.
   Buffer C: 10 mM Tris-HCl, pH 8.0, 100 mM NaCl.
2. Microgranular DEAE-cellulose 52 is obtained from Whatman. Desalting resin, such as Sephadex G-25, is from Pharmacia LKB (Uppsala, Sweden). Both resins are stored in water containing 0.1% Na azide. Sephadex resins may be reused after extensive washing with large volumes of water, whereas DEAE-cellulose requires regeneration before reuse (*see* Note 2).
3. A 500-mL sintered glass funnel (no. 1 sinter) is useful for washing resin during equilibration and regeneration. This is usually connected to water suction.

4. For desalting and ion-exchange chromatography (column volumes of 300 mL), sintered glass columns (45 × 3 cm in diameter) are required. Smaller columns could be used for smaller column volumes.
5. A gradient maker (2 × 500 mL) is used to generate salt gradients.
6. A calibrated conductivity meter is required.

3. Methods

3.1. Sample Preparation

Protein samples require desalting before being applied to the ion-exchange column. This may be achieved either by dialysis or by gel filtration on a Sephadex G-25 column *(3)*. Pharmacia produces a PD10 column for rapid desalting of small volumes.

3.2. Resin Equilibration

1. Before use, wash 100–200 g resin in 1–2 L water followed by removal of fines. This is achieved by stirring the resin with a glass rod, allowing the bulk of the resin to settle, and then aspirating supernatant, which contains the fines. Repeat this process three times. It is sometimes convenient to do this procedure in a graduated cylinder.
2. Wash the defined resin in 500 mL buffer A (*see* Note 3) to achieve equilibration. Again, following settlement of the resin, aspirate off the supernatant. Repeat this process twice (i.e., a total of three washes). Gently stir the resin into a slurry, and transfer it into a sintered glass column that contains a small volume of water (this prevents air bubbles from being trapped in the sinter). Pass buffer B through the column until it is completely equilibrated (*see* Note 4).

3.3. Chromatography

1. Apply the desalted sample to the equilibrated resin, and collect the effluent. This should be assayed for the protein of interest in case it has not bound (*see* Note 5).
2. Then elute bound protein by applying a gradient (2 × 500 mL) of 0–100 mM NaCl in 10 mM Tris buffer, pH 8.0 (*see* Note 1).
3. Collect fractions in a fraction collector, and assay for the protein of interest, protein concentration *(4)*, and conductivity. The conductivity meter should be carefully calibrated (Note: this property is temperature-dependent), so that conductivity measurements may be expressed as NaCl concentrations. In this way, the reproducibility of each chromatography experiment may be assessed. These measurements can also be made on-line, using on-line UV and conductivity detectors connected to a PC or chart recorder.
4. Pool appropriate fractions, and concentrate for further study or purification (*see* Note 6).

3.4. Resin Regeneration

The resin may usually be regenerated by washing with 10 mM Tris-HCl, pH 8.0, buffer containing 2M NaCl after use. Three to four uses of a batch of resin are possible in most cases before extensive regeneration is neccessary (*see* Note 2). However, owing to column compaction as a consequence of high-salt concentration gradients, in some cases, only single usage of a packed resin is possible before extensive regeneration is required.

1. Place the resin in a beaker containing 15 vol of 0.5M HCl for 30 min, and then decant the supernatant.
2. Wash the resin extensively on a glass sinter until the pH of the wash-through reaches an intermediate value of 4.0, then place the resin in a beaker containing 15 vol of 0.5M NaOH for 30 min, and decant the supernatant. Again, the resin is extensively washed with water until a pH of 8.0–7.0 is achieved. The resin may now be stored in 0.1% Na azide until required.

4. Notes

1. For proteins not previously purified by DEAE-cellulose chromatography, buffer selection may be helped if the pI of the protein of interest is known. In general, proteins will bind to anion exchangers at pH values above their pIs, while binding to cation exchangers, such as CM-cellulose, at pH values below their pIs.

   Phosphate, Tris, and other common buffers are used in ion-exchange chromatography in concentrations of 10–50 mM. Urea (desalted on a mixed-bed column of Dowex or else molecular biology grade) may be included in buffers at concentrations up to 8M if separation of denatured samples is required (e.g., chromatography of CNBr-generated peptides). Care should be taken that the urea does not precipitate in the column during chromatography (especially at low temperatures).

2. Fibrous DEAE-cellulose is exposed to 0.5M HCl for 60 min rather than 30 min. Regeneration of CM-cellulose is achieved by a similar process, except that the resin is washed first with 0.5M NaOH for 30 min, followed later by 0.5M HCl. The intermediate pH is 8, whereas the final pH is 5.5–6.5.

3. Concentrated buffer is used to equilibrate and charge the exchange groups, because much lower volumes are required than if more dilute buffer is used. This results in more rapid equilibration.

4. Equilibration is assessed by measuring the pH and conductivity of buffer B and the column eluate. When these values are the same, the column is equilibrated.

5. For proteins not previously purified by this method, it is often neccessary to develop a method. The main variables one can use to do this include

choice of ion-exchange group, pH (*see* Note 1), salt concentration, and gradient steepness. In general, for example, around 70% of rat liver cytosolic proteins will bind to DEAE-cellulose at pH 7.5. If the protein of interest has an alkaline pI, then passage through such a column at this pH might be a useful first step in the purification. Conversely, if the protein of interest binds to DEAE-cellulose at pH 7.5, then it is worth repeating the experiment at progressively higher pH values, e.g., 8.5 and 9.5. Only highly acidic proteins will bind at pH 10.0. It will also be advantageous to assess a range of salt gradients. Generally, shallower gradients (e.g., 0–100, 300–400 mM) give better resolution. However, steep gradients (e.g., 0–1M) can be used in initial experiments to identify elution positions of proteins of interest. It is also possible to use stepwise washes (e.g., 100 mM NaCl followed by 200 mM NaCl) rather than continuous gradients. The best approach to developing a new method, therefore, is to assess the binding (i.e., binding vs nonbinding) of the protein of interest on a small scale on both CM-cellulose and DEAE-cellulose at a range of pH values in the alkaline and acidic ranges, respectively. Then develop chromatograms where the protein of interest has bound with a steep gradient. Finally, shallower gradients are assessed at the most promising pH values. It is also possible to use nonlinear gradients for more difficult separations.
6. A major factor in the success of chromatography is the pooling of fractions. In general, fractions are never pooled in a new purification until SDS-PAGE analysis *(5,6)* has been carried out on aliquots taken from active fractions. This gives a good indication of fraction purity. It is best to accept some losses of the protein of interest if this means removing significant contaminants. Carrying out the chromatography at different pH values (*see* Note 5) usually gives quite different chromatograms and this is often useful for displacing contaminant peaks.

References

1. Himmelhoch, S. R. (1971) Ion-exchange chromatography. *Methods Enzymol.* **22,** 273–290.
2. Walsh, G. and Headon, D. R. (1994) Downstream processing of protein products, in *Protein Biotechnology,* Wiley, Chichester, UK, pp. 39–117.
3. Boyer, R. F. (1993) Gel exclusion chromatography, in *Modern Experimental Biochemistry,* Benjamin Cummings, New York, pp. 81–89.
4. Lowry, O. H., Rosebrough, N. J., Farr, A. L., and Randall, R. J. (1951) Protein measurement with the folin-phenol reagent. *J. Biol. Chem.* **193,** 267–275.
5. Laemmli, U. K. (1970) Cleavage of structural proteins during the assembly of the head of bacteriophage T4. *Nature* **227,** 680–685.
6. Hames, B. D. and Rickwood, D. (1990) *Gel Electrophoresis of Proteins–A Practical Approach,* Oxford, University Press, Oxford, UK.

CHAPTER 15

Hydrophobic Interaction Chromatography

Paul O'Farrell

1. Introduction

During the development of affinity chromatography, spacer arms consisting of alkyl chains of 2–10 carbons were used to hold the affinity ligand at some distance from the solid matrix so as to minimize ligand/support interactions. Control experiments were carried out to determine the effect of the spacer itself where the spacer alone was attached to the matrix. It was found that columns prepared from such material in fact retained certain proteins. These observations lead to the first systematic studies of hydrophobic interaction chromatography (HIC).

Although the nature of the hydrophobic interaction is still not fully understood, it is clear that it has its origins largely in the properties of the solvent. When a nonpolar solute molecule is dissolved in a polar solvent, such as water, the solvent is forced to form a structural cavity around it. This causes a decrease in entropy and is therefore thermodynamically unfavorable. When two such solute molecules come into contact they may aggregate as a result of the consequent favorable increase in entropy, since less solvent molecules are required for structure formation around the dimer. Thus hydrophobic interaction is not an attractive force as such, rather it is forced on a nonpolar solute by the structure of the polar solvent. It follows that the addition of modifiers to the solvent which change its ability to form structure will have an effect on hydrophobic interactions. Chaotropic salts decrease hydrophobic interactions (because there is less overall structure in the solvent in the presence of such agents, the energetic advantage gained by the aggregation of nonpolar solute molecules is reduced), while so-called "structure-forming" salts increase the strength of the interaction (see Note 1).

It has been estimated that as much as 50% of the solvent accessible surface of a soluble protein can consist of nonpolar atoms *(1)*, but this percentage varies over a wide range from one protein to another. This characteristic represents a general difference between proteins which can be exploited for their separation and which provides a third dimension in addition to size exclusion and ion exchange as chromatographic methods.

Although the exact nature of hydrophobic interaction is not fully understood, nevertheless its exploitation in column chromatography is a straightforward matter. In general, protein is loaded onto the matrix in high-salt buffer, and eluted by decreasing the salt concentration, either by a gradient or in a stepwise fashion. The high salt concentration of the loading buffer means that it is not necessary to remove salt present in the protein sample by dialysis or gel filtration. Consequently, HIC is an excellent choice of procedure following initial purification by ion-exchange chromatography or if the protein of interest is in the supernatant fraction after partial purification by precipitation with ammonium sulfate; it is necessary only to add enough salt to ensure binding to the column. In addition, the presence of salts in the sample solution can exert a stabilizing effect and prevent loss of activity.

If a decreasing salt gradient is used for elution, the gradient in the column is stabilized by density effects. However a number of problems also exist. Proteins may tend to aggregate during elution; conditions are, after all, similar to those used for fractionation by salting out. Because of this and other factors, such as slow association/dissociation, peaks can be broad and separation is not always ideal. Such effects are generally reduced if a step elution method is used.

The method presented here describes the final stage in the purification of aspartate aminotransferase (AAT) from *Thermus aquaticus*. The extract applied to the column had already been partially purified by a combination of ammonium sulfate precipitation, protamine sulfate precipitation, and anion- and cation-exchange chromatography *(2)*. The protein concentration was approx 150 µg/mL and the AAT specific activity was 25 U/mg.

2. Materials

1. Buffer 1: 10 mM Tris-HCl, pH 8.0, 0.6M NaCl.
2. Buffer 2: 10 mM Tris-HCl, pH 6.5 (see Note 2).

Hydrophobic Interaction Chromatography 153

3. Phenyl Sepharose CL-4B (Pharmacia, Uppsala, Sweden) (*see* Note 3). The ligand density is 40 µmol/mL of gel.
4. Chromatography column: 10 cm long × 1.5 cm in diameter, total bed vol 18 mL.

3. Method

Previous steps in the purification schedule had been carried out at 4°C, the HIC step was carried out at room temperature (*see* Note 4).

1. Wash the gel, which is supplied preswollen in 20% ethanol, with at least 10 vol of water to ensure that all the ethanol is removed; this is important since ethanol strongly interferes with hydrophobic interactions.
2. Allow the gel to settle and decant the supernatant.
3. Prepare a slurry of gel in buffer 1 with a ratio of 75% settled gel to 25% buffer.
4. Pour the slurry into the column and, after fitting the column top piece, equilibrate by pumping buffer 1 through the column at 80 mL/h for 1.5–2 h.
5. Pool active fractions from the previous purification step (chromatography on DEAE-cellulose, DE-52 in the present case) and adjust to approx 0.6M NaCl.
6. Dialyze this material overnight against buffer 1 (*see* Note 5).
7. Apply the resultant solution to the column at 60 mL/h and collect 3 mL fractions.
8. Remove unbound proteins by washing the column with buffer 1 at 60 mL/h.
9. When the column effluent is free of protein, determined by measuring the absorbance at 280 nm, elute the AAT by changing the running buffer to buffer 2 (*see* Note 6). Assay fractions for AAT activity and pool active fractions.
10. Clean the column by passing 3–4 bed volumes of distilled water through it and then re-equilibrate with loading buffer (*see* Note 7).

4. Notes

1. The Hofmeister series lists ions in the order from those which favor hydrophobic interactions to those which decrease the strength of the interaction:
   a. Anions: PO_4^{3-}, SO_4^{2-}, CH_3COO^-, Cl^-, Br^-, NO_3^-, ClO_4^-, I^-, SCN^-.
   b. Cations: NH_4^+, Rb^+, K^+, Na^+, Cs^+, Li^+, Mg^{2+}, Ca^{2+}, Ba^{2+}.
2. Both buffers contained 4 µM pyridoxal-5'-phosphate and 2 mM 2-oxoglutarate but these were added specifically to stabilize the AAT and, of course, would not be used in other applications. Other additives may be needed in particular cases and it is necessary to confirm that they do not interfere with chromatographic behavior.
3. A wide variety of hydrophobic resins are currently commercially available. In addition there are a number of products which combine hydrophobic characteristics with others such as size exclusion or affinity. The most commonly used purely hydrophobic matrices are substituted with alkyl

(generally octyl) or phenyl ligands. The choice of which of these matrices to use is empirical and the preferred method for a particular separation should be determined by small-scale screening experiments. Reference *3* gives excellent advice on developing and interpreting such screening experiments. In general, a weakly hydrophobic resin should be used with an uncharacterized protein, since a strongly hydrophobic protein may not be eluted easily from a strongly binding resin (e.g., octyl-substituted resins).

4. As HIC is entropy based, it can be expected that temperature will have an effect on strength of binding. In general, hydrophobic interactions increase in strength with increasing temperature; there can be a 20–30% reduction in binding when the temperature is changed from 20 to 4°C *(4)*. However, temperature also affects protein conformation and solubility and consequently can affect a protein's interaction with a hydrophobic matrix. It is difficult to predict the effect of temperature on a particular separation, but it should be remembered that changes in temperature can cause changes in elution patterns and that in some cases a protein which binds to a particular hydrophobic matrix at room temperature may not bind in the cold room.

5. Generally, the strength of the interaction between proteins and hydrophobic ligands decreases with increasing pH. This is presumably because of an increase in the hydrophilicity of proteins due to the titration of charged groups. The effect of pH is different for different proteins and thus it may be expected that elution profiles may be improved by carrying out elution at different pH values. However, the effect of changing the pH is not great at moderate pH values, and for most applications it is probably better simply to work at a pH where the protein of interest is known to be stable.

6. Elution can be achieved in a number of ways and can be either via a gradient or a stepwise method. Although gradient elution should in principle give better separation of components, as noted in the Introduction, peaks can be broad and the elution volume may therefore be larger than desired. This may be of little consequence if HIC is to be followed by another adsorption procedure since the next step will concentrate the sample. However, if HIC is to be followed by, for example, size exclusion chromatography, or if the protein of interest is unstable in dilute solution, a stepwise method which will result in a smaller elution volume may be indicated.

A variety of salts may be used for loading and elution *(5)*. In general the effects of various salts on HIC mimic their effects on salting out. Those salts which are most effective in salting out ("structure-forming" salts) are most effective at binding proteins to hydrophobic matrices, whereas salts used for salting in decrease the interaction. Thus, elution may be achieved by decreasing the concentration of structure forming salt, or increasing the concentration of a chaotropic agent. In fact, a combination of the two methods has been shown to result in better separations and sharper peaks *(6)*.

If a protein is very strongly bound, then alcohols (e.g., ethylene glycol, gradient of 0–80%) or detergents (e.g., Triton X-100, 1% [w/v]) may be used to effect elution. However, such agents can have deleterious effects on protein structure and in addition, detergents can prove very difficult to remove from hydrophobic matrices. If it is found to be necessary to resort to such extremes, it is probably best to consider using a more weakly hydrophobic ligand.

7. For more stringent cleaning, 4 bed volumes of 0.5–1.0M NaOH, followed by 2–3 bed vol of distilled water may used. The column can be stored filled with 0.01M NaOH; after storage, 2–3 bed vol of water are passed through the column which is then ready to be equilibrated with the loading buffer.

References

1. Lee, B. and Richards, E. M. (1971) The interpretation of protein structures: estimation of static availability. *J. Mol. Biol.* **55,** 379–400.
2. Walker J. M. and Wang, Y.-X. (1993) Purification of aspartate aminotransferase from *Thermus aquaticus. Biochem. Mol. Biol. Inter.* **29,** 867–873.
3. Pharmacia Biotech (1993) *Hydrophobic Interaction Chromatography–Principles and Methods*, Uppsala, Sweden.
4. Kennedy, R. M. (1990) Hydrophobic chromatography. *Methods Enzymol.* **182,** 339–343.
5. Hjerten, S. (1973) Some general aspects of hydrophobic interaction chromatography. *J. Chromatog.* **87,** 325–331.
6. El Rassi, Z., De Ocampo, L. F., and Bacolod, M. D. (1990) Binary and ternary salt gradients in hydrophobic interaction chromatography. *J. Chromatog.* **499,** 141–152.

CHAPTER 16

Affinity Chromatography

Paul Cutler

1. Introduction
1.1. General Principles

Affinity chromatography is a method of selectively and reversibly binding proteins to a solid support matrix based on the exploitation of known biological affinities between molecules. The interaction between the target protein and the matrix is not based on general properties, such as the isoelectric point (pI) or hydrophobicity, which are more commonly used in adsorption chromatography, but on individual structural properties, such as the interaction of antibodies with antigens, enzymes with substrate analogs, nucleic acids with binding proteins, and hormones with receptors. In order to exploit the interaction for the purposes of purification, one of the components, the ligand, must be immobilized onto a solid matrix in a manner that renders it stable and active. Once produced, the matrix can be used to purify its specific target from a suitable biological extract *(1)*.

Affinity chromatography is a particularly powerful technique, since it offers the potential of purifying target proteins that exist in very low titer from complex mixtures with a very high degree of purification in a single step. Since the aim is to purify material on the basis of biological function, it is even possible to separate active and inactive forms of the same material selectively.

Affinity matrices can be formed from ligands that are either mono-specific or group-specific. Mono-specific ligands recognize a single form of a protein, such as a receptor with affinity for a specific hormone or an

Table 1
Some Commonly Employed
Group-Specific Affinity Ligands

Ligand	Target protein
5'AMP, ATP	Dehydrogenases
NAD, NADP	Dehydrogenases
Protein A	Antibodies
Protein G	Antibodies
Lectins	Polysaccharides, glycoproteins
Histones	DNA
Heparin	Lipoproteins, DNA, RNA
Gelatin	Fibronectin
Lysine	rRNA, dsDNA, plasminogen
Arginine	Fibronectin
Benzamidine	Serine proteases
Polymyxin	Endotoxins
Calmodulin	Kinases
Cibacron Blue	Kinases, phosphatases, dehydrogenases, albumin

enzyme recognizing an inhibitor. These matrices are often "tailor-made" for particular separations. Group-specific ligands include enzyme cofactors, plant lectins, and protein A from *Staphylococcus aureus* (Table 1). Group-specific matrices owing to their generic nature, are frequently available commercially from a range of suppliers, often in prepacked columns.

Originally, in order to purify a protein by affinity chromatography, it was necessary to find a naturally occurring ligand. In recent years, the definition of affinity chromatography has expanded to include the separation of proteins by specific interactions other than purely biological interactions. Included in this category are dye affinity ligands, such as Cibacron Blue *(2)*, and immobilized metal ion-affinity matrices *(3)*. With the advent of monoclonal antibodies (MAbs) and recombinant protein technology, it is possible to manipulate a biological affinity. By using hybridoma technology, it is possible to produce economically viable amounts of MAbs where the target protein is the antigen. Immobilizing the antibody on a solid support creates an immunopurification matrix *(4)*. It is also possible with recombinant DNA technology to produce a target protein in cell culture which has a specific affinity region, the tag, engineered into the product *(5,6)*. The tag is used to purify the protein on

Affinity Chromatography

a relatively inexpensive and well-defined affinity matrix. Enzymic cleavage points may also be inserted between the protein and the tag sequences so that once purified from the culture, the tag can be removed. An example of this is a fusion containing the target protein with a hexahistidine sequence, which shows preferential binding to immobilized metal affinity matrices *(7)*.

Two schools of thought exist regarding where an affinity chromatography step should lie in a purification scheme. One suggests that owing to the expense of producing affinity matrices, the material should be partially purified prior to the affinity step to protect the matrix from proteases, and so forth. Conversely, the affinity matrix can come first to purify rapidly what may be an unstable protein to near homogeneity in a single high-yielding step, while dramatically reducing the volume of material to be processed. With the increasing availability of affinity matrices manufactured using recombinant technologies to produce less costly ligands, the latter school of thought appears to be gaining popularity. However, the actual position of the affinity chromatography step within a purification scheme must be decided on a case-by-case basis.

1.2. Affinity Matrices

1. Ligand: The immobilized affinity ligand must be able to form a reversible complex with the target protein. The binding constant (the inverse of the dissociation constant) should be high enough to enable stable complex formation ideally at physiological conditions. However, the affinity must be sufficiently weak to facilitate the elution of the protein under relatively mild conditions, thereby avoiding denaturation of either the ligand or target protein. Binding constants of 10^5–$10^{11} M$ are usually considered a good working range *(see* Note 1).

    A detailed description of the immobilization of the ligand to generate an affinity support is beyond the scope of this chapter. However, reviews exist describing immobilization techniques *(see* ref. *8* and Note 2). In addition to immobilization in a form that retains biological function of the ligand, good chemical stability must be maintained in order to withstand the harsh elution and cleaning regimes used in affinity chromatography processes. Leaching of the ligand from the matrix must be minimized to avoid contamination of the purified protein.

2. Matrix: The matrix of choice should show the physical and chemical properties required for adsorption chromatography techniques *(9,10)*. The matrix typically constitutes macroporous beads (50–400 µm) to produce a large surface area, which maximizes the capacity of the activated matrix

and prevents any size exclusion effects. It must be hydrophilic, but uncharged to prevent nonspecific ionic binding. It must demonstrate good mechanical rigidity and chemical stability under the conditions used for activiation, derivatization, elution, and regeneration. Although the most commonly used matrix support is agarose, a range of other matrix supports, including silica and polyacrylamide, are available. Selection of the matrix should be made based on the physicochemical properties required.
3. Spacers: One of the common problems encountered when derivatizing a matrix for affinity chromatography is the steric hindrance observed when the ligand is too close to the solid phase. To minimize this affinity matrices are often activated with a spacer arm, commonly a six-carbon hydrophilic chain to distance the ligand from the solid phase. The length of the spacer arm should be optimized for each individual matrix. Although longer spacer arms may allow greater ligand availability, they can themselves lead to steric hindrance and confer unwanted hydrophobicity onto the support.
4. Activation of the matrix: The protein ligand is almost always covalently attached to the solid support. The matrix is activated with a reagent, such as cyanogen bromide, which reacts with the ε-amino groups of lysine residues in protein ligands, resulting in the protein being covalently attached. Other chemistries enable immobilization of proteins via carboxyl, hydroxyl, and thiol functionalities (*see* ref. *8* and Note 2).

1.3. Practical Aspects

Purification by affinity chromatography is a relatively straightforward technique because of the selective binding of the target protein (Fig. 1). The crude extract is passed through the column, the target material binds, and the matrix is then washed to remove the nonbound fraction. This may involve more than one wash step if nonspecific interactions or selective binding of impurities to the bound target protein are suspected. The conditions of the mobile phase are then modified usually by changes in pH or ionic strength, or by inclusion of either a general chaotropic agent or a competing soluble ligand to facilitate elution of the protein. The eluted protein is collected and the column regenerated under suitable conditions.

1. Sample preparation: Efficient binding is facilitated by ensuring that the sample and the column are equilibrated in a buffer that is optimal for binding; this requires a defined pH, ionic strength, and so forth. Since most biological affinities in mammalian systems occur at physiological pH, buffers such as phosphate-buffered saline (PBS) are common. The buffer should contain any elements, such as cofactors, required to maintain activity of the target protein. The sample can be conditioned for binding to the matrix by such methods as dialysis, desalting, ultrafiltration or simply by

Affinity Chromatography

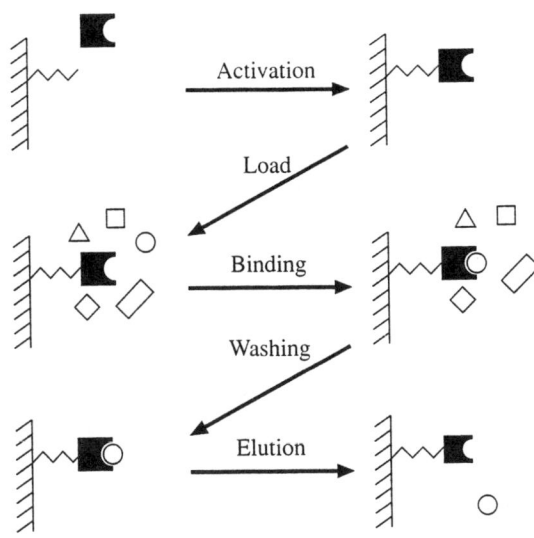

Fig. 1. Principle of affinity chromatography. The ligand is immobilized onto a solid support matrix. The crude extract is passed through the column. The target molecule for which the ligand possesses affinity is retained, whereas all other material is eluted. The bound target protein is eluted by alteration of the mobile-phase conditions.

pH adjustment. If crude material is to be loaded, then some form of pretreatment may be required, such as filtration or centrifugation. It may be necessary to include protease inhibitors, such as leupeptin or phenylmethylsulfonyl fluoride (PMSF), to prevent proteolytic degradation of the target (see Note 3 and Chapter 9).
2. Column: The column size and the degree of ligand substitution determine the capacity of the column for the target protein. The affinity constant of the ligand for the target protein under the operating conditions will also dictate column performance. This can be determined empirically by overloading the column or in a more sophisticated manner by producing a breakthrough curve to calculate the adsorption isotherm (Fig. 2). Capacities of 1–20 mg protein bound/mL matrix are common. In addition to standard chromatographic columns, other affinity separation methodologies exist, including activated filters (11) and stired tanks or expanded beds (see ref. 12 and Note 4).

Under normal conditions short, wide columns are used to facilitate rapid separations. However, where the affinity is low and the material is only retarded, separation is enhanced by use of a thinner, longer column (13).

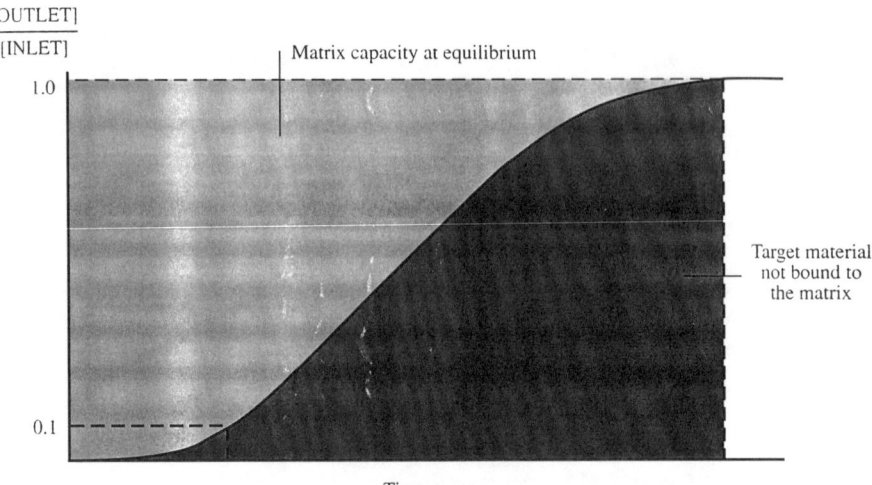

Fig. 2. Breakthrough curves can be used to establish the effective capacity of an affinity matrix. By specifically detecting or assaying the target protein appearing in the flowthrough from the column, the breakthrough of material can be plotted. Eventually, as the matrix reaches equilibrium (theoretical maximum capacity), the inlet concentration and outlet concentration are equivalent. The effective, or dynamic capacity of the matrix is reached sooner and is often taken when the ratio of outlet to inlet is 0.1.

Columns should be packed in accordance with the standard procedures described by the manfacturers. Briefly, the matrix should be packed in a glass or acrylic column with flow adaptors to ensure a minimum of dead space. For small-scale separations, cheaper disposable columns can be used under gravimetric flow.

The flow rate used for adsorption depends on the physical properties of the matrix support and the binding constant of the ligand for the target protein. Maximum adsorption usually occurs at a low flow rate, although flow rates tend to be a compromise between optimal binding and process time, which affects cost and target protein stability. In practice, linear flow rates of 30–50 cm/h are used for low-pressure preparative separations.

3. Loading: Loading volume is generally not critical as long as the total amount of target protein does not exceed the capacity of the column. The passage of proteins through the column is monitored most commonly using a UV detector. Following loading, the column is washed with equilibration buffer to remove any unbound material. Once baseline has been achieved on the detector, a more stringent wash can be made if necessary

to remove any weakly binding material that may have bound nonspecifically. This wash is designed so as not to cause elution of the target protein. Often the wash buffers can contain low levels of detergents or moderate concentrations of salt.

4. Elution: The protein of interest is eluted by weakening the ligand protein interaction. This can be done either nonspecifically, for example, by using changes in pH, ionic strength, and so forth, or by the addition of a specific solute in the eluant to remove the target protein selectively (e.g., a competing ligand). In extreme circumstances chaotropic agents may be used, although these significantly increase the likelihood of denaturation. Owing to the selectivity of binding, it is common for affinity elutions to be performed in a stepwise mode (*see* Note 1). However, in certain circumstances, particularly where group-specific ligands are employed, a gradient elution may be used. This has been used successfully to separate antibodies of differing subclasses on protein A affinity matrices.

The conditions for elution are one of the primary reasons why affinity chromatography fails to yield active protein (*see* Note 5). The need to use harsh conditions to maximize recovery must be balanced against potential denaturation and inactivation of the target protein. A knowledge of the physiochemical properties of the target protein is invaluable when designing an elution buffer. In general, the pH of the buffer should not be near the pI of the protein. It is advisable to avoid buffering systems with chelation effects (e.g., citrate) for purification of metalloproteins.

The purified protein should be placed in an environment that promotes its stability as soon as possible after elution. Frequently in the case of pH elution, this involves titration to near neutrality. The material may be placed in a suitable buffer by using buffer-exchange techniques, such as desalting, ultrafiltration, or dialysis. Treatment of the eluted protein will be largely influenced by subsequent purification steps and the use for which the protein is designed (*see* Note 6).

5. Purity analysis: The performance of the affinity chromatography step is usually determined by comparison of the purity before and after purification by polyacrylamide gel electrophoresis (PAGE). Other techniques for assessing purity include gel filtration and reverse-phase high-performance liquid chromotography (HPLC). Where possible, specific assays should be used to detect recovery (% yield), such as Western blotting techniques or, in the case of enzymes, calculating the activity per mass of total protein using techniques, such as enzyme-linked immunosorbent assay (ELISA).

It is important to analyze the flowthrough to ensure maximum recovery. Several methods exist for analyzing the flowthrough material, such as rechromatography, to establish that the capacity of the column has not been

exceeded. This, however, will not reflect the inability to bind owing to denaturation of the ligand or inappropriate binding conditions.
6. Matrix regeneration and storage: It is often the case that the harshest conditions a solid-phase chromatography matrix is exposed to are those used for regeneration and sanitization. Where a labile protein is used as a ligand, the affinity matrix cannot withstand treatment with stringent agents, such as sodium hydroxide. Commonly used regeneration regimes involve the use of chaotropic agents, such as $6M$ guanidine hydrochloride or $3M$ sodium thiocyanate.

Affinity matrices should be stored under conditions that prevent bacterial and fungal growth. Matrices are typically stored in the presence of antimicrobial agents, such as 20% (v/v) ethanol, 0.02% (w/v) Thimerasol, or 0.01% (w/v) sodium azide at standard cold-room/refrigerator temperatures ($< 8°C$).

2. Materials

1. Protein A Sepharose 4B (Pharmacia, Uppsala, Sweden).
2. Buffer A: 100 mM Tris-HCl, pH 7.5, $3M$ NaCl.
3. Buffer B: 50 mM Tris-HCl, pH 7.5, $1.5M$ NaCl.
4. Buffer C: $0.1M$ citrate, pH 3.0.
5. Buffer D: $0.1M$ citrate, pH 2.0.

3. Methods

Immunoglobulin Gs (IgGs) bind to protein A from *S. aureus* via the immunoglobulins Fc region. The affinity of the antibody for protein A is both species- and subclass-dependent. A range of protein A affinity matrices can be bought commercially. The crude antibody extract can be prepared by dilution or by dialysis into the binding buffer. The matrix is equilibrated with the binding buffer before the extract is loaded. The unbound material is washed away with the buffer prior to elution with a low-pH buffer. An example purification strategy is given (*see* ref. *14* and Fig. 3):

1. Dilute 5 mL of goat polyclonal antiserum raized against human albumin with 5 mL of buffer A, and retitrate to pH 7.5.
2. Pack a 5 mL (1.0 × 6.4 cm) column with protein A Sepharose 4B, and equilibrate in buffer B. Apply the diluted serum to the column at a flow rate of 50 cm/h. Wash the column with 5 column volumes (25 mL) of buffer B.
3. Program the fraction collector to collect 1 mL fractions. Exchange the inlet and outlet tubing of the column to reverse the flowthrough the column.
4. Elute the antibody with 5 column volumes of buffer C. Wash the column with 5 column volumes of buffer D, followed by 2 column volumes of $6M$ guanidine hydrochloride.

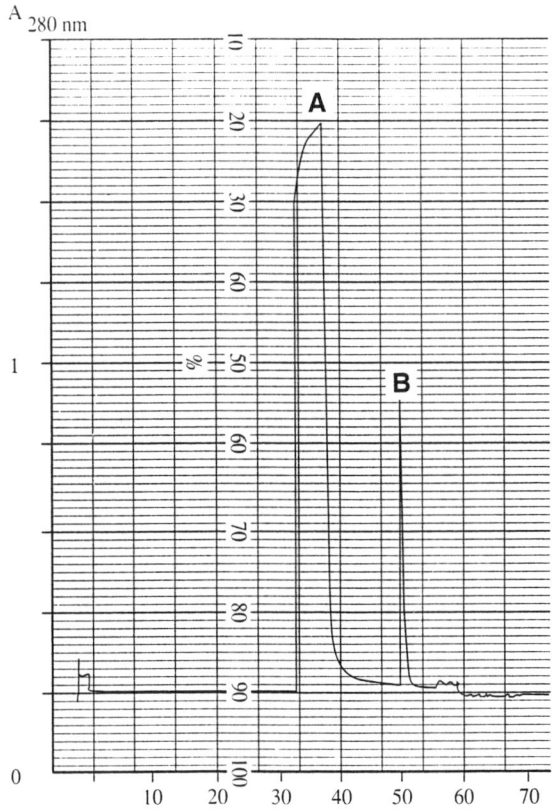

Fig. 3. Purification of goat immunoglubin from serum by protein A affinity chromotography. The majority of the protein elutes in the flowthrough fraction (Peak A). The bound immunoglobulin is eluted by lowering the pH of the mobile phase (Peak B).

5. Pool fractions containing antibody, and adjust the pH to pH 7.5 with $1M$ untitrated Tris base. Analyze the antibody for purity by SDS-PAGE (Fig. 4).

4. Notes

1. It is a commonly held belief that the higher the capacity of a column, the better. Although a high capacity is clearly advantageous, it should be remembered that on adsorption, the protein of interest is being removed from a dilute solution and concentrating on the solid phase. On elution, the protein is often released in a relatively high concentration form in a dissociating buffer. This can lead to insolubility of the protein and dramatic losses. This concentration can be partly modulated by the elution flow rate.

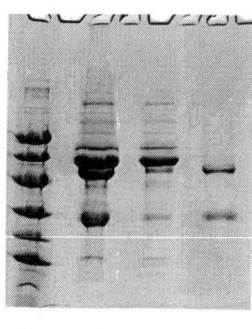

Fig. 4. Analysis of the purification of immunoglobulins from goat serum by sodium dodecyl sulfate polyacrylamide gel electrophoresis (SDS-PAGE). The crude serum (track 2) was purified by protein A affinity chromatography. The unbound material (track 3) was eluted in the column flowthrough. The purified antibody (track 4) was eluted from the column by lowering the pH of the mobile phase. The SDS-PAGE is run under reducing conditions resulting in the antibody appearing as its constituent "heavy" chains (~50 kDa) and "light" chains (~25 kDa). The molecular weights are compared to standard marker proteins (track 1).

In general, low flow rates promote elution in a concentrated form, whereas higher flow rates produce more dilute solutions, which in certain cases may be advantageous. Where the binding constant is too high to facilitate elution of the protein in an active form, it may be necessary to attenuate the ligand affinity via chemical modification. This is, however, a relatively complicated process requiring extensive knowledge of the interaction between ligand and target molecule.

2. Knowledge of the nature of the interaction between ligand and target protein is invaluable when selecting the chemistry by which a ligand is to be immobilized. A common immobilization strategy is to immobilize proteins via lysine residues using cyanogen bromide-activated supports. If it is known that a lysine residue on the ligand is involved in the interaction that is being exploited, then it is clear that the matrix may show poor performance. A recent advance is "site-directed immobilization" *(15)*. Many ligands are glycoproteins with defined regions of oligosaccharide commonly linked to asparagine residues. Use of hydrazine chemistry, which reacts specifically with the *cis* diols of the sugar moities allows the ligand to be selectively immobilized. This has been used successfully in immunopurification matrices where antibodies are immobilized via their oligosaccharide-containing Fc regions, leaving their antigen binding regions

Affinity Chromatography

    free to interact with their target protein. Ligands have also been engineered to include defined biotinylated residues, which can then be immobilized on streptavidin-activated matrices in an oriented manner.

3. An important factor in protein purification is the stability of the target protein. The target protein may be susceptible to inactivation or degradation owing to the physicochemical conditions employed during purification, proteolytic degradation by enzymes that may be inadvertently activated during purification, and the harsh conditions that may be used for elution of the bound material. All of these must be considered when planning a purification protocol.

4. Affinity chromatography has evolved to fit into the format of general protein purification and as such has been predominantly performed in columns. Since methodologies are required for preliminary purification of crude extracts, the use of the matrix in "batch mode" has begun to be developed. The advantage of batch mode is that the matrix can be applied to very crude cell cultures or tissue extracts with only minimal need for any form of clarification (removal of cell debris, etc.). At lab scale, matrices, such as silica, are popular owing to their density allowing rapid recovery of the matrix from the extract by centrifugation. At analytical scale, the methodology can be applied to separations in microcentrifuge tubes. Once the target protein has bound to the ligand, the matrix can be washed, and either eluted in batch mode or packed into a column format. On a larger scale, the method has been performed with purpose-designed equipment, such as "fluidized beds," e.g., Streamline® (Pharmacia).

5. If a highly concentrated eluate is required, then reverse elution can often be advantageous. When a column is loaded to under capacity, the protein occupies ligands at the top of the column preferentially. By eluting in the reverse mode, the material elutes in a tighter band. This is also advantageous if there is a heterogeneous pool of proteins with differing affinities, such as an extract of serum containing polyclonal antibodies with differing affinities for the immunogen that is acting as a ligand. The antibodies with the highest avidity for the epitope rest at the top of the column, so reverse elution aids recovery.

6. Purification is always a compromise between purity and recovery. The more steps that are to be used in the purification, the purer the end product, but the lower the recovery will be. Losses are incurred during the chromatography and in the manipulation of the protein between chromatography steps. The eluted protein fractions commonly undergo buffer exchange and concentration prior to subsequent steps. By selecting elution conditions that are consistent with subsequent chromatography steps, such as high salt for hydrophobic chromatography or defined pH changes for ion-

exchange chromatography, the target may be eluted in a form requiring minimal conditioning prior to further purification and, thereby maximizing the recovery.

References

1. Ostrove, S. (1990) Affinity chromatography: general methods, in *Methods in Enzymology*, vol. 182 (Deutscher, M. P., ed.), Academic, San Diego, CA, pp. 357–371.
2. Clonis, Y. D. (1988) The application of reactive dyes in enzyme and protein downstream processing. *Crit. Rev. Biotechnol.* **7(4),** 263–280.
3. Sulkowski, E. (1989) The saga of IMAC and MIT. *Bioessays* **10(5),** 170–175.
4. Desai, M. A. (1990) Immunoaffinity adsorption: process scale isolation of therapeutic-grade biochemicals. *J. Chem. Tech. Biotechnol.* **48,** 105–126.
5. Sherwood, R. (1991) Protein fusions: bioseparation and application. *TIBTECH* **9,** 1–3.
6. Sassenfield, H. M. (1990) Engineering proteins for purification. *TIBTECH* **8,** 88–93.
7. Schmitt, J., Hess, H., and Stunnenberg, H. G. (1993) Affinity purification of histidine tagged proteins. *Mol. Biol. Rep.* **18,** 223–230.
8. Dean, P. D. G., Johnson, W. S., and Middle, F. A. (1985) Activation procedures, in *Affinity Chromatography* (Rickwood, D. and Hames, B. D., eds.), IRL Press, Oxford, UK, pp. 31–59.
9. Groman, E. V. and Wilchek, M. (1987) Recent developments in affinity chromatography supports. *TIBTECH* **5,** 220–224.
10. Narayanan, S. U. and Crane, L. J. (1990) Affinity chromatography supports: a look at performance requirements. *TIBTECH* **8,** 12–16.
11. Bamford, C. H., Al-Lamee, K. A., Purbrick, M. D., and Wear, T. J. (1992) Studies of a novel membrane for affinity separations. *J. Chromatogr.* **606,** 19–31.
12. Chase, H. A. (1994) Purification of proteins by adsorption chromatography in expanded beds. *TIBTECH* **12,** 296–303.
13. Ohlson, S., Lundblad, A., and Zopf, D. (1988) Novel approach to affinity chromatography using "weak" monclonal antibodies. *Anal. Biochem.* **169,** 204–208.
14. Perry, M. and Kirby, H. (1990) Monoclonal antibodies and their fragments, in *Protein Purification Applications* (Harris, E. L. V. and Angal, S., eds.), IRL Press, Oxford, UK, pp. 147–164.
15. O'Shannessy, D. J. and Quarles, R. H. (1985) Specific conjugation reactions of the oligosaccharide moieties of immunoglobulins. *J. Appl. Biochem.* **7,** 347–355.

Chapter 17

Dye-Ligand Affinity Chromatography

D. Margaret Worrall

1. Introduction

Dye-ligand affinity is based on the ability of reactive dyes to bind proteins in a selective and reversible manner *(1,2)*. The dyes are generally either monochloro-triazine or dichloro-triazine compounds (two example structures are shown in Fig. 1) and were originally developed in the textile industry. The reactive chloro group allows easy immobilization of the triazine dye to a support matrix, such as Sepharose or agarose, by nucleophilic displacement of the chlorine atom.

The initial discovery of the ability of these dyes to bind proteins came from the observation that Blue Dextran (a conjugate of Cibacron Blue FG-3A), used as a void volume marker on gel-filtration columns, could retard the elution of certain proteins *(3)*. A number of studies have been carried out on the specificity of the dyes for particular proteins, mostly using the prototype Cibacron Blue dye. The dyes appear to be most effective at binding proteins and enzymes that utilize nucleotide cofactors, such as kinases and dehydrogenases. It has been proposed that the aromatic triazine dye structure resembles the nucleotide structure of NAD and that the dye interacts with the dinucleotide fold in these proteins *(4)*. In many cases, bound proteins can be eluted from the columns by a substrate or nucleotide cofactor in a competitive fashion *(5),* and dyes have been shown to compete for substrate binding sites in free solution *(6)*. However, other proteins with no nucleotide binding sites, such as human serum albumin, can also bind dye ligands tightly. It seems likely that these dyes can bind proteins by both electrostatic interactions and/or by more specific "pseudoaffinity" interactions with ligand bind-

From: *Methods in Molecular Biology, Vol. 59: Protein Purification Protocols*
Edited by: S. Doonan Humana Press Inc., Totowa, NJ

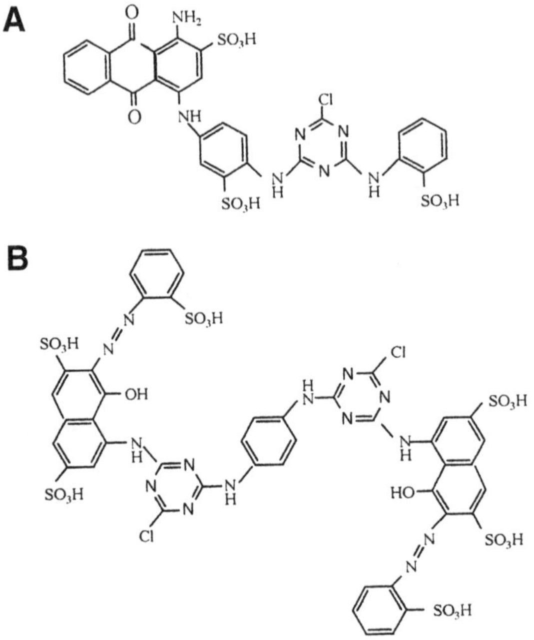

Fig. 1. **(A)** Structure of Cibacron Blue F3G-A (Procion Blue H-B, Reactive Blue 2) and **(B)** structure of Procion Red HE-3B (Reactive Red 120).

ing sites. In practical terms, the degree of purification achieved with dye-ligand chromatography is generally better than that obtained with less specific techniques, such as ion-exchange or gel-filtration chromatography. A further advantage is that the reactive dyes are relatively inert and unaffected by enzymes in crude cellular extracts.

Individual dyes show differences in binding profiles, and it is generally useful to carry out a screening procedure to determine the best dye ligand for a given purification protocol. It has been reported that NAD-utilizing enzymes bind Cibacron Blue, whereas NADP-utilizing enzymes preferentially bind to Procion Red HE-3B *(7)*. In general, however, it is not possible to predict which ligand will give the best purification, and a detailed method for screening is given in Section 3.1.

Elution of the bound protein is generally carried out by increasing ionic strength or by using a competing ligand, such as the substrate or cofactor. The optimized conditions for purification of a mammalian bifunctional enzyme CoA synthase, using both of these methods, are described.

Dye-Ligand Chromatography

Table 1
Commercially Available Immobilized Dyes

Procion dye	Amicon	Sigma
Blue H-B (Cibacron Blue)	Matrex Gel Blue A	Reactive Blue 2
Blue MX-R		Reactive Blue 4
Red HE-3B	Matrex Gel Red A	Reactive Red 120
Yellow H-A	Matrex Gel Orange A	Reactive Yellow 3
Yellow MX-3R	Matrex Gel Orange B	
Green H-4G		Reactive Green 5
Green H-E4BD	Matrex Gel Green A	Reactive Green 19
Brown MX-5BR		Reactive Brown

2. Materials

1. Dye screening kits are available from Sigma (St. Louis, MO) and Amicon (Beverly, MA) containing a range of immobilized dyes as described in Table 1. Both of these suppliers use agarose as the support matrix.
2. Blue Sepharose CL-6B (Cibacron Blue F3G-A linked to Sepharose CL-6B) and Red Sepharose (Procion Red HE-3B-Sepharose CL-6B) are available from Pharmacia (Piscataway, NJ) (*see* Note 1).
3. Equilibration buffer: 20 mM Tris-HCl, pH 8.0, 0.5 mM DTT (*see* Notes 2 and 5).
4. Elution buffer: 20 mM Tris-HCl, pH 8.0, 1M KCl, 0.5 mM DTT.
5. Blue Sepharose affinity elution buffer: 0.1 mM Coenzyme A, 0.1M KCl in equilibration buffer.

All buffers contain analytical-grade reagents. DTT is added freshly before use.

3. Methods
3.1 Screening Dye Ligands

1. Prepacked columns are available to screen for binding of the protein of interest. Otherwise, pour individual columns of 1-mL packed bed volume of a range of adsorbants, such as those listed in Table 1. Wash the columns with 10 mL of equilibration buffer.
2. Adjust the pH of the protein sample to be applied to the columns to that of the equilibration buffer. The ionic strength of the protein sample should also be close to that of the equilibration buffer and preferably not exceeding a total of 0.05M. This can be achieved by simple dilution. The protein concentration should be about 10–20 mg/mL, and if necessary, the sample should be centrifuged or filtered to remove any particulate matter.

3. Load 1 mL of the sample to the columns under gravity, and wash through with 5 mL of the equilibration buffer, collecting the unbound proteins in one fraction. Elute the bound protein with 5 mL elution buffer and collect in a fresh tube.
4. Assay both fractions from each column for total protein using a quantitative assay, such as the Bradford assay *(8)*. Assay for the protein of interest using the most quantitative method available. Calculate the purification factor achieved in each case.
5. The best binding dye will be the one that effectively binds all the target protein, but allows much of the contaminants to pass through. Further development of this step will involve optimization of the elution conditions to remove contaminants by step or gradient elution. A dye that does not bind the protein of interest, but adsorbs some of the contaminating proteins can also be useful as a negative binding purification step (*see* Note 4).

3.2. Optimization of Purification

Further small-scale columns are recommended for optimization of purification conditions prior to scaling up.

1. For a negative binding step, vary the composition of the equilibration buffer with respect to both ionic strength and pH. Lowering of the ionic strength and lowering pH will generally increase the amount of total protein binding to the immobilized dyes. As in the dye screening step, collect the unbound material, and assay for total protein and protein of interest. Calculate the specific activity and the recovery of the protein of interest. The best buffer conditions will give the highest specific activity in the unbound fraction, provided a good yield of the protein is also achieved.
2. For a positive binding purification step, the equilibration buffer conditions should similarly be varied in order to maximize binding of the protein of interest and minimize binding of contaminant proteins.
3. The capacity of the dye for the protein of interest can be determined by frontal analysis. This is achieved by continuous loading of the sample solution onto the column until the protein of interest is detected in the eluate. This occurs when the target protein is displaced by proteins with higher affinities for the immobilized dye. Frontal analysis can be useful for examining the relative affinities of more than one protein in a mixture of proteins for a particular matrix. Optimum loading is 80% of the sample volume required for frontal detection.
4. Elution conditions can now be optimized. Wash the column with 5 vol of equilibration buffer to remove nonbinding proteins fully. Apply increasing salt concentrations to the column up to $1M$ NaCl (or KCl) in a stepwise

fashion with increases of 200 mM/step and with 2 column volumes/step. Alternatively, a salt gradient can be applied, and a 0–1M NaCl linear gradient could be tried in the first instance. If gradient programming is available as in FPLC, then nonlinear gradients can be used later to optimize purification and speed of elution.

5. The ability of specific ligands, such as a substrate, cofactor, or inhibitor, to elute the protein of interest can also be screened for, and this "affinity elution" can often provide a more powerful purification step.

   As before, a small-scale column is loaded with sample and washed with equilibration buffer. Having already assessed the ionic strength conditions necessary for elution as described in the previous step, this information is used to optimize affinity elution conditions. First wash the column with buffer of an ionic strength just below that required to elute the protein in order to remove contaminants (try 20 mM < eluting buffer). Check that the eluate does not contain the protein of interest, and if so, lower the ionic strength of this wash solution. To affinity elute, lower the salt concentration of the buffer further (by 5–10 mM depending on the ionic strength of the specific ligand) and spike this solution with the specific ligand. In cases where the specific ligand will elute many proteins (e.g., ATP), better purification may be attained by applying a concentration gradient of the affinity ligand.

6. When scaling up of the purification step, keep the proportion of sample to column volume constant, although column size should be increased with larger diameters rather than larger column lengths. The linear flow rate should remain constant so that the sample protein will have the same contact time with the dye (*see* Notes 5 and 6).

3.3. Example Purification: Isolation of Coenzyme A Synthase

The bifunctional enzyme CoA synthase isolated and partially purified from pig liver was further purified to near homogeneity by dye-ligand chromatography on two types of immobilized dyes *(9)*. The crude protein solution was loaded onto a Red Sepharose column equilibrated with 10 vol of equilibration buffer. The column was further washed with 2 column vol of equilibration buffer, and a linear gradient of 0–1M KCl in 20 mM Tris-HCl, pH 8.0, was applied to elute the protein. The fractions containing CoA synthase (Fig. 2A) were pooled and diluted with 2 vol of 0.5 mM DTT in order to reduce the ionic strength and allow binding to Blue Sepharose. This material was loaded onto a Blue Sepharose Cl-6B column equilibrated with 10 mM Tris-HCl, pH 8.0. The column was then washed with 0.12M KCl, in 10 mM Tris-HCl, pH 8.0, which does not

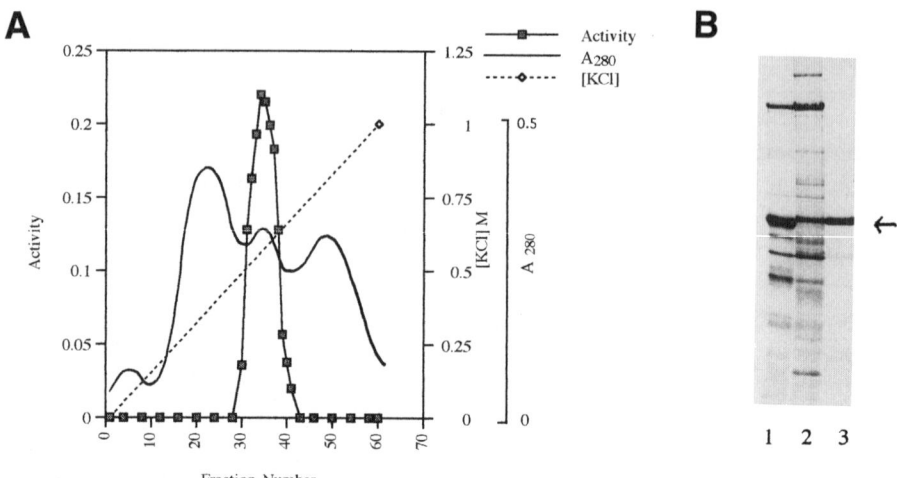

Fig. 2. Purification of CoA synthase on Procion Red Sepharose and affinity elution from Blue Sepharose. **(A)** Elution profile of CoA synthase from Red Sepharose CL-6B (Procion Red HE-3B) using a 0–1 M KCl gradient. Fractions 30–40 were pooled, diluted with 2 vol 0.5 mM DTT, and applied to a Blue Sepharose CL-6B column. Affinity elution was carried out with 0.1 mM Coenzyme A. **(B)** SDS-PAGE of purification steps; lane 1, crude fraction (CoA synthase band of 60 kDa not visible); lane 2, pooled Red Sepharose fractions; lane 3, Blue Sepharose affinity-eluted protein.

elute the CoA synthase, but does elute most of the contaminating proteins. The enzyme was then eluted with a solution of 0.11 M KCl, 10 mM Tris-HCl, pH 8.0, containing 0.1 mM Coenzyme A. As can be seen in Fig. 2B, the combination of these two steps can purify this protein from a crude extract where it is not visible on SDS-PAGE to near homogeneity.

4. Notes

1. Immobilization of dyes onto a support matrix can be readily carried out in the laboratory, and this may be desirable for cost-saving measures, particularly for large-scale column purification. The reactive dyes can be purchased relatively cheaply from Sigma and Polysciences. Detailed methods for coupling the dyes to the support matrix are given elsewhere *(10)*. The major drawback from homemade columns is usually leakage of the dye from the matrix even after extensive washing. Novel purpose-made dye ligands have recently been synthesized that reduce leakage problems and have higher protein binding capacity *(11)*.

2. The composition of the equilibration and elution buffers can be varied greatly. The examples given use Tris buffers at pH 8.0, which is relatively high, and so will reduce the amount of total protein binding. If binding of the protein of interest is low or weak in these conditions, it may be desirable to decrease the pH, in which case phosphate buffers may be more suitable.
3. The addition of divalent and trivalent metal ions (e.g., Mg^{2+}, Ca^{2+}, Zn^{2+}, Cu^{2+}, Mn^{2+}, Fe^{3+}) to the equilibration buffer can promote binding of certain proteins *(12)*. Similarly, addition of EDTA to elution buffers can increase desorption of proteins.
4. Negative binding steps have proven to be very effective in the purification of proteins from plasma and from cell-culture extracts, which contain large amounts of fetal calf serum. This is owing to the high affinity of serum albumin, which is the major contaminating protein in such preparations, to Cibacron Blue and procion Red HE-3B dyes *(13)*.
5. Regeneration of dye-ligand adsorbants can be carried out using either high-salt ($2M$ NaCl) and/or high-pH ($1M$ NaOH) solutions, and this is generally quite effective. However, if the column capacity is noticeably decreasing with continued use, then it may be improved by using chaotropic agents, such as $8M$ urea or $6M$ guanidinium hydrochloride, to denature and so remove residual protein.
6. HPLC-grade support matrices for dye-ligand chromatography have been used for purification of a number of proteins by immobilization of dyes onto silica-based matrices *(14)*. Prepacked columns of Procion Blue MX-R, Procion Red HE-3B, and Cibacron Blue immobilized to silica (300 and 500 Å) are now commercially available from Serva Fine Chemicals.

References

1. Stellwagen, E. (1990) Chromatography on immobilized reactive dyes. *Methods Enzymol.* **182,** 343–357.
2. Scopes, R. K. (1986) Strategies for enzyme isolation using dye-ligand and related adsorbents *J. Chromatogr.* **376,** 131–140.
3. Kopperschlager, G., Freyer, R., Diezel, W., and Hoffman, E. (1968) Some kinetic and molecular properties of yeast phosphofructokinase. *FEBS Lett.* **1,** 137–141.
4. Thompson, S. T., Cass, K. H., and Stellwagon, E. (1975) Blue Dextran Sepharose: an affinity column for the dinucleotide fold in proteins. *Proc. Natl. Sci. Acad. USA* **72,** 669–672.
5. Baird, J., Sherwood, R., Carr, R. H. G., and Atkinson, A. (1976) Enzyme purification by substrate elution chromatography from Procion dye–polysaccharide matrices. *FEBS Lett.* **70,** 61–66.
6. Reuter R., Metz, P., Lorenz, G., and Kopperschlager, G. (1990) Interaction of bacterial glucose-6-phosphate dehydrogenase with triazine dyes: a study by means of affinity partitioning and kinetic analysis. *Biomed. Biophys. Acta* **49,** 151–160.

7. Watson, D. H., Harvey, M. J., and Dean, P. D. G. (1978) The selective retardation of NADP$^+$-dependent dehydrogenases by immobilised Procion HE-3B. *Biochem. J.* **173,** 591–596.
8. Bradford, M. M. (1976) A rapid and sensitive method for the quantification of microgram quantities of protein using the principle of protein–dye binding. *Anal. Biochem.* **72,** 248–254.
9. Worrall, D. M. and Tubbs, P. K. (1983) A bifunctional enzyme complex in CoA biosynthesis; purification of phosphopantetheine adenylyltransferase and dephospho-CoA kinase. *Biochem. J.* **215,** 153–157.
10. Lowe, C. R. and Pearson, J. C. (1984) Affinity chromatography on immobilised dyes. *Methods Enzymol.* **104,** 97–113.
11. Lowe, C. R., Burton, S. J., Burton, N., Sterart, D. J., Purvis, D. R., Pitfield, I., and Eapen, S. (1990) New developments in affinity chromatography. *J. Mol. Recognit.* **3,** 117–122.
12. Hughes, P. (1989) Metal enhanced interactions of proteins with triazine dyes and applications to downstream processing, in *Protein–Dye Interactions: Developments and Applications* (Vijayalakshima, M. A. and Bertrand, O., eds.), Elsevier, London, pp. 207–215.
13. Travis, J., Bowden, D., Tewksbury, D., Johnson, D., and Pannell, R. (1976) Isolation of albumin from whole human plasma and fractionation of albumin-depleted plasma. *Biochem. J.* **157,** 301–306.
14. Groman, E. V. and Wilchek, M. (1987) Recent developments in affinity chromatography supports. *Trends Biotechnol.* **5,** 220–224.

CHAPTER 18

Lectin Affinity Chromatography

Iris West and Owen Goldring

1. Introduction

Lectins are glycoproteins or proteins that have a selective affinity for a carbohydrate, or a group of carbohydrates. Many purified lectins are readily available, and these may be immobilized to a variety of chromatography supports.

Immobilized lectins are powerful tools which can be used to separate and isolate glycoconjugates *(1–3)*, polysaccharides, soluble cell components, and cells *(4)* that contain specific carbohydrate structures. Glycopeptide mixtures can be fractionated by sequential chromatography on different lectins *(5)*. Immobilized lectins can also be used to remove glycoprotein contaminants from partially purified proteins. In addition, immobilized lectins can be used to probe the composition and structure of surface carbohydrates *(6)*. Subtle changes in glycoprotein structure owing to disease may be reflected in altered affinity for lectins *(7)*. The predominant lectins used for affinity chromatography are listed in Table 1.

In principle, the mixture of glycoconjugates to be resolved is applied to the immobilized lectin column. The glycoconjugates are selectively adsorbed to the lectin and components without affinity for the lectin are then washed away. The adsorbed carbohydrate containing components are dissociated from the lectin by competitive elution with the hapten sugar, i.e., the best saccharide inhibitor. This may be a simple monosaccharide or an oligosaccharide *(24,25)*.

There are many matrices suitable for coupling to lectins, and a number of reliable coupling methodologies have been developed. Many supports

Table 1
Common Lectins used in Affinity Chromatography

Lectin	Specificity	Useful eluants	Uses	Refs.
Concanavalin A	α-D-Mannopyranosyl with free hydroxyl groups at C3, C4, and C6	0.01–0.5 M Methyl α-D-mannoside	Separation of glycoproteins	8
		D-Mannose	Purification of glycoprotein enzymes	9
			Partial purification of IgM	10
		D-Glucose	Separation of lipoproteins	11
Lens culinaris	α-D-Glucopyranosyl residues	Methyl-α-D-glucoside	Purification of gonadotrophins	12
	α-D-Mannopyranosyl	0.1 M Na Borate, pH 6.5	Purification of HeLa cells	13
	Binds less strongly than Con A	0.15 M Methyl α-D-mannoside	Isolation of mouse H antigens	14
			Purification of detergent-solubilized glycoproteins	15
			Schistosoma mansoni: surface membrane isolation	16
Tritium vulgaris	N-acetyl-D-glucosamine	0.1 M N-acetyl-glucosamine	Biochemical characterization of H4G4 antigen from HOON pre-B leukemic cell line	17
			Purification and analysis of RNA polymerase transcription factors	18
Ricinus communis	α-D-Galactopyranosyl residues	0.15 M D-galactose	Fractionation of glycopeptide binding proteins	19
Jacalin	D-Galactopyranosyl residues	0.1 M melibiose in PBS	Separating IgA1 and IgA2	20,21
			Purification of C1 inhibitor	22
Bandeira simplicifolia	α-D-Galactopyranosyl and N-acetyl-D-galactosamyl	PBS	Resolving mixtures of nucleotide sugars	23

Table 2
Commonly Available Activated Matrices

Commercial name	Supplier[a]
Affi 10, Affi 15	Bio-Rad
Affi Prep 10 and 15 *(26)*	Bio-Rad
CNBr agarose	Sigma
Epoxy-activated agarose	Sigma
CDI agarose	Sigma
Reacti 6X	Pierce
Polyacrylhydrazide agarose	Sigma

[a]Locations: Bio-Rad (Hercules, CA); Sigma (St. Louis, MO); Pierce (Rockford, IL).

are now supplied preactivated. These are simple to use. Table 2 lists a few of the most popular activated matrices (*see* Note 1).

The lectin must be chosen for its affinity for a carbohydrate or sequence of sugars. Many of the lectins are not specific for a single sugar and may react with many different glycoproteins. Concanavalin A (Con A) has a very broad specificity, and will bind many serum glycoproteins, lysosomal hydrolases, and polysaccharides. The binding efficiency of Con A is reduced by the presence of detergents 0.1% sodium dodecyl sulfate (SDS), 0.1% sodium deoxycholate, and 0.1% cetyltrimethyl-ammonium bromide (CTAB) *(27)*. *Lens culinaris* lectin has similar carbohydrate specificities to Con A, but binds less strongly. The lectin retains its binding efficiency in the presence of sodium deoxycholate. It is therefore useful for the purification of solubilized membrane glycoconjugates. Wheat germ lectin can interact with mucins that contain N-acetylglucosamine residues. It retains its binding efficiency in the presence of 1% sodium deoxycholate. Jacalin is a lectin isolated from the seeds of *Artocarpus integrifolia*, which has an affinity for D-galactose residues *(22)*. It is reported to separate IgA_1 from IgA_2 subclass antibodies. It is therefore useful for removing contaminating IgA from preparations of purified IgG.

Two simple procedures (Sections 3.1. and 3.2.) for the immobilization of lectins are given, one for Con A onto carbonyldiimidazole (CDI)-activated agarose, and the other for Con A onto Affigel 15. Once the Con A has been immobilized to the matrices, the purification and elution procedures are common. A single protocol (Section 3.3.) is given detailing this procedure.

2. Materials
2.1. Immobilization of Con A on CDI-Activated Agarose

1. CDI-activated agarose (Sigma). The activated gel is stable in the acetone slurry for 1–2 yr at 4°C.
2. Affinity-purified Con A.
3. Protective sugar: methyl-α-D-mannoside.
4. Coupling buffer: $0.1 M$ Na_2CO_3, pH 9.5.
5. $0.1 M$ Tris (hydroxymethyl) methylamine (Tris) pH 9.5.
6. Phosphate-buffered saline (PBS).
7. $0.1 M$ Sodium acetate, pH 4.0.
8. $0.1 M$ $NaHCO_3$, pH 8.0.

2.2. Immobilization of Con A on Affigel 15

1. Affigel 15.
2. Affinity-purified Con A.
3. Protective sugar: methyl-α-D-mannoside.
4. Coupling buffer (PBS): 8.0 g NaCl, 0.2 g KH_2PO_4, 2.9 g $Na_2HPO_4 \cdot 12H_2O$, 0.2 g KCl made up to 1 L with distilled water.
5. Equilibration buffer: 5 mM Sodium acetate buffer containing $0.1 M$ NaCl, 1 mM $MnCl_2$, 1 mM $CaCl_2$, 1 mM $MgCl_2$, and 0.02% sodium azide (*see* Note 4).

2.3. Purification of Glycoproteins on Immobilized Con A

1. Equilibration buffer: 5mM sodium acetate buffer, pH 5.5, containing $0.1 M$ NaCl, 1 mM $CaCl_2$, 0.02% sodium azide.
2. $0.01 M$ Methyl-α-mannoside in acetate buffer.
3. $0.3 M$ Methyl-α-mannoside in acetate buffer.
4. 5 mM Sodium acetate buffer containing $1 M$ NaCl.

3. Methods
3.1. Immobilization of Con A (see Note 3) on CDI-Activated Agarose

1. Place the gel in a sintered glass funnel. Apply gentle suction (a water pump is adequate).
2. Wash the gel free of acetone with 10 vol of ice-cold distilled water (*see* Note 2). Do not dry cake.
3. Dissolve Con A at 20 mg/mL in $0.1 M$ Na_2CO_3 buffer, pH 9.5, containing the protective sugar $0.5 M$ methyl-α-D-mannoside (*see* Note 3). Keep at 4°C.
4. Gently break up the gel cake, then add the cake directly to the coupling buffer containing the ligand, 1 vol of gel to 1 vol of buffer.

5. Incubate for 48 h at 4°C with gentle mixing.
6. Transfer the gel to a column (*see* Note 9), and then wash the gel with 0.1M Tris to block any remaining unreacted sites.
7. Wash the gel with 3 vol of 0.1M sodium acetate buffer, pH 4.0, and then with 3 vol of 0.1M NaHCO$_3$, pH 8.0.
8. Repeat step 7.
9. Wash the gel with PBS containing 0.02% sodium azide, and then store the gel at 4°C in buffer containing 0.02% sodium azide.

3.2. Immobilization of Con A (see Note 3) on Affigel 15 (see Notes 5, 7, and 8)

1. Place the gel in a sintered glass funnel. Apply gentle suction to remove the isopropyl alcohol. Do not let the gel dry out.
2. Wash the gel with 3 vol of cold distilled water (*see* Note 2).
3. Drain off excess water.
4. Dissolve Con A at 10 mg/mL in PBS containing 0.3M methyl-α-D-mannoside (*see* Note 3).
5. Transfer the gel to the coupling buffer, 1 vol gel:1 vol coupling buffer (*see* Note 6).
6. Mix gently for 2 h at 4°C.
7. Transfer the gel to a column (*see* Note 9) and wash with 5 vol of PBS.
8. Wash with equilibration buffer, and then store the column at 4°C in the presence of 0.2% sodium azide.

3.3. Purification of Glycoproteins on Immobilized Con A

1. Equilibrate the column with 5mM sodium acetate buffer containing 0.1M NaCl, 1 mM MnCl$_2$, 1 mM CaCl$_2$, 1 mM MgCl$_2$, 0.02% sodium azide (*see* Note 10), flow rate 0.1 mL/min.
2. Either dissolve solid glycoprotein in a little equilibration buffer, or if the glycoprotein is a solution, dilute 1:1 with equilibration buffer. Remove any particulate matter.
3. Apply to the column, and wash with equilibration buffer to remove nonbound components. Monitor effluent at 280 nm until the A_{280} is 0 (*see* Note 11).
4. Elute with 0.01M methyl-α-D-mannoside in acetate buffer (5 vol). Collect 1-mL fractions, and store on ice.
5. Elute with 0.3M methyl-α-D-mannoside in acetate buffer (*see* Notes 12 and 13). Collect 1-mL fractions, and store on ice.
6. Wash column with acetate buffer containing 1M NaCl, and then regenerate the column by extensive washing with equilibration buffer.
7. Pooled bound fractions can either be concentrated by ultrafiltration or dialyzed against a buffer to remove the hapten sugar.

Table 3
Common Lectins and Their Protective Saccharides

Lectin	Hapten saccharide	References
Con A	Methyl-α-D-mannoside	28
L. culinaris	Methyl-α-D-mannoside	—
T. vulgaris	Chitin oligosacchrides	29
R. communis	Methyl-β-galactoside	18
Jacalin	D-Galactose	22

4. Notes

1. When using CNBr-activated matrices, charged groups may be introduced during the coupling step. This may lead to nonspecific adsorption owing to ion-exchange effects.
2. Adequate washing of the gel prior to coupling is required. Many proteins are sensitive to the acetone or alcohols used to preserve the activated gels.
3. Coupling methods are similar for different lectins; however, the lectin binding site must be protected by the presence of the hapten monosaccharide. Table 3 lists the common lectins and the saccharides used to protect the active site during immobilization.
4. Con A requires Mn^{2+} and Ca^{2+} to preserve the activity of the binding site.
5. Affigel 10 and 15 are N-hydroxysuccinimide esters of crosslinked agarose. Both are supplied as a slurry in isopropyl alcohol. The gels can be stored without loss of activity for up to a year at –20°C.
6. Buffers containing amino groups should not be used for coupling. Suitable buffers include HEPES, PBS, and bicarbonate.
7. Affigel 10 couples neutral or basic proteins optimally. The coupling buffers used should be at a pH at, or below, the pI of the ligand. To couple acidic proteins, coupling buffers should contain 80 mM $CaCl_2$. Affigel 15 couples acidic proteins optimally. The pH of coupling buffers should be above or near the pI of the ligand. To couple basic proteins, include 0.3M NaCl in the coupling buffer.
8. Steps 1–4 should be completed in 20 min to ensure optimum coupling.
9. The size of column depends on the amount of sample to be purified. Preliminary experiments can easily be carried out using Pasteur pipets, with supports of washed glass wool. Alternatively, the barrels of disposable plastic syringes can be used. Long, thin columns will give the best resolution of several components. However, where the requirement is to remove only contaminants from the protein, a short, fat column will elute the proteins in a smaller volume.
10. Equilibration buffers used will depend on the stability of the glycoconjugate being purified, and where protease activity is suspected, the appropriate

protease inhibitors should be included. Some glycoconjugates are sensitive to acetate *(18)*. Phosphate buffers will precipitate out any Ca^{2+} added as calcium phosphate. The optimal pH for the equilibration buffer should be established to ensure maximum adsorption to the affinity column. Binding may be enhanced in certain cases by the inclusion of NaCl in the buffer *(30)*.

11. Sparingly reactive glycoconjugates may be retarded rather than bound by the column *(31)*. Washing should continue for at least five bed volumes. The nonbound fractions can be retained and reapplied to the column on a subsequent occasion, when it is suspected that the load applied to the column is saturating.
12. Elution can be achieved either as a "step," i.e., a change to a buffer containing the hapten sugar, or using a gradient of the hapten sugar. Different inhibiting sugars can be used for elution, e.g., D-glucose, α-D-mannose, methyl-α-D-glucoside.
13. Occasionally, glycoconjugates may bind very tightly to the matrix and may not elute even with $0.5M$ hapten sugar. Sodium borate buffers offer an alternative: these may elute components with high affinity for the lectin without denaturing it. The eluant temperature can be raised above 4°C, dependent on the protein stability. Detergents, e.g., 0–5% SDS together with $6M$ urea will often strip components from the lectin, but may irreversibly denature the lectin. Occasionally, it may be expedient to use stringent denaturing conditions to desorb bound fractions with a high yield. Glycoproteins have been eluted from Con A Sepharose by heating in media containing 5% SDS and $8M$ urea *(32)*. $1M$ NH_4OH has also been used as an eluant *(13)* with high yield, but irreversibly denaturing the lectin.

References

1. Kennedy, J. F. and Rosevear, A. (1973) An assessment of the fractionation of carbohydrates on Concanavalin A Sepharose by affinity chromatography. *J. Chem. Soc. Perkin Trans.* **19,** 2041–2046.
2. Fidler, M. B., Ben-Yoseph, Y., and Nadler, H. L. (1979) Binding of human liver hydrolases by immobilised lectins. *Biochem. J.* **177,** 175.
3. Kawai, Y. and Spiro, R. G. (1980) Isolation of major glycoprotein from fat cell plasma membranes. *Arch. Biochem. Biophys.* **199,** 84–91.
4. Sharma, S. K. and Mahendroo, P. P. (1980) Affinity chromatography of cells and cell membranes. *J. Chromatogr.* **184,** 471–499.
5. Cummings, R. D. and Kornfeld, S. (1982) Fractionation of asparagine linked oligosaccharides by serial lectin affinity chromatography. *J. Biol. Chem.* **257,** 11,235–11,240.
6. Boyle, F. A. and Peters, T. J. (1988) Characterisation of galactosyltransferase isoforms by ion exchange and lectin affinity chromatography. *Clin. Chim. Acta* **178,** 289–296.
7. Pos, O., Drechoe, A., Durand, G., Bierhuizen, M. F. A., Van der Stelt, M. E., and Van Dijk, W. (1989) Con A affinity of rat Â–1-acid glycoprotein (rAGP): changes

during inflammation, dexamethasone or phenobarbital immunoelectrophoresis (CAIE) are not only a reflection of biantennary glycan content. *Clin. Chim. Acta* **184,** 121–132.
8. Nikawa, T., Towatari, T., and Katunuma, N. (1992) Purification and characterisation of Cathepsin J from rat liver. *Eur. J. Biochem.* **204(1),** 381–393.
9. Esmann, M. (1980) Concanavalin A Sepharose purification of soluble Na,K-ATPase from rectal glands of the spiny dogfish. *Anal. Biochem.* **108,** 83–85.
10. Weinstein, Y., Givol, D., and Strausbauch, D. (1972) The fractionation of immunoglobulins with insolubilised concanavalin A. *J. Immunol.* **109,** 1402–1404.
11. Yamaguchi, N., Kawai, K., and Ashihara, T. (1986) Discrimination of gamma-glutamyltranspeptidase from normal and carcinomatous pancreas. *Clin. Chim. Acta* **154,** 133–140.
12. Kinzel, V., Kübler, D., Richards, J., and Stöhr, M. (1976) *Lens Culinaris* lectin immobilised on Sepharose: binding and sugar specific release of intact tissue culture cells. *Science* **192,** 487–489.
13. Matsuura, S. and Chen, H. C. (1980) Simple and effective solvent system for elution of gonadotropins from concanavalin A affinity chromatography. *Anal. Biochem.* **106,** 402–410.
14. Kvist, S., Sandberg-Tragardh, L., and Östberg, L. (1977) Isolation and partial characterisation of papain solubilised murine H-2 antigens. *Biochemistry* **16,** 4415–4420.
15. Dawson, J. R., Silver, J., Shepherd, L. B., and Amos, B. D. (1974) The purification of detergent solubilised HL-A antigen by affinity chromatography with the haemagglutin from *Lens culinaris*. *J. Immunol.* **112,** 1190–1193.
16. Pujol, F. H. and Cesari, I. M. (1993) *Schistosoma Mansoni:* surface membrane isolation with lectin coated beads. *Membr. Biochem.* **10(3),** 155–161.
17. Gougos, A. and Letarte, M. (1988) Biochemical characterisation of the 44g4 antigen from the HOON pre-B leukemic cell line. *J. Immunol.* **141(6),** 1934–1940.
18. Dulaney, J. T. (1979) Eractors affecting the binding of glycoproteins to concanavalin A and *Ricinus communis* agglutin and the dissociation of their complexes. *Anal. Biochem.* **99,** 254.
19. Gregory, R. L., Rundergren, J., and Arnold, R. R. (1987) Separation of human IgA$_1$ and IgA$_2$ using jacalin-agarose chromatography. *J. Immunol. Methods* **99,** 101–106.
20. Jackson, S. F. and Tjian, R. (1989) Purification and analysis of RNA polymerase 11 transcription factors by using wheat germ agglutinin affinity chromatography. *Proc. Natl. Acad. Sci. USA* **86(6),** 1781–1785.
21. Pilatte, Y., Hammer, C. H., and Frank, M. M. (1983) A new simplified procedure for C1 inhibitor purification. *J. Immunol. Methods* **120,** 37–43.
22. Roque-Barreira, M. C. and Campos-Neto, A. (1985) Jacalin: an IgG-binding lectin. *J. Immunol.* **134,** 1740.
23. Blake, D. A. and Goldstein, I. J. (1982) Resolution of carbohydrates by lectin affinity chromatography. *Methods Enzymol.* **83,** 127–132.
24. Debray, H., Pierce-Cretel, A., Spik, G., and Montreuil, J. (1983) Affinity of ten insolubilised lectins towards various glycoproteins with the N-glycosylamine linkage and related oligosaccharides, in *Lectins in Biology, Biochemistry, Clinical Biochemistry*, vol. 3 (Bog-Hanson, T. C. and Spengler, G. A., eds.), De Gruyter, Berlin, p. 335.

25. Debray, H., Descout, D., Strecker, G., Spik, G., and Montreuil, J. (1981) Specificity of twelve lectins towards oligosaccharides and glycopeptides related to N-glycosylglycoproteins. *Eur. J. Biochem.* **117,** 41–55.
26. Matson, R. S. and Siebert, C. J. (1988) Evaluation of a new N-hydroxysuccinimide activated support for fast flow immunoaffinity chromatography. *Prep. Chromatogr.* **1,** 67–91.
27. Lotan, R. and Nicolson, G. L. (1979) Purification of cell membrane glycoproteins by lectin affinity chromatography. *Biochem. Biophys. Acta* **559,** 329–376.
28. Cook, J. H. (1984) Turnover and orientation of the major neural retina cell surface protein protected from tryptic cleavage. *Biochemistry* **23,** 899–904.
29. Mintz, G. and Glaser, L. (1979) Glycoprotein purification on a high capacity wheat germ lectin affinity column. *Anal. Biochem.* **97,** 423–427.
30. Lotan, R., Beattie, G., Hubbell, W., and Nicolson, G. L. (1977) Activities of lectins and their immobilised derivatives in detergent solutions. Implications on the use of lectin affinity chromatography for the purification of membrane glycoproteins. *Biochemistry* **16(9),** 1787.
31. Narashimhan, S., Wilson, J. R., Martin, E., and Schacter, H. (1979) A structural basis for four distinct elution profiles on concanavalin A Sepharose affinity chromatography of glycopeptides. *Can. J. Biochem.* **57,** 83–97.
32. Poliquin, L. and Shore, G. C. (1980) A method for efficient and selective recovery of membrane glycoproteins from concanavalin A Sepharose using media containing sodium dodecyl sulphate and urea. *Anal. Biochem.* **109,** 460–465.

CHAPTER 19

Immunoaffinity Chromatography

George W. Jack and David J. Beer

1. Introduction

Immunoaffinity chromatography (IAC) harnesses the specificity and avidity of the interaction between an antigen and its antibody to purify the antigen. The technique may employ either polyclonal or monoclonal antibodies (MAbs) *(1)*, but MAbs are preferred for a variety of reasons. Polyclonal sera are never specific for a single antigen, but reflect the variety of immunological challenges sustained by the animal since birth. Furthermore, the best polyclonal sera are generated by immunization with a highly purified antigen, so if a purification method already exists for the protein, why raise antisera to develop a second purification method? Perhaps the main disadvantage of polyclonal sera lies in the wide range of avidity of the antibodies they contain, because although ensuring that such antibodies will capture their antigen, the dissociation of the antigen–antibody complex may prove impossible under conditions that retain the activity of both.

Conversely, MAbs may be generated even when pure antigen is not available and produced on an appropriate scale by growing as ascites or in cell culture. A unique and consistent product may be obtained by this means, whereas every animal immunized with a given antigen will produce a different set of antibodies, each with their own properties. Each MAb exhibits a unique association constant (K_a) for its antigen, so that from a panel of MAbs one can be chosen for use in IAC with sufficient avidity to ensure good binding of the antigen, while still allowing the MAb–antigen complex to be dissociated under relatively mild conditions.

A rapid ELISA plate technique has been developed (2) to determine if a given MAb will withstand a range of potential elution conditions.

The technique of IAC comprises first the immobilization of an MAb on an insoluble support matrix followed by packing the sorbent as a column on which to perform adsorption/desorption chromatography. A wide range of matrices for immobilization is available commercially (3), but probably the most commonly employed are the Sepharoses produced by Pharmacia (Uppsala, Sweden) and the Affigel series from Bio-Rad (Hercules, CA). The Affigels are crosslinked agaroses with spacer arms for carbodiimide-induced coupling or derivatized with N-hydroxy-succinimide esters for spontaneous coupling of proteins through free amino groups via a 10- or 15-atom long spacer arm; see Fig. 1. Pharmacia markets a range of activated and derivatized Sepharoses, also agarose-based, the most commonly used of which is CNBr-activated Sepharose; see Fig. 2 for the coupling reaction. CNBr-activated agarose (4) may be synthesized in-house, but since the production of it and a range of other activated matrices involves handling highly toxic chemicals and may result in varying degrees of substitution between batches, it is probably best brought in as the prepared derivative.

The degree of substitution of the matrix influences the capacity of the sorbent for antigen (5), but in general, substitution levels over 10 mg MAb/mL of matrix result in inefficient utilization of the MAb. The choice of the MAb to be used in IAC must to some extent be determined by trial and error even after MAbs unstable in commonly used elution buffers have been eliminated. If the K_as of a panel of MAbs are known, the number to be tested can be restricted to those with K_as in the range 5×10^{-7} to $5 \times 10^{-8} M$, since this has been shown to be optimal for IAC (6). Pure MAb should be used to minimize the amount of nonfunctional immobilized protein. This will reduce the nonspecific binding capacity of the sorbent, since by virtue of their amphoteric nature, immobilized proteins may act as ion exchangers, whereas any exposed hydrophobic sequences may result in nonspecific hydrophobic interactions. The selection of appropriate buffers for the binding and washing of the MAb–antigen complex will minimize contamination of the eluted product.

The nature of the chemistry used to immobilize the MAb must also be determined by trial and error, since all chemistries will not necessarilly result in an active sorbent, even though coupling of the MAb to the matrix has been achieved. A further constraint on the choice of coupling chemistry is the

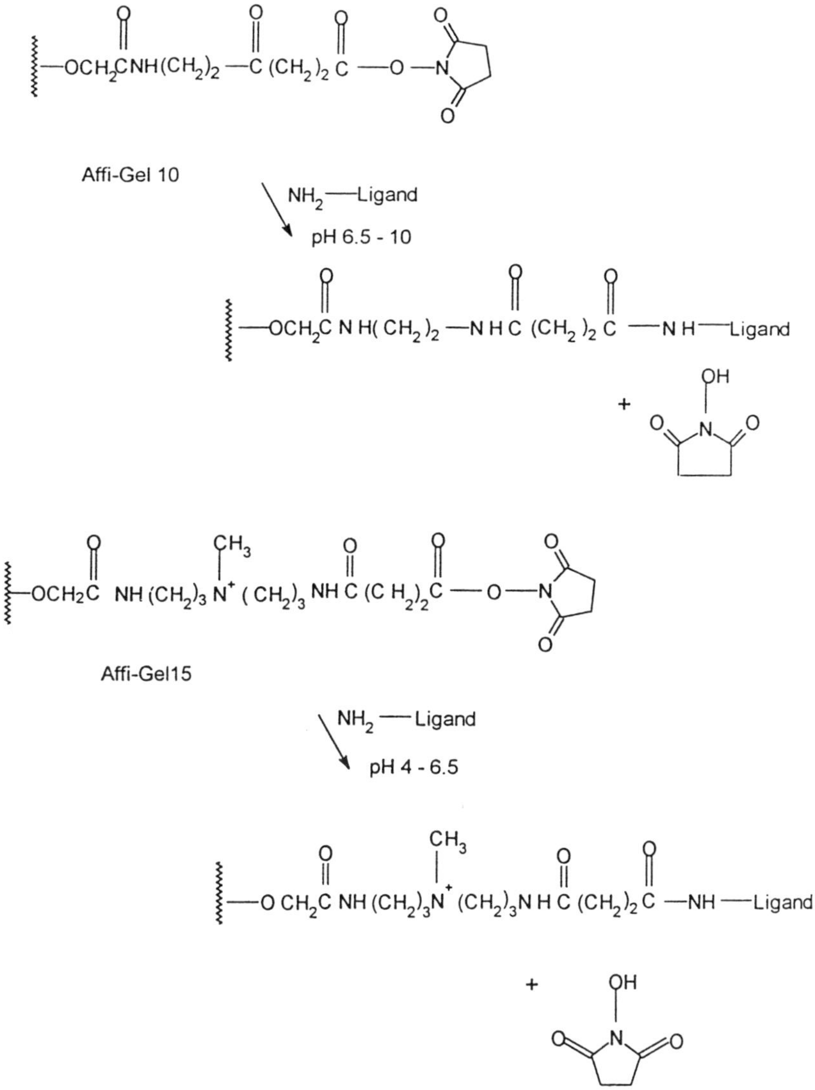

Immunoaffinity Chromatography

Fig. 1. The reactions of Affigels 10 and 15 with an amine containing ligand.

known leakage of ligand from sorbents generated by some chemistries. Although all IAC sorbents are known to leak ligand to some extent, this can be minimized by using those chemistries that have been demonstrated to leak least (7,8). The coupling of MAbs to Affigel 10 has been shown (9) to produce the most stable sorbents and, hence, is described in this example.

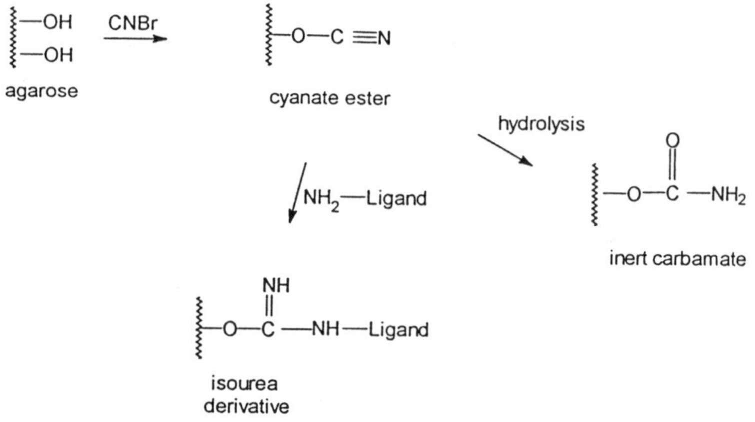

Fig. 2. The reaction of cyanogen bromide with an agarose matrix and the subsequent coupling of an amine containing ligand.

Among the advantages of IAC is its ability to produce virtually pure protein in a single step, thereby avoiding multistep procedures, which invariably result in low overall yields. Because of the high specificity and avidity of the adsorption stage, very dilute antigen solutions in the presence of large amounts of extraneous protein may be processed successfully, since the sorbent will only bind the antigen of the immobilized MAb. The capacity of sorbents for antigen is, however, low with capacities in excess of 1 mg antigen/mL of gel being something of a rarity. This factor, together with the high cost of producing IAC columns, has encouraged the use of relatively small columns operated in a cycling mode and automated by microprocessor or computer control of pumps, multiway solenoid valves, and fraction collectors (5,6). Commercially produced instruments to run this sort of chromatography are available from Amicon (Beverly, MA), Pharmacia, Sepracor (Marlborough, MA), Bio-Rad, and PerSeptive (Cambridge, MA). The use of short, fat columns at relatively high flow rates enables several cycles to be completed per day. The inclusion of protective devices such as pressure and level sensors in automated systems allows for unattended operation 24 h/d ensuring high productivity from the system despite a low yield/cycle. Furthermore, should the column become contaminated and lose activity, the replacement cost is not prohibitive.

For small-scale laboratory investigative use, several cycles of chromatography may be run per day using relatively unsophisticated equipment, while the high resolving power of the technique ensures high product quality in a single step.

2. Materials

Use Analar- or even Aristar-grade reagents and distilled water for all solutions. All solutions must be sterilized either by filtration (0.2-μm filter) or autoclaving (121°C for 30 min) where appropriate.

1. The apparatus required for running laboratory-scale IAC is commonly available and includes a glass column with adjustable flow adaptors, a peristaltic pump, a flowthrough UV-absorbance monitor and recorder, and a fraction collector. For synthesis of the IAC sorbent, a sintered glass funnel of porosity 2 to British Standard 1752, i.e., 40–90 μm, with a Buchner flask attached to a vacuum pump is required.
2. Coupling buffer: $0.1M$ phosphate/citrate, pH 6.4. Dissolve Na_2HPO_4 (14.2 g) in water and adjust to pH 6.4 with solid citric acid then make to 1 L with water. Sterilize by autoclaving.
3. Blocking buffer: $1.0M$ ethanolamine pH 8.5. Dilute 61 mL ethanolamine in water, and adjust the pH to 8.5 by the careful addition of $6M$ HCl in a fume hood. Make to 1 L with water. Sterilize by filtration.
4. Binding buffer: borate/salt, pH 8.5. Dissolve boric acid (60 g), NaCl (140 g), and NaOH (10 g) in water and make to 1 L. Sterilize by autoclaving. For use, dilute 200 mL buffer concentrate with 800 mL sterile water.
5. Wash buffer (acid): acetate/salt, pH 4.0. Dissolve sodium acetate trihydrate (13.6 g) and NaCl (29.2 g) in water and adjust to pH 4.0 with glacial acetic acid, and then with water to 1 L. Sterilize by autoclaving.
6. Column cleaning buffer: $8M$ urea/acetate, pH 4.0. Dissolve Aristar-grade urea (480 g) and sodium acetate trihydrate (13.6 g) in water, adjust to pH 4.0 with glacial acetic acid, and then to 1 L with water. Sterilize by filtration.

3. Method

1. Take 10 mL of a 50% (v/v) suspension of Affigel 10 (*see* Note 1), place in a sintered glass funnel, and suck through the suspending isopropanol. Ensure that the gel is not allowed to go dry during washing.
2. Wash the gel rapidly with 200 mL of cold water/mL of gel in several aliquots to remove the isopropanol.
3. Suck the liquid through between successive additions of water. Rapidly wash the gel with 100 mL of coupling buffer, and then immediately transfer the gel to a flask containing approx 40 mg of MAb in coupling buffer.
4. Stir the gel gently to obtain an even suspension, and then set the flask to shake at 4°C for 4 h.

5. Take a sample of the supernatant above the gel beads, and assay it for protein. This cannot be done by a simple A_{280nm} measurement owing to the absorbance of the leaving group from the matrix. If more than 80% of the MAb is no longer in solution, proceed; if not, allow the reaction to continue at 4°C.
6. When coupling is complete, add to the suspension 0.1 mL of blocking buffer/mL of suspension, and shake at room temperature for 30 min.
7. Return the gel to the sintered funnel, and wash it with alternate 100-mL lots of binding buffer, and wash buffer (acid). Five cycles is adequate.
8. Pack the gel in a 1.6-cm diameter glass column with the flow adaptors in contact with the gel bed. At a flow rate of 40 mL/h, wash the column with 3 column volumes (15 mL) of binding buffer followed by an equal volume of wash buffer. Repeat the process three times and then wash the the gel with 1 column (5 mL) of urea/acetate before equilibrating the gel with 3 column vol of binding buffer.
9. If the purification protocol is known, subject the column to five blank cycles with no sample application before the first purification cycle is run.
10. Equilibrate the extract to be chromatographed with the binding buffer either by dialysis or by buffer exchange through Sephadex G-25. Filter-sterilize the extract (*see* Note 2).
11. Pump the extract through the IAC sorbent at the above flow rate, and then wash the column with binding buffer until the UV absorbance of the column effluent falls back to the baseline level. Use a sensitive range on the absorbance monitor, such as 0–0.1 absorbance units, for full-scale recorder deflection (*see* Note 3).
12. Wash the column with 2 column volumes of a solution of low ionic strength. This could be the binding buffer omitting the salt, or even water will suffice.
13. Stop the flow through the column, and alter the configuration of the plumbing to reverse the direction of flow through the column. This may be also achieved by the use of a number of three-way valves. Elute the column with 3 column volumes of a solution (*see* Notes 4 and 5), which will dissociate the MAb–antigen complex.
14. Wash the column with 1 column volume of urea/acetate followed by 2 column volumes of binding buffer. Again, reverse the direction of flow through the column, and equilibrate the sorbent with 3 column volumes of binding buffer before applying the next sample of extract. The entire chromatographic process should take between 2 and 3 h.

4. Notes

1. Affigels 10 and 15 are supplied as 50% (v/v) suspensions in isopropanol and may be stored at –20°C for extended periods without loss of binding capac-

ity if kept dry. Adherence to the manufacturer's recommendations should result in successful coupling of MAb to matrix. If the pI of the MAb is known, then a rational choice can be made between the 10 and 15 series gels, since at neutral pH, Affigel 10 couples proteins of neutral to alkaline pI, whereas Affigel 15 couples proteins of acidic pI with high efficiency. In general, however, MAbs couple more efficiently to Affigel 10.

2. All proteins are susceptible to proteolysis even when afforded the stabilizing influence of immobilization. An IAC column run with nonsterile buffers and extracts will rapidly lose its antigen binding capacity because of bacterial or fungal proteolysis of the immobilized antibody. The use of sterile buffers and extracts is essential for the long-term activity of an IAC column. The choice of sterilization method for buffers is also important, since some buffers will change pH (e.g., carbonate/bicarbonate buffers) or decompose (buffers containing urea) on autoclaving, and for those, filtration is the sterilization method of choice. If operated carefully under sterile conditions, columns have been used for over 50 cycles of operation spread over many months, but such longevity is enhanced if the column is operated in a "clean room" environment. The ease with which a column can become contaminated and inactivated is exacerbated by the inability to sanitize the column owing to the protein nature of the ligand. This reinforces the recommendation to perform multiple cycles on a small column since you have less MAb to lose following accidental contamination.

3. All IAC columns will bind protein nonspecifically, although appropriate selection of the buffer for the absorption stage can reduce this phenomenon quite markedly. In the purification of human thyrotropin *(10)*, altering the binding buffer from Tris to borate, and then adding NaCl to the borate sequentially reduced the level of contamination of the thyrotropin by lutropin by approx 40-fold. Since protein binds nonspecifically to IAC columns by a mixture of ionic and hydrophobic interactions, washing the sorbent with solutions of both high and low ionic strength will minimize these opportunistic contaminants.

4. The nature of the eluting agent to be used in IAC must be determined empirically in each instance. A change to more acidic conditions is one of the most commonly used elution strategies, but to be successful requires some knowledge of the pH stability of the protein under purification; human thyrotropin, for example, starts losing activity below pH 3.5 *(10)*. Immobilized MAbs have much greater pH stability than antibodies in free solution and can withstand acid pH down to pH 2.0, but many appear susceptible to inactivation above pH 10.0. They are also able to withstand the effects of chaotropes and protein denaturants. Protein denaturants, such as $8M$ urea or $6M$ guanidine hydrochloride at neutral pH have been used as eluents if

the antigen retains its activity on removal of the denaturant. High concentrations of chaotropic agents, such as thiocyanate and iodide ions and $MgCl_2$ at 2.5M, are commonly used as eluents. A commonly used ploy with chaotropes is to pump just <1 column volume of eluent onto the column and then stop pumping for 30 min before continuing the elution. The above are only a small selection of the compounds that have been employed as eluents in IAC (2). The nature of the eluent requires carefull consideration in the case of potential therapeutic agents so as not to compromise their administration to humans.

5. Whenever practical, reverse-flow elution is to be preferred, since it can result in elution volumes down to one-third of those obtained by unidirectional flow through the sorbent. It also has the advantage of clearing debris away from the net or sinter of the normal inlet flow adaptor of the column.

6. Leakage of antibody occurs from all IAC columns. The extent to which such leakage results in product contamination can be minimized both by careful column washing and by choosing a coupling chemistry known to generate a stable linkage. A regime as outlined in Section 3. for washing a new gel will produce a sorbent from which antigen may be eluted with very low, yet still detectable, contamination. During thyrotropin purification, the first three cycles on a newly synthesized sorbent yielded thyrotropin contaminated with MAb at 24, 12, and then <6 ng/mL, showing that leakage from a new gel decreases with use (10). Using more sensitive assays for MAbs (11), Affigel 10-based sorbents have been shown to leak <0.3 ng MAb/mL of eluent (9). If minimum contamination levels are sought from the first cycle of operation, the precycling protocol described in Section 3. should be employed as well as after the column has been unused for some time.

    Leakage results largely from solvolysis of the matrix–ligand bond. However, the antibody is bound to the gel by multipoint attachment, and more than one bond must hydrolyze before leakage occurs. Mechanical fragmentation and chemical degradation of the matrix do not appear to contribute significantly (9) to leakage. The presence of proteases in the extracts applied to IAC sorbents can induce leakage of antibody fragments (9) and can be minimized by treating the extract with protease inhibitors prior to the chromatography. The surest way to achieve minimal leakage is to test a range of immobilization chemistries and determine which results in least leakage in the case of the MAb of choice. In general, however, the order of stability of coupling chemistries is N-hydroxysuccinimide > CNBr > 1,1' carbonyldiimidazole > tresyl > hydrazido, for crosslinked agarose as matrix (9).

7. MAbs are produced by cell lines derived from a cancerous cell, and as a consequence, oncogenic nucleic acid may be present in crude MAb prepa-

rations. Before use in IAC sorbents, MAbs should be purified both to eliminate oncogenic material and to diminish the amount of nonfunctional protein immobilized. This may be achieved by a variety of means, but the most common methods currently in use are chromatography on either ion exchangers or immobilized protein A or G *(12)*. Both protein A and protein G are receptors for the Fc region of antibodies derived from staphylococcal and streptococcal strains respectively. Both may be purchased already immobilized from suppliers, such as Bio-Rad and Pharmacia, with instructions for use. Each sorbent does not bind all subclasses of mouse and rat antibodies, but one or the other should purify almost any MAb. The product, however, is likely to contain small amounts of protein A or G leaking off the matrix.

As an alternative, a two-step procedure of cation- and anion-exchange chromatography will purify virtually all MAbs to homogeneity. A 0–0.3M NaCl gradient in 20 mM acetate, pH 5.0, on S-Sepharose followed by a 0–0.3M NaCl gradient in 20 mM Tris-HCl, pH 7.6, on Q-Sepharose should provide pure MAb.

8. Of necessity, this is a brief overview of IAC. For a more comprehensive treatment of the subject, the reader is directed to recent books *(3)* and reviews *(5,13–15)* of what has become an extensive literature.

References

1. Köhler, G. and Milstein, C. (1975) Continuous culture of fused cells secreting antibody of predefined specificity. *Nature* **250**, 495–497.
2. Bonde, M., Frøkier, H., and Pepper, D. S. (1991) Selection of monoclonal antibodies for immunoaffinity chromatography: model studies with antibodies against soy bean trypsin inhibitor. *J. Biochem. Biophys. Methods* **23**, 73–82.
3. Hermanson, G. T., Mallia, A. K., and Smith, P. K. (1992) *Immobilized Affinity Ligand Techniques.* Academic, San Diego, CA.
4. Kohn, J. and Wilchek, M. (1982) A new approach (cyano-transfer) for cyanogen bromide activation of Sepharose at neutral pH, which yields activated resins, free of interfering nitrogen derivatives. *Biochem. Biophys. Res. Commun.* **107**, 878–884.
5. Chase, H. A. (1984) Affinity separations utilizing immobilised monoclonal antibodies—a new tool for the biochemical engineer. *Chem. Eng. Sci.* **39**, 1099–1125.
6. Jack, G. W., Blazek, R., James, K., Boyd, J. E., and Micklem, L. R. (1987) The automated production by immunoaffinity chromatography of the human pituitary glycoprotein hormones thyrotropin, follitropin and lutropin. *J. Chem. Tech. Biotechnol.* **39**, 45–58.
7. Bessos, H., Appleyard, C., Micklem, L. R., and Pepper, D. S. (1991) Monoclonal antibody leakage from gels: effect of support, activation and eluent composition. *Prep. Chromatogr.* **1**, 207–220.
8. Ubrich, N., Hubert, P., Regnault, V., Dellacherie, E., and Rivat, C. (1992) Compared stability of Sepharose-based immunosorbents prepared by various methods. *J. Chromatogr.* **584**, 17–22.

9. Beer, D. J. (1993) *The Stability of Immunoaffinity Chromatography Matrices.* PhD Thesis, The Open University, Milton Keynes, UK.
10. Jack, G. W. and Blazek, R. (1987) The purification of human thyroid-stimulating hormone by immunoaffinity chromatography. *J. Chem. Tech. Biotechnol.* **39,** 1–10.
11. Beer, D. J., Yates, A. M., and Jack, G. W. (1994) Development of enzyme-linked immunosorbent assays for the detection of monoclonal antibody leakage from immunoaffinity chromatography matrices. *J. Immunol. Methods* **173,** 103–109.
12. Pharmacia LKB Biotechnology. (1993) *Monoclonal Antibody Purification Handbook,* Uppsala, Sweden.
13. Jack, G. W. (1990) The use of immunoaffinity chromatography in the preparation of therapeutic products, in *Focus on Laboratory Methods in Immunology,* vol. II (Zola, H., ed.), CRC, Boca Raton, FL, pp. 127–146.
14. Yarmush, M. L., Weiss, A. M., Antonsen, K. P., Odde, D. J., and Yarmush, D. M. (1992) Immunoaffinity purification: basic principles and operational considerations. *Biotechnol. Adv.* **10,** 413–446.
15. Jack, G. W. (1994) Immunoaffinity chromatography. *Mol. Biotechnol.* **1,** 59–86.

CHAPTER 20

Immobilized Metal Ion Affinity Chromatography

Tai-Tung Yip and T. William Hutchens

1. Introduction

Immobilized metal ion affinity chromatography (IMAC) *(1,2)* is also referred to as metal chelate chromatography, metal ion interaction chromatography, and ligand-exchange chromatography. We view this affinity separation technique as an intermediate between highly specific, high-affinity bioaffinity separation methods, and wider spectrum, low-specificity adsorption methods, such as ion exchange. The IMAC stationary phases are designed to chelate certain metal ions that have selectivity for specific groups (e.g., His residues) in peptides (e.g., *3–7*) and on protein surfaces *(8–13)*. The number of stationary phases that can be synthesized for efficient chelation of metal ions is unlimited, but the critical consideration is that there must be enough exposure of the metal ion to interact with the proteins, preferably in a biospecific manner. Several examples are presented in Fig. 1. The challenge to produce new immobilized chelating groups, including protein surface metal-binding domains *(14,15)* is being explored continuously *(16)*. Table 1 presents a list of published procedures for the synthesis and use of stationary phases with immobilized chelating groups. This is by no means exhaustive, and is intended only to give an idea of the scope and versatility of IMAC.

The number and spectrum of different proteins (Fig. 2, p. 200) and peptides (Fig. 3, p. 201) characterized or purified by use of immobilized metal ions are increasing at an incredible rate. The three reviews listed *(8,19,20)* barely present the full scope of activities in this field.

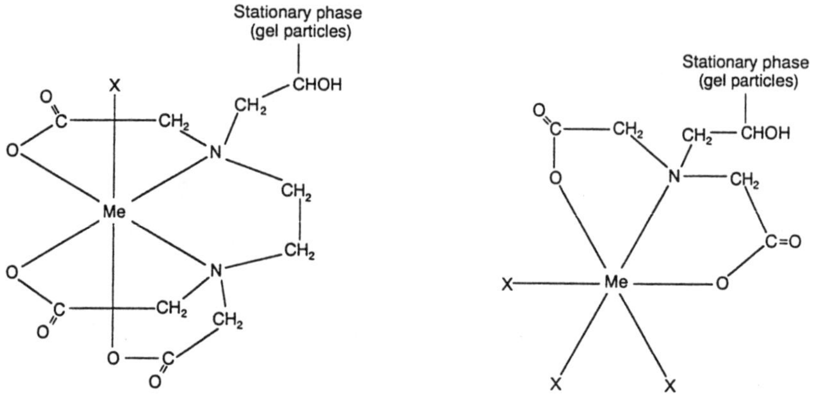

A) Tris(carboxymethyl)ethylenediamine (TED)

B) Iminodiacetate

Agarose-GHHPHG
Agarose-GHHPHGHHPHG
Agarose-GHHPHGHHPHGHHPHG
Agarose-GHHPHGHHPHGHHPHGHHPHGHHPHG

C) Immobilized metal-binding peptides

Fig. 1. Schematic illustration of several types of immobilized metal-chelating groups, including, iminodiacetate (IDA), tris(carboxymethyl) ethylenediamine (TED), and the metal-binding peptides (GHHPH)$_n$G (where n = 1, 2, 3, and 5) *(14,15)*.

Beyond the use of immobilized metal ions for protein purification are several analytical applications, including mapping proteolytic digestion products *(5)*, analyses of peptide amino acid composition (e.g., *5,6*), evaluation of protein surface structure (e.g., *8–11*), monitoring ligand-dependent alterations in protein surface structure (Fig. 4, p. 202) *(12,16)*, and metal ion exchange or transfer (e.g., *14,21*).

The versatility of IMAC is one of its greatest assets. However, this feature is also confusing to the uninitiated. The choice of stationary phases and the metal ion to be immobilized is actually not complicated. If there is no information on the behavior of the particular protein or peptide on IMAC in the literature, use a commercially available stationary phase (immobilized iminodiacetate), and pick the relatively stronger affinity transitional metal ion, Cu(II), to immobilize. If the interaction with the sample is found to be too strong, try other metal ions in the series, such as Ni(II) or Zn(II), or try an immobilized metal chelating group with a

Table 1
Immobilized Chelating Groups and Metal Ions
Used for Immobilized Metal Ion Affinity Chromatography

Chelating group	Suitable metal ions	Refs.	Commercial source[a]
IDA	Transitional	1,2	Pharmacia LKB Pierce Sigma Boehringer Mannheim TosoHaas
2-Hydroxy-3[N-(2-pyridylmethyl) glycine]propyl	Transitional	3	Not available
α-Alkyl nitrilotriacetic acid	Transitional	4	Not available
Carboxymethylated aspartic acid	Ca(II)	13	Not available
TED	Transitional	2	Not available
$(GHHPH)_nG$[b]	Transitional	14,15	Not available

[a]Locations: Pharmacia LKB, Uppsala, Sweden; Pierce, Rockford, IL; Sigma, St. Louis, MO; Boehringer Mannheim, Mannheim, Germany; TosoHaas, Philadelphia, PA.

[b]Letters represent standard one-letter amino acid codes (G = glycine; H = histidine; P = proline). The number of internal repeat units is given by n (n = 1, 2, 3, and 5).

lower affinity for proteins (2,22). An important contribution to the correct use of IMAC for protein purification is a simplified presentation of the various sample elution procedures. This is especially important to the first-time user. There are many ways to decrease the interaction between an immobilized metal ion and the adsorbed protein. Two of these methods are efficient and easily controlled; they will be presented in detail in this chapter. Interpretation of IMAC results for purposes other than separation (i.e., analysis of surface topography and metal ion transfer) has been discussed elsewhere and is beyond the scope of this chapter.

2. Materials

The following list of materials and reagents is only representative. Other stationary phases, metal ions, affinity reagents, and mobile phase modifiers are used routinely.

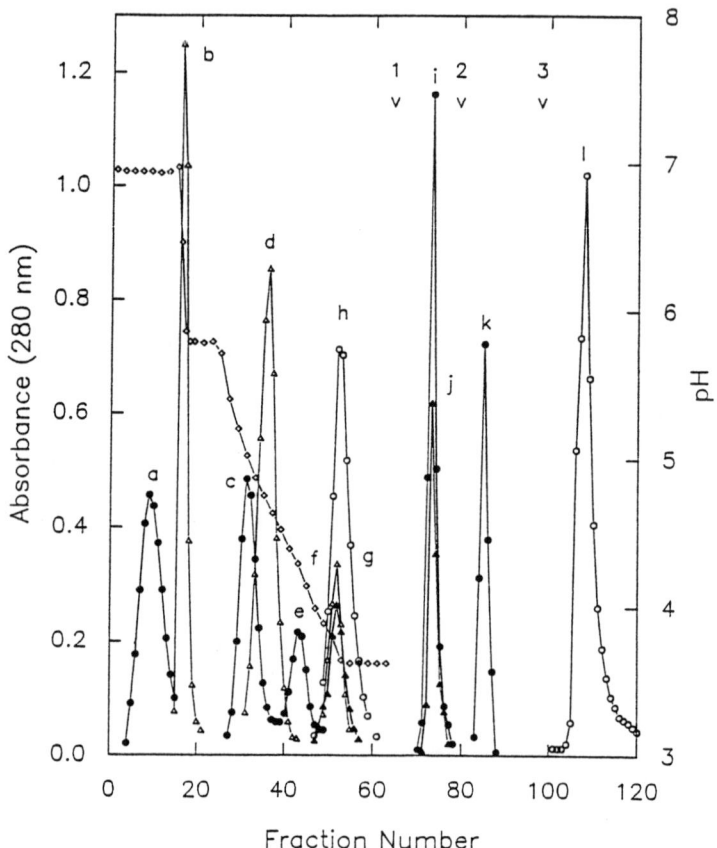

Fig. 2. Protein elution from immobilized (IDA) Cu(II) ions as a function of decreasing pH and increasing imidazole concentration. Proteins were eluted in the following order: chymotrypsinogen (a), chymotrypsin (b), cytochrome c (c), lysozyme (d), ribonuclease A (e), ovalbumin (f), soybean trypsin inhibitor (g), human lactoferrin (h), bovine serum albumin (i), porcine serum albumin (j), myoglobin (k), and transformed (DNA-binding) estrogen receptor (l). Open triangles represent pH values of collected fractions. Arrows 1–3 mark the introduction of 20, 50, and 100 mM imidazole, respectively, to elute high-affinity proteins resistant to elution by decreasing pH. Protein elution was evaluated by absorbance at 280 nm. In the case of the [^{3}H]estradiol-receptor complex, receptor protein elution was determined by liquid scintillation counting. Except for the estrogen receptor (l), protein recovery exceeded 90%. Only 50–60% of the DNA-binding estrogen receptor protein applied at pH 7.0 was eluted with 100–200 mM imidazole. Reproduced with permission from ref. *17*.

Immobilized Metal Ion Interaction

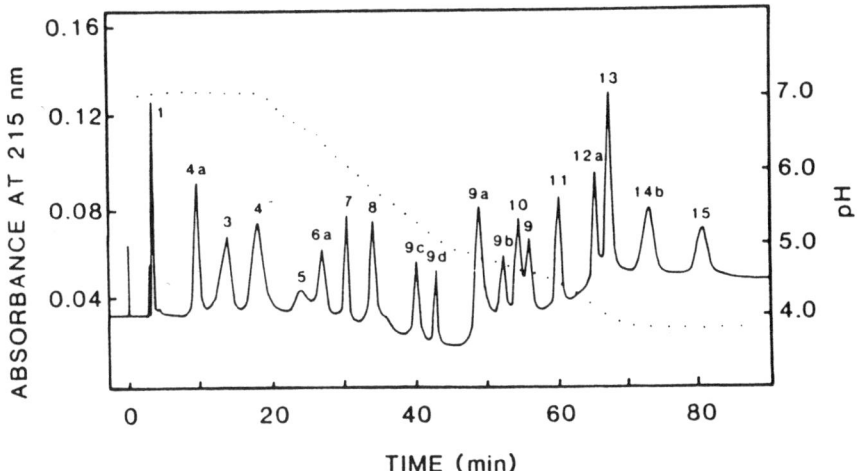

Fig. 3. Separation of bioactive peptide hormones by IMAC. The pH-dependent separation of a synthetic peptide hormone mixture (19 peptides) was accomplished using a TSK chelate-5PW column (8 × 75 mm, 10-µm particle diameter) loaded with Cu(II). A 20-µL sample (1–4 µg of each peptide) was applied to the column equilibrated in 20 mM sodium phosphate containing 0.5M NaCl, pH 7.0. After 10 min of isocratic elution, pH-dependent elution was initiated with a 50-min gradient to pH 3.8 using 0.1M sodium phosphate containing 0.5M NaCl at a flow rate of 1 mL/min. Peptide detection during elution was by UV absorance at 215 nm (0.32 AUFS). The pH profile of effluent is indicated by the dotted line. Sample elution peaks were identified as: 1, neurotensin; 4a, sulfated [leu^5] enkephalin; 3, oxytocin; 4, [leu^5] enkephalin; 5, mastoparan; 6a, tyr-bradykinin; 7, substance P; 8, somatostatin; 9c, [Asu^{1,7}] eel calcitonin; 9d, eel calcitonin (11–32); 9a, [Asu^{1,7}] human calcitonin; 9b, human calcitonin (17–32); 10, bombesin; 9, human calcitonin; 11, angiotensin II; 12a, [Trp (for) ^{25,26}] human GIP (21–42); 13, LH-RH; 14b, human PTH (13–34); 15, angiotensin I. Reproduced with permission from ref. *18*.

2.1. Stationary Phase for IMAC

2.1.1. Conventional Open Column Stationary Phases (Agarose)

One example is Chelating Sepharose Fast Flow (Pharmacia), which uses the iminodiacetate (IDA) chelating group. Another example is Tris(carboxymethyl)ethylenediamine (TED) (*see* Table 1 and Fig. 1). This stationary phase is used for proteins whose affinity for IDA-metal groups is too high *(2,22)*.

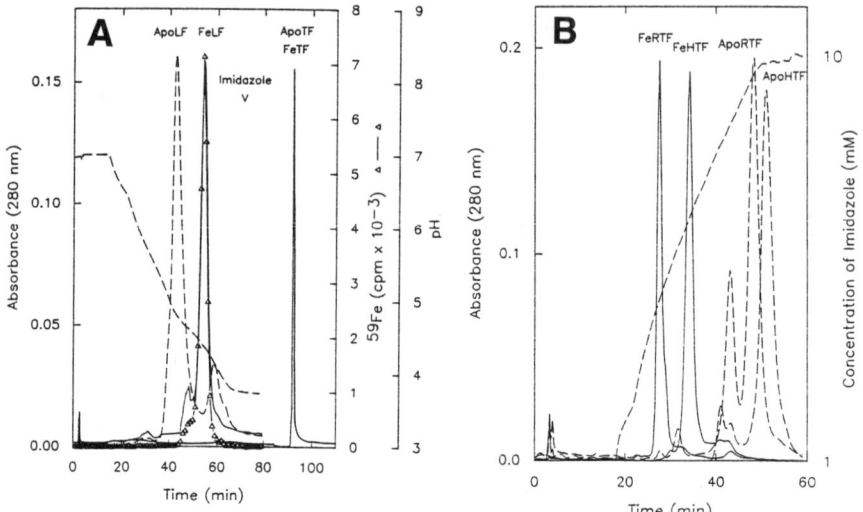

Fig. 4. **(A)** Separation of apolactoferrin (dashed line) and iron-saturated (solid line) human lactoferrins on high-performance IDA-Cu(II) columns with a phosphate-buffered gradient of decreasing pH in the presence of 3M urea. Both apotransferrin (ApoTF) and iron-saturated human serum transferrins (FeTF) were eluted only on introduction of 20 mM imidazole (imidazole-labeled arrow). UV absorbance was monitored at 280 nm. The elution of iron-saturated human lactoferrin was determined by measuring protein-bound ^{59}Fe radioactivity (open triangles). **(B)** Separation of iron-free (dashed line) and iron-saturated (solid line) transferrins on a high-performance IDA-Cu(II) affinity column using an imidazole elution gradient protocol. The apo and holo forms of human serum transferrin (hTF) and rabbit serum transferrin (rTF) are shown. The Fe and Apo prefixes to these abbreviations designate iron-saturated and metal-free transferrins, respectively. Reproduced with permission from ref. *12*.

2.1.2. High-Performance Stationary Phases (Rigid Polymer)

One example is TSK Chelate 5PW (TosoHaas) (IDA chelating group).

2.1.3. Immobilized Synthetic Peptides as Biospecific Stationary Phases

Stationary phases of this type are designed based on the known sequence of protein surface metal-binding domains *(14,15)*. The metal-binding sequence of amino acids is first identified (e.g., *14,23*). The synthetic protein surface metal-binding domain is then prepared by solid-phase methods of peptide synthesis *(14)* and verified to have metal-binding properties in solution *(14,24)*. Finally, the peptides are immobilized (e.g., to agarose) using chemical coupling procedures consistent with the retention of metal-binding properties *(15)* (Table 1 and Fig. 1).

The procedures outlined in this chapter emphasize the specific use of agarose-immobilized IDA metal chelating groups. In general, however, these procedures are all acceptable for use with a wide variety of different immobilized metal chelate affinity adsorbents.

2.2. Metal Ion Solutions in Water

1. 50 mM CuSO$_4$.
2. 50 mM ZnSO$_4$.
3. 50 mM NiSO$_4$.

2.3. Buffers

1. Buffer 1: 20 mM sodium phosphate (7.8 mM NaH$_2$PO$_4$, 12.2 mM Na$_2$HPO$_4$), 0.5M sodium chloride, pH 7.0.
2. Buffer 2: 0.1M sodium acetate, 0.5M sodium chloride, pH 5.8. Use 0.1M acetic acid and adjust the pH with 2–5M sodium hydroxide.
3. Buffer 3: 0.1M sodium acetate, 0.5M sodium chloride, pH 3.8. Use 0.1M acetic acid and adjust the pH with 2–5M sodium hydroxide.
4. Buffer 4: 50 mM sodium dihydrogen phosphate, 0.5M sodium chloride. Add concentrated HCl until pH is 4.0.
5. Buffer 5: 20 mM imidazole (use the purest grade or pretreat with charcoal), 20 mM sodium dihydrogen phosphate (or HEPES), 0.5M sodium chloride. Adjust to pH 7.0 with HCl.
6. Buffer 6: 2 mM imidazole, 20 mM sodium dihydrogen phosphate (or HEPES), 0.5M sodium chloride. Adjust to pH 7.0 with HCl.
7. Buffer 7: buffer 1 containing 50 mM EDTA.
8. Buffer 8: 20 mM sodium phosphate (3.2 mM NaH$_2$PO$_4$, 16.8 mM Ha$_2$HPO$_4$), 0.5M sodium chloride, 3M urea, pH 7.5.
9. Buffer 9: 50 mM sodium dihydrogen phosphate, 0.5M sodium chloride, 3M urea. Add concentrated HCl until pH is 3.8.
10. Milli-Q (Millipore) water or glass-distilled, deionized water.

Urea (1–3M) and ethylene glycol (up to 50%) have been found useful as additives to the abovementioned buffers (*see* Notes 7 and 8).

2.4. Columns and Equipment

1. 1-cm Inner diameter columns, 5–10-cm long for analytical and micropreparative scale procedures.
2. 5–10-cm Inner diameter columns, 10–50 cm long for preparative scale procedures.
3. Peristaltic pump.

4. Simple gradient forming device to hold a vol 10–20X column bed vol (if stepwise elution is unsuitable).
5. Flow-through UV detector (280 nm) and pH monitor.
6. Fraction collector.

3. Methods

3.1. Loading the Immobilized Metal Ion Affinity Gel with Metal Ions and Column Packing Procedures (Agarose-Based IDA Chelating Gel)

1. Suspend the IDA gel slurry well in the bottle supplied. Pour an adequate portion into a sintered glass funnel. Wash with 10 bed vol of water to remove the alcohol preservative.
2. Add 2–3 bed vol of 50 mM metal ion solution in water. Mix well.
3. Wash with 3 bed vol of water (use buffer 3 for IDA-Cu^{2+}) to remove excess metal ions.
4. Equilibrate gel with 5 bed vol of buffer 1.
5. Suspend the gel and transfer to a suction flask. Degas the gel slurry.
6. Add the gel slurry to column with the column outlet closed. Allow the gel to settle for several minutes; then open the outlet to begin flow.
7. When the desired volume of gel has been packed, insert the column adaptor. Pump buffer through the column at twice the desired end flow rate for several minutes. Readjust the column adapter until it just touches the settled gel bed. Re-equilibrate the column at a linear flow rate (volumetric flow rate/cross-sectional area of column) of approx 30 cm/h.

3.2. Elution of Adsorbed Proteins

3.2.1. pH Gradient Elution (Discontinuous Buffer System)

1. After sample application, elute with 5 bed vol of buffer 1.
2. Change buffer to buffer 2, and elute with 5 bed vol (*see* Note 1).
3. Elute with a linear gradient of buffer 2 to buffer 3. Total gradient vol should be equal to 10–20 bed volume.
4. Finally, elute with additional buffer 3 until column effluent pH is stable and all protein has eluted (recovery should exceed 90%).

3.2.2. pH Gradient Elution (Continuous Buffer System)

1. After sample application, elute with 2.5 bed vol of buffer 1.
2. Start a linear pH gradient of buffer 1 to buffer 4. Total gradient vol should be equal to approx 15 bed vol (*see* Note 2).
3. Elute with additional buffer 4 until the column effluent pH is stable.
4. If the total quantity of added protein is not completely recovered, elute with a small vol (<5 bed vol) of buffer 4 adjusted to pH 3.5.

Immobilized Metal Ion Interaction 205

3.2.3. Affinity Gradient Elution with Imidazole

1. Equilibrate the column first with 5 bed vol of buffer 5. Now, equilibrate the column with 5–10 bed vol of buffer 6 (*see* Note 3).
2. After sample application, elute with 2.5 bed vol of buffer 6.
3. Now elute with a linear gradient of buffer 6 to buffer 5. Total imidazole gradient vol should equal 15 bed volumes (*see* Notes 4 and 5).

3.3. Evaluation of Metal Ion Exchange or Transfer from the Stationary Phase to the Eluted Peptide/Protein

1. Use trace quantities of radioactive metal ions (e.g., ^{65}Zn) to label the stationary phase (i.e., immobilized) metal ion pool (*see* Section 3.1.). After elution of adsorbed proteins (*see* Section 3.2.), determine the total quantity of radioactive metal ions transferred to eluted proteins from the stationary phase (by use of a gamma counter).
2. To avoid the use of radioactive metal ions, the transfer of metal ions from the stationary phase to apo (metal-free) peptides present initially in the starting sample may be determined by either of two methods of soft ionization mass spectrometry. Both electrospray ionization mass spectrometry *(25)* and matrix-assisted UV laser desorption time-of-flight mass spectrometry *(14,23–25)* have been used to detect peptide-metal ion complexes (Fig. 5). Both techniques are rapid (<10 min), sensitive (picomoles), and are able to address metal-binding stoichiometry.

3.4. Column Regeneration

1. Wash with 5 bed vol of buffer 7.
2. Wash with 10 bed vol of water. The column is now ready for reloading with metal ions.

3.5. High-Performance IMAC

For example, use a TSK chelate 5PW column (7.5 mm id × 750 mm) 10-μm particle size.

1. High-performance liquid chromatography (HPLC) pump system status.
    a. Flow rate: 1 mL/min.
    b. Upper pressure limit: 250 psi.
2. Metal ion loading.
    a. Wash the column with 5 bed vol of water.
    b. Inject 1 mL of 0.2M metal sulfate in water.
    c. Wash away excess metal ion with 3 bed vol of water; for Cu(II), wash with 3 bed vol of buffer 3.
3. pH Gradient elution: phosphate buffers pH 7.0 and pH 4.0.
    a. 100% buffer 1, 5–10 min.

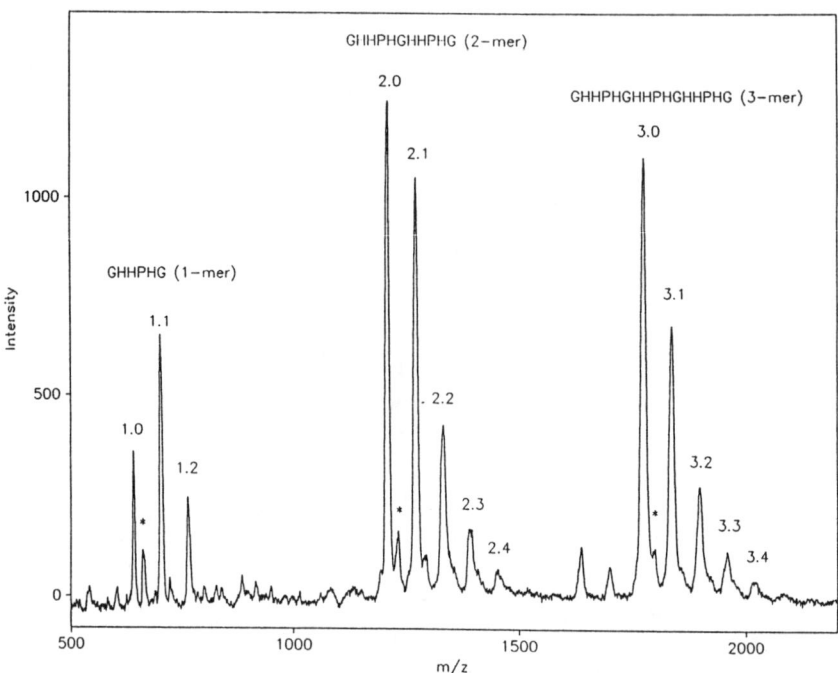

Fig. 5. Matrix-assisted UV laser desorption time-of-flight mass spectrometry (LDTOF) of a mixture of three synthetic metal-binding peptides (1-, 2-, and 3-mer) *after* elution from a column of immobilized GHHPHGHHPHG (2-mer) loaded with Cu(II) ions *(14)*. The synthetic peptide-metal ion affinity column used for metal ion transfer was prepared by coupling GHHPHGHHPHG (2-mer) to Affi-10 (Bio-Rad, Hercules, CA). Cu(II) ions were loaded as described in Section 3.1. The column was equilibrated with 20 mM sodium phosphate buffer (pH 7.0) with 0.5M NaCl. An equimolar mixture of the three different synthetic peptides (free of bound metal ions) was passed through the column unretained. Flow-through peptide fractions were anaylzed directly by LDTOF *(23–25)*. The metal ion-free peptides GHHPHG (1-mer peak 1.0), GHHPHGHHPHG (2-mer peak 2.0), and GHHPHGHHPHGHHPHG (3-mer peak 3.0) are observed along with peptides with 1, 2, 3, or 4 bound Cu(II) ions (e.g., 3.1, 3.2., 3.3., 3.4). The small peaks marked by an asterisk indicate the presence of a peptide-sodium adduct ion. A detailed description of these results is provided in ref. *14* (reproduced with permission).

    b. 0–80% buffer 4, duration 25 min.
    c. 80–100% buffer 4, duration 20 min.
    d. 100% buffer 4 until column effluent pH is constant or until all proteins have been eluted.

4. pH Gradient elution in 3*M* urea: phosphate buffers pH 7.5 to pH 3.8.
   a. 100% buffer 8, 5 min.
   b. 0–10% buffer 9, duration 10 min.
   c. 10–80% buffer 9, duration 18 min.
   d. 80–100% buffer 9, duration 25 min.
   e. 100% buffer 9 until eluent pH is constant or all proteins have been eluted.
5. Imidazole gradient elution: 1 m*M* imidazole (5% buffer 5 in buffer 1) to 20 m*M* imidazole (100% buffer 5) (*see* Notes 5 and 6).
   a. 5–10% buffer 5, immediately after sample injection, duration 10 min.
   b. 10–100% buffer 5, duration 30 min.
   c. 100% buffer 5 until all samples have been eluted.

4. Notes

1. The discontinuous buffer pH gradient is ideal for the pH 6.0–3.5 range. We have observed that acetate is also a stronger eluent than phosphate.
2. The phosphate buffer pH gradient is good for the pH 7.0–4.5 range. It has the advantage of UV transparency and is particularly suitable for peptide analysis *(18)*.
3. HEPES can be used instead of phosphate for both discontinuous buffer pH gradients and in the imidazole gradient elution mode. HEPES is a weaker metal ion "stripping" buffer than phosphate. HEPES is also good for preserving the metal binding capacity of some carrier proteins such as transferrin *(12)*.
4. For well-characterized proteins, a stepwise gradient of either pH or imidazole can be used to eliminate the need for a gradient-forming device.
5. For the affinity elution method with imidazole, the imidazole gradient actually formed must be monitored by, for example, absorbance at 230 nm or by chemical assay, if reproducible results are desired. Even when the column is presaturated with concentrated imidazole, and then equilibrated with buffers containing a substantial amount (2 m*M*) of imidazole, the immobilized metal ions can still bind additional imidazole when the affinity elution gradient is introduced. As a result, when simple (nonprogrammable) gradient-forming devices (typical for open-column chromatography) are used, a linear imidazole gradient is not produced; a small imidazole elution front (peak) at the beginning of the gradient will cause some proteins to elute "prematurely." The multistep gradient described for use with the HPLC systems is designed to overcome this problem. However, we emphasize that this particular program is custom designed only for the high-performance immobilized metal ion column of given dimensions and capacity. The program must be adjusted for other column types.
6. The imidazole gradient of up to 20 m*M* is only an example. Quite often, much higher concentrations (up to 100 m*M*) are required to elute higher affinity proteins *(9,10,17)* *(see* Fig. 2).

7. To facilitate the elution of some proteins, mobile phase modifiers (additives), such as urea, ethylene glycol, detergents, and alcohols, can be included in the column equilibration and elution buffers *(10,17,18,26)*.
8. To ensure reproducible column performance for several runs without a complete column regeneration in between, low concentrations of free metal ions can be included in the buffers to maintain a fully metal-charged stationary phase. This will not affect the resolution and elution position of the proteins *(7,17,27)*. On the other hand, if free metal ions are not desired in the protein eluent, a separate metal ion scavenger column (e.g., a blank or metal-free IDA-gel or TED-gel column) may be connected in series.
9. Batch-type (i.e., nonchromatographic) equilibrium binding assays have been described *(17,28)*. This is useful in screening different types of immobilized metal ion–protein interaction variables. These variables include the selection of appropriate immobilized metal ion type *(17)*, the effects of temperature and mobile phase conditions *(17,26)*, and free metal ions *(7)* on protein-immobilized metal ion interaction capacity and affinity.
10. Immobilized metal ions may be useful for the reversible site- or domain-specific immobilization of functional receptor proteins or enzymes. Data collected with the estrogen receptor protein suggest that receptor immobilization on IDA-Zn(II) *(29)* and IDA-Cu(II) (unpublished) does not impair receptor function (ligand-binding activity).

Acknowledgment

This work was supported, in part, by the US Department of Agriculture, Agricultural Research Service Agreement No. 58-6250-1-003. The contents of this publication do not necessarily reflect the views or policies of the US Department of Agriculture, nor does mention of trade names, commercial products, or organizations imply endorsement by the US Government.

References

1. Porath, J., Carlsson, J., Olsson, I., and Belfrage, G. (1975) Metal chelate affinity chromatography, a new approach to protein fractionation. *Nature* **258,** 598,599.
2. Porath, J. and Olin, B. (1983) Immobilized metal ion affinity adsorption and immobilized metal ion affinity chromatography of biomaterials. Serum protein affinities for gel-immobilized iron and nickel ions. *Biochemistry* **22,** 1621–1630.
3. Monjon, B. and Solms, J. (1987) Group separation of peptides by ligand-exchange chromatography with a Sephadex containing N-(2-pyridyl-methyl)glycine. *Anal. Biochem.* **160,** 88–97.
4. Hochuli, E., Dobeli, H., and Schacher, A. (1987) New metal chelate adsorbent selective for proteins and peptides containing neighbouring histidine residues. *J. Chromatogr.* **411,** 177–184.

5. Yip, T.-T. and Hutchens T. W. (1989) Development of high-performance immobilized metal affinity chromatography for the separation of synthetic peptides and proteolytic digestion products, in *Protein Recognition of Immobilized Ligands*. UCLA Symposia on Molecular and Cellular Biology, vol. 80 (Hutchens, T. W., ed.), Alan R. Liss, New York, pp. 45–56.
6. Yip, T. T., Nakagawa, Y., and Porath, J. (1989) Evaluation of the interaction of peptides with Cu(II), Ni(II), and Zn(II) by high-performance immobilized metal ion affinity chromatography. *Anal. Biochem.* **183,** 159–171.
7. Hutchens, T. W. and Yip, T. T. (1990) Differential interaction of peptides and protein surface structures with free metal ions and surface-immobilized metal ions. *J. Chromatogr.* **500,** 531–542.
8. Sulkowski, E. (1985) Purification of proteins by IMAC. *Trends Biotechnol* **3,** 1–7.
9. Hutchens, T. W. and Li, C. M. (1988) Estrogen receptor interaction with immobilized metals: differential molecular recognition of Zn^{2+}, Cu^{2+}, and Ni^{2+} and separation of receptor isoforms. *J. Mol. Recog.* **1,** 80–92.
10. Hutchens, T. W., Li, C. M., Sato, Y., and Yip, T.-T. (1989) Multiple DNA-binding estrogen receptor forms resolved by interaction with immobilized metal ions. Identification of a metal-binding domain. *J. Biol. Chem.* **264,** 17,206–17,212.
11. Hemdan, E. S., Zhao, Y.-J., Sulkowski, E., and Porath, J. (1989) Surface topography of histidine residues: a facile probe by immobilized metal ion affinity chromatography. *Proc. Natl. Acad. Sci. USA* **86,** 1811–1815.
12. Hutchens T. W. and Yip, T.-T. (1991) Metal ligand-induced alterations in the surface structures of lactoferrin and transferrin probed by interaction with immobilized Cu(II) ions. *J. Chromatogr.* **536,** 1–15.
13. Mantovaara-Jonsson, T., Pertoft, H., and Porath, J. (1989) Purification of human serum amyloid ccmponent (SAP) by calcium affinity chromatography. *Biotechnol. Appl. Biochem.* **11,** 564–571.
14. Hutchens, T. W., Nelson, R. W., Li, C. M., and Yip, T.-T. (1992) Synthetic metal binding protein surface domains for metal ion-dependent interaction chromatography. I. Analysis of bound metal ions by matrix-assisted UV laser desorption time-of-flight mass spectrometry. *J. Chromatogr.* **604,** 125–132.
15. Hutchens, T. W. and Yip, T.-T. (1992) Synthetic metal binding protein surface domains for metal ion-dependent interaction chromatography. II. Immobilization of synthetic metal-binding peptides from metal-ion transport proteins as model bioactive protein surface domains. *J. Chromatogr.* **604,** 133–141.
16. Hutchens, T. W. and Yip, T.-T. (1990) Model protein surface domains for the investigation of metal ion-dependent macromolecular interactions and metal ion transfer. *Methods: A Companion to Methods in Enzymology* **4,** 79–96.
17. Hutchens, T. W. and Yip, T.-T. (1990) Protein interactions with immobilized transition metal ions: quantitative evaluations of variations in affinity and binding capacity. *Anal. Biochem.* **191,** 160–168.
18. Nakagawa, Y., Yip, T. T., Belew, M., and Porath, J. (1988) High performance immobilized metal ion affinity chromatography of peptides: analytical separation of biologically active synthetic peptides. *Anal. Biochem.* **168,** 75–81.

19. Fatiadi A. J. (1987) Affinity chromatography and metal chelate affinity chromatography. *CRC Critical Rev. Anal. Chem.* **18,** 1–44.
20. Kagedal, L. (1989) Immobilized metal ion affinity chromatography, in *High Resolution Protein Purification* (Ryden, L. and Jansson, J.-C., eds.), Verlag Chemie Inst., Deerfield Beach, FL, pp. 227–251.
21. Muszynska, G., Zheo., Y.-J., and Porath, J. (1986) Carboxypeptidase A: a model for studying the interaction of proteins with immobilized metal ions. *J. Inorg. Biochem.* **26,** 127–135.
22. Yip, T.-T. and Hutchens, T. W. (1991) Metal ion affinity adsorption of a ZN(II)-transport protein present in maternal plasma during lactation: structural characterization and identification as histidine-rich glycoprotein. *Protein Express. Purification* **2,** 355–362.
23. Hutchens, T. W., Nelson, R. W., and Yip, T.-T. (1992) Recognition of transition metal ions by peptides: identification of specific metal-binding peptides in proteolytic digest maps by UV laster desorption time-of-flight spectrometry. *FEBS Lett.* **296,** 99–102.
24. Hutchens, T. W., Nelson, R. W., and Yip, T.-T. (1991) Evaluation of peptide-metal ion interactions by UV laser desorption time-of-flight mass spectrometry. *J. Mol. Recog.* **4,** 151–153.
25. Hutchens, T. W., Nelson, R. W., Allen, M. H., Li, C. M., and Yip, T.-T. (1992) Peptide metal ion interactions in solution: detection by laser desorption time-of-flight mass spectrometry and electrospray ionization mass spectrometry. *Biol. Mass Spectrom.* **21,** 151–159.
26. Hutchens, T. W. and Yip, T.-T. (1991) Protein interactions with surface-immobilized metal ions: structure-dependent variations in affinity and binding capacity constant with temperature and urea concentration. *J. Inorg. Biochem.* **42,** 105–118.
27. Figueoroa, A., Corradini, C., Feibush, B., and Karger, B. L. (1986) High-performance immobilized metal ion affinity chromatography of proteins on iminodiacetic acid silica-based bonded phases. *J. Chromatogr.* **371,** 335–352.
28. Hutchens, T. W., Yip, T.-T., and Porath, J. (1988) Protein interaction with immobilized ligands. Quantitative analysis of equilibrium partition data and comparison with analytical affinity chromatographic data using immobilized metal ion adsorbents. *Anal. Biochem.* **170,** 168–182.
29. Hutchens, T. W. and Li, C. M. (1990) Ligand-binding properties of estrogen receptor proteins after interaction with surface-immobilized Zn(II) ions: evidence for localized surface interactions and minimal conformational changes. *J. Mol. Recog.* **3,** 174–179.

Chapter 21

Chromatography on Hydroxyapatite

Shawn Doonan

1. Introduction

Hydroxyapatite (HT/HTP) has achieved only limited popularity as a chromatographic material for the purification of proteins. This is for a variety of reasons, including difficulties in predicting its chromatographic behavior, its relatively low capacity, and the fact that its handling properties are not ideal. For most applications, other materials with superior chromatographic properties are to be preferred. That being said, because its modes of protein adsorption and desorption are different from those of techniques such as ion-exchange chromatography (*see* next paragraph), it is sometimes possible to achieve fractionation using HT/HTP when other techniques have failed; the example described in Section 3. is a case in point. Aficionados of HT/HTP may consider this relegation of the material to "last chance" status as unjustified, but it is nonetheless the case that most practitioners of protein purification will try something else first.

HT/HTP is a crystalline form of calcium phosphate with the molecular formula $Ca_{10}(PO_4)_6(OH)_2$. The mechanism of interaction of proteins with the material has been investigated and reviewed by Gorbunoff *(1)*. When the HT/HTP is equilibrated with phosphate buffer (the most common mode of use), it then appears that positively charged proteins interact nonspecifically with the general negative charge on the column produced by immobilized phosphate ions. The proteins can then be eluted by increasing the phosphate concentration, addition of a salt, such as sodium chloride, or by use of Ca^{2+} or Mg^{2+}, which complex with phosphate ions on the column and effectively neutralize it. In the case of negatively charged proteins, interaction is a balance between electrostatic repulsion by the negative charge on the column and specific complexation between protein carboxylic acid groups and column calcium sites; the latter effect will depend on the disposition of carboxylate side chains in the protein rather than simply on their number. Elution is effected by ions that complex more strongly with calcium than do carboxylic acids (e.g., phosphate or fluoride); Cl⁻ is not effective because it does not complex with calcium.

These considerations have led to the formulation of set of guidelines for the use of HT/HTP to fractionate an unknown mixture of proteins *(1)*. The column should be equilibrated with dilute phosphate buffer (e.g., 1 mM, pH 6.8) if the objective is to retain basic proteins and some of the acidic ones from a mixture; alternatively equilibration can be in an unbuffered NaCl solution (1 mM) if the protein of interest is acidic and is to be retained on the column. Elution should then carried out by using an initial wash with 5 mM MgCl$_2$ (or a gradient of 1–5 mM) to remove basic proteins, a wash with 1M NaCl (or a gradient of 0.01–1M) to elute proteins with isoelectric points around neutrality, and finally a gradient of 0.1–0.3M phosphate to elute acidic proteins. Clearly for particular protein mixtures, it may be possible to use a more simple elution schedule depending on the composition, but the procedures outlined above provide a basis for initial examination of an unknown protein sample from which simpler protocols can be derived. Some typical uses of HT/HTP chromatography are given in refs. *1–5*.

The particular application described in Section 3. concerns the separation of the cytosolic and mitochondrial isoenzymes of fumarase *(5)*. These two proteins are products of the same gene and differ only in a small number (so far undefined) of amino acid residues at the N-terminus. They are not separable by conventional ion-exchange chromatography, but can be easily resolved by chromatography on HT/HTP.

2. Materials
1. HT/HTP from Bio-Rad (Watford, UK) (*see* Note 1).
2. Buffer A: 20 mM potassium phosphate, pH 7.0, containing 10 µM phenylmethanesulfonyl fluoride (PMSF) (*see* Note 2).
3. Buffer B: 350 mM potassium phosphate, pH 7.0, containing 10 µM PMSF.

3. Method
1. Suspend about 25 g of HT/HTP in 100 mL of buffer A. Stir gently, allow to settle, and then remove fines by aspiration. Repeat the last step (*see* Note 3).
2. Pack the HT/HTP into a column (2.0 × 10 cm), and wash with five bed volumes of buffer A at a flow rate of 30 mL/h (*see* Note 4).

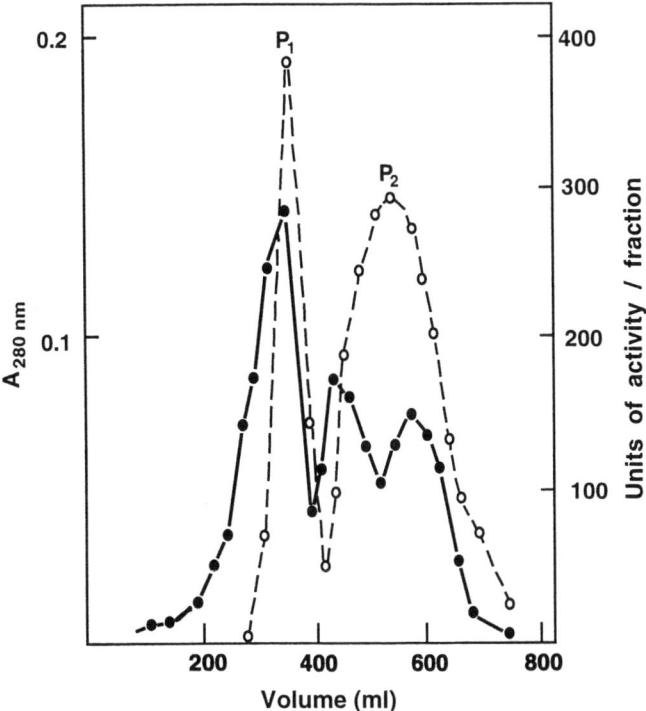

Fig. 1. Separation of cytosolic and mitochondrial fumarases from baker's yeast by HT/HTP chromatography. Full and dashed traces show absorbance at 280 nm and enzyme activity, respectively. (Reprinted from ref. 5 with kind permission from Elsevier Sci., Ltd.)

3. Apply the protein sample previously dialyzed in buffer A, and continue washing the column with the same buffer until no more protein is eluted as judged from A_{280} measurements.
4. Elute bound protein by application of a linear gradient formed from 400 mL each of buffers A and B.
5. Regenerate the column by washing with two column volumes of buffer B, followed by five column volumes of buffer A. Store in buffer A containing 0.02% sodium azide or in buffer A saturated with toluene.

The elution profile obtained is shown in Fig. 1. The source of the partially purified mixture of fumarases is described in Note 5.

4. Notes

1. HT/HTP suitable for chromatography can be prepared in the laboratory. The procedure involves mixing calcium chloride and sodium phosphate to

form the amorphous compound brushite, and then converting this to crystalline hyroxyapatite by boiling in sodium hydroxide or ammonia. A detailed procedure is given by Bernardi *(6)*. Although this product may be superior to commercial HT/HTPs *(7)*, it is likely that the occasional user will prefer to purchase the material.

There are several sources of supply, and the nature of the material varies from one to another. BDH (Poole, Dorset UK) markets a spheroidal HT/HTP that has similar properties to that used here, but may have greater mechanical strength. Other products consist of matrices with surface coatings of HT/HTP. For example, HA-Ultrogel from IBF (Luton, UK) consists of coated agarose. The binding capacities of these products vary from 2.5–10 mg BSA/mL of packed column. It is not possible to predict which of these commercial materials will perform best in any particular application; the important point is to stick to one product, and if following a procedure from the literature, to use the same product as described in the original work. For HPLC applications, Bio-Rad also markets a high-performance HT/HTP (Bio-Gel HPHT) packed in a stainless-steel column.

2. PMSF was included in this particular application because it was vital to the objective of the investigation to preclude the possibility of proteolysis *(5)*. The necessity to add protease inhibitors and which inhibitors to add will depend on the problem at hand (*see* Chapter 9).

    Stock solutions of PMSF (50–100 mM) should be made in dry propan-2-ol, methanol, or ethanol; in this form, they are stable for some months at 4°C. The half-life in aqueous solutions is short, and hence fresh PMSF should be added to protein solutions or to buffers daily.

3. Crystalline HT/HTPs are not very stable mechanically, and it is important to avoid fragmentation owing to rough handling and to be sure to remove fines before packing columns. Failure to do so will result in high back-pressures and low flow rates. Similarly, the flow rates of columns of HT/HTP decrease with repeated use, and when this occurs, it is necessary to repack them. Absorption of carbon dioxide and consequent formation of a plug at the top of the column is one reason for development of back-pressure; this can be avoided by boiling buffers before use to remove dissolved CO_2 or, more simply, by replacement of the top 1–2 cm of the column bed if plugging occurs.

4. Use of high flow rates, particularly with crystalline HT/HTP, will lead to compaction of the material and necessitate repacking of the column.

5. The fumarase isoenzymes were extracted from 1 kg of baker's yeast, and then partially purified by precipitation with 60% saturation ammonium sulfate, followed by ion-exchange chromatography on carboxymethyl cellulose CM-23 at pH 5.8. It was this fraction that was subjected to chroma-

tography on HT/HTP with the results shown in Fig. 1. Peak 1 eluted from the column at about 120 mM phosphate buffer and contained mainly the cytosolic fumarase *(8)*. Peak 2 eluted at about 250 mM phosphate buffer and contained mainly the mitochondrial form. The separation was not complete, so fractions from the front of peak 1 and from the back of peak 2 were combined to give the respective pools for the two isoenzymes; the central fractions were rejected. Final purification of the separated isoenzymes was achieved by affinity chromatography using pyromellitic acid-Sepharose 4B *(9)*. Apart from HT/HTP chromatography, the only other technique that has been found to be effective in separating fumarase isoenzymes is chromatofocusing, which was used for the isoenzymes from rat liver *(10)*.

References

1. Gorbunoff, M. J. (1990) Protein chromatography on hydroxyapatite columns. *Methods Enzymol.* **182,** 329–339.
2. Gorbunoff, M. J. (1980) Purification of ovomucoid by hydroxyapatite chromatography. *J. Chromatogr.* **187,** 224–228.
3. Mizuuchi, K., O'Dea, M. H., and Gellert, M. (1978) DNA gyrase: subunit structure and ATPase activity of the purified enzyme. *Proc. Natl. Acad Sci. USA* **75,** 5960–5963.
4. Peng, H. and Marians, K. J. (1993) *Escherichia coli* topoisomerase IV: purification, characterisation, subunit structure and subunit interactions. *J. Biol. Chem.* **268,** 24,481–24,490.
5. Boonyarat, D. and Doonan, S. (1988) Purification and structural comparisons of the cytosolic and mitochondrial fumarases from baker's yeast. *Int. J. Biochem.* **20,** 1125–1132.
6. Bernardi, G. (1971) Chromatography of proteins on hydroxyapatite. *Methods Enzymol.* **22,** 325–339.
7. Roe, S. (1989) Separation based on structure, in *Protein Purification Methods: A Practical Approach* (Harris, E. L. V. and Angal, S., eds.), IRL, Oxford, UK, pp. 238–242.
8. Hiraga, K., Inoue, I., Manaka, H., and Tuboi, S. (1984) Chromatographic differentiation of the mitochondrial and cytosolic fumarases of rat liver and baker's yeast and differential induction of two fumarases of baker's yeast. *Biochem. Int.* **9,** 455–461.
9. Beeckmans, S. and Kanarek, L. (1977) A new purification procedure for fumarase based on affinity chromatography: isolation and characterisation of pig-liver fumarase. *Eur. J. Biochem.* **78,** 437–444.
10. O'Hare, M. C. and Doonan, S. (1985) Purification and structural comparisons of the cytosolic and mitochondrial fumarases from rat liver. *Biochim. Biophys. Acta* **827,** 127–134.

CHAPTER 22

Affinity Precipitation Methods

Jane A. Irwin and Keith F. Tipton

1. Introduction
1.1. Overview

Affinity chromatography (see Chapter 16) is a powerful protein purification technique, that exploits the specific interaction between a biological ligand (e.g., a substrate, coenzyme, hormone, antibody, or nucleic acid) or its synthetic analog and its complementary binding site on a protein. One of the variations on this technique (see refs. *1–3* for reviews) was that of affinity precipitation. As in affinity chromatography, the protein binds to a specific ligand, but the latter is free in solution, rather than bound to an insoluble support. Ligand binding gives rise to the precipitation of the protein from solution, which is then followed by centrifugation. The pellet contains the protein of interest and the ligand, whereas the other components of the mixture remain in the supernatant, allowing easy separation.

There are two main approaches to affinity precipitation. The first of these is called the "bis-ligand" or "homobifunctional ligand" approach. The ligand is bifunctional, bearing two identical ligands connected by a spacer arm. If the spacer is long enough, each ligand can bind to a ligand binding site on a different protein molecule. Oligomeric proteins can bind two or more bis-ligands, with the consequent formation of crosslinked lattices. When the lattice becomes large enough, it will precipitate from solution.

The second approach to affinity precipitation differs from this in that the affinity ligand has two functions, one of which binds the target protein and a second to promote the precipitation of the aggregate. In some cases, precipitation is induced by the addition of a third component, e.g., a lectin or metal ion. This technique is referred to in the literature as affinity precipitation, but in contrast to the first approach, the precipitation of the complex does not occur as a direct consequence of the affinity interaction between the protein and its ligand, but as a result of adding a third component. Table 1 gives a list of examples of both "bis-ligand" and this second form of affinity precipitation.

This chapter is confined to a discussion of the "bis-ligand" form of affinity precipitation, that can be described as "true" affinity precipitation, occurring solely as a result of the direct interaction between the ligand and the protein, and not as a result of a third component of the system.

1.2. Bis-Coenzymes and Their Applications

Larsson and Mosbach were the first to synthesize a bifunctional affinity precipitation reagent *(4)*. This affinity precipitation reagent was N_2,N_2'-adipodihydrazido-bis-(N^6-carbonylmethyl)NAD$^+$ (bis-NAD$^+$). It consisted of two molecules of the NAD$^+$ derivative N^6-carboxymethyl-NAD$^+$, linked by an adipic acid dihydrazide spacer arm (Fig. 1). This compound was used to affinity precipitate purified lactate dehydrogenase from bovine heart in up to 90% yield. Bovine liver GDH and yeast alcohol dehydrogenase were also precipitated by this technique *(5–7)*. The latter enzyme also required high ionic strengths for precipitation to occur. Further applications of bis-NAD$^+$ are mentioned in Table 1, and some limitations on its use are described in Note 1.

An alternative synthesis of a range of bis-NAD$^+$ affinity reagents has been developed *(28)*. It involves the carbodiimide-mediated condensation of two molecules of N^6-(2-aminoethyl)NAD$^+$ with different dicarboxylic acids (Fig. 2). The synthesis of bis-coenzyme derivatives has not been limited only to NAD$^+$; a synthesis of a bis-ATP derivative, N_2,N_2'-adipodihydrazido-bis-(N^6-carbonylmethyl)ATP *(10)* is described here.

1.3. Structural and Kinetic Requirements for Affinity Precipitation with Bis-Coenzymes

The technique of bis-ligand affinity precipitation with bis-NAD$^+$ only works under certain conditions *(29)*, which may be summarized as follows:

1. The enzyme of interest has to contain more than one coenzyme binding site.
2. The bis-ligand has to have a strong affinity for the enzyme.

Table 1
Published Examples of Protein Purification by Affinity Precipitation

Bis-ligands[a]

Protein	Bis-ligand	Refs.
LDH	Bis-NAD$^+$	4–8
GDH	Bis-NAD$^+$	5–7,9
YADH	Bis-NAD$^+$	5,7
Isocitrate dehydrogenase	Bis-NAD$^+$	7
PFK	Bis-ATP	10
LDH (rabbit)	Bis-Cibacron Blue F3G-A	11,12
Bovine serum albumin	Bis-Cibacron Blue F3G-A	11,12
LDH (rabbit)	Methoxylated p-sulfonated isomer of Procion Blue H-B	13,14
GDH (recombinant)	EGTA (Zn)$_2$	15
Human hemoglobin, sperm whale myoglobin	Cu(II)$_2$EGTA, Cu(II)$_2$ polyethylene glycol-(iminodiacetic acid)$_2$	16

Hetero-bifunctional ligands[b]

Protein	Ligand	Carrier	Precipitant	Refs.
Trypsin	p-Aminobenzamidine	N-acryloyl-aminobenzoic acid	Low pH	17
Trypsin	Soybean trypsin inhibitor	Chitosan	High pH	18
Wheat germ agglutinin	N-acetyl-D-glucosamine	Chitosan	High pH	19
Protein A	IgG	Hydroxymethyl cellulose acetate succinate	Low pH	20
Recombinant protein A	IgG	Eudragit S100	Low pH	21
IgG	Protein A	Galactomannan	Potassium borate	22
LDH	Cibacron Blue	Dextran	Concanavalin A	23
Trypsin	Soybean trypsin inhibitor	Alginate	Ca^{2+} ions	24
Endo-polygalacturonase	Alginate	Alginate	Ca^{2+}, low pH	25
Protein A	IgG	N-isopropylacrylamide polymer	Temperature increase	26
Trypsin	p-Aminobenzamidine	N-isopropylacrylamide polymer	Temperature increase	27

[a]These do not require a second component to effect affinity precipitation.
[b]These require a ligand carrier and a third component to promote the precipitation of the ligand–protein complex.

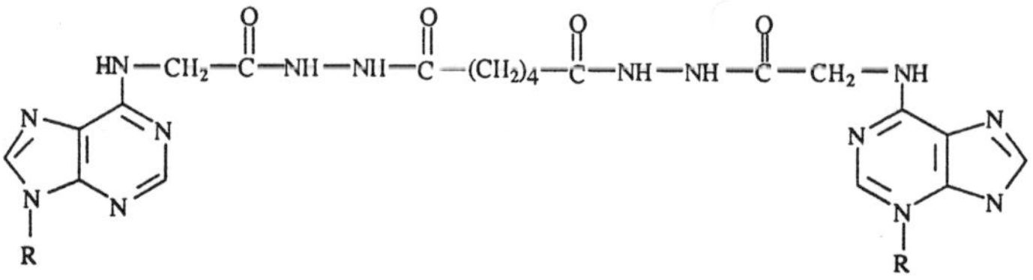

Fig. 1. Structure of bis-NAD$^+$, first synthesized by Larsson and Mosbach *(4)*, and used as an affinity ligand for several dehydrogenases. R represents nicotinamide mononucleotide.

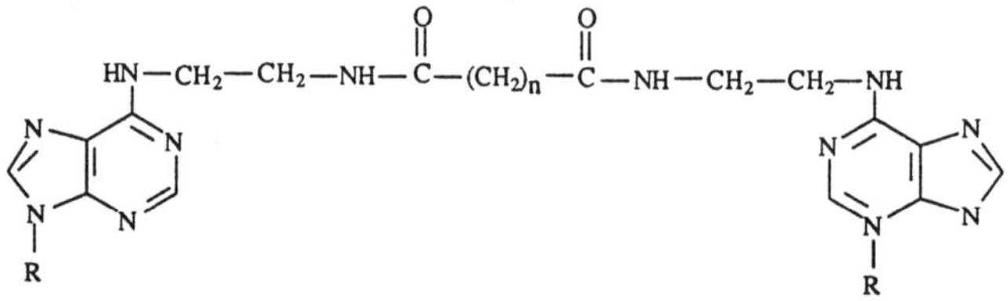

Fig. 2. General structure of bis-NAD$^+$ derivatives, based on N^6-(2-aminoethyl)-NAD$^+$ as a starting material. These include N,N'-bis(N^6-ethylene-NAD$^+$) glutaramide ($n = 3$); N,N'-bis(N^6-ethylene-NAD$^+$) adipamide ($n = 4$), and N,N'-bis(N^6-ethylene-NAD$^+$)pimelamide ($n = 5$). R represents nicotinamide mononucleotide phosphoribose.

3. The spacer connecting the two ligands has to be long enough to bridge the distance between two ligand binding sites on two different enzyme molecules. In the case of N_2, N_2'-adipodihydrazido-bis-(N^6-carbonylmethyl)-NAD$^+$ (bis-NAD$^+$), the spacer length was approx 1.7 nm, which permitted easy simultaneous access for a molecule of bis-NAD$^+$ to two different molecules of a dehydrogenase.

A further limitation is the ratio of coenzyme derivative to enzyme subunit. If this is low, a lattice will not form since there are not enough crosslinks; if it is too high, each dehydrogenase subunit can be occupied by a separate molecule of bis-NAD$^+$ and no crosslinks will form. Maximum crosslinking occurs at an optimum ratio of NAD$^+$ eq/enzyme subunit (assuming two NAD$^+$ Eq/bis-NAD$^+$), which in the case of tetramers has been found to be approximately unity. This can vary; for example, the hexameric mammalian glutamate dehydrogenase (GDH) precipitated over a broad range of NAD$^+$ equivalents to active site ratios (0.3:10), with up to 70% precipitation occurring at a ratio as low as 0.16, although

Affinity Precipitation Methods

this probably involves the existence of higher GDH polymers *(5)*. The ratio of approximately unity for tetramers is similar to the behavior observed in immunoprecipitation, in which two antigen molecules/antibody give rise to the optimum precipitation of immune complexes *(30)*.

Adding bis-NAD^+ alone to a crude extract containing many different dehydrogenases will not in itself cause specific affinity precipitation, since it is a "general ligand," binding to most NAD^+-dependent dehydrogenases. In addition to being nonspecific, bis-NAD^+ forms a weak binary complex with several dehydrogenases, e.g., lactate and horse liver alcohol dehydrogenases. The addition of a substrate analog (usually a competitive inhibitor, relative to the enzyme's second substrate) strengthens the binding interaction. Since NADH binds, on average, one order of magnitude more tightly to the active sites of many dehydrogenases than NAD^+, it displaces bis-NAD^+ and is commonly used to dissolve crosslinked aggregates.

The specificity of bis-NAD^+ of affinity precipitation for any given dehydrogenase is conferred by a property described by O'Carra as the "locking-on" effect *(31)*. This was originally developed to increase the strength of enzyme binding to an immobilized ligand by adding analogs of substrates specific to the enzyme under study. In the case of lactate dehydrogenase, for example, the strength of adsorbtion to an immobilized NAD^+ derivative was increased by adding oxalate, a structural analog of lactate, to the irrigating buffer. The enzyme was eluted by simply leaving out the oxalate.

This "locking-on" property does not occur with all enzymes. It is confined to those with sequential mechanisms (*see* refs. *3* and *31* for more details). Most coenzyme-dependent enzymes, including the majority of dehydrogenases, have ordered sequential kinetic mechanisms, in which the coenzyme binds before the second substrate. If this is the case, the "locking-on" effect occurs. The addition of an unreactive, competitive inhibitor of the second substrate will displace the coenzyme binding equilibrium by ternary complex formation (sometimes described as an "abortive" complex) and thereby increase the strength of binding. In affinity precipitation, the bis-coenzyme is "locked" into place by the substrate analog, provided it is saturating. Some enhancement of bis-ligand binding can occur for an enzyme with a random sequential mechanism, provided that the equilibrium of the reaction under the conditions employed in affinity precipitation is such that the binding of the substrate analog then favors coenzyme binding.

The identity of the substrate analog will determine which dehydrogenase is precipitated from a crude extract. For example, adding bis-NAD$^+$ and glutarate will lead to the precipitation of GDH from a crude extract; addition of pyrazole or oxalate would favor alcohol dehydrogenase or lactate dehydrogenase precipitation, respectively. This is the key to making the technique specific for one enzyme.

In the case of the only reported bis-ATP derivative, which was used to affinity precipitate bovine heart phosphofructokinase, precipitation was not accomplished with the aid of a second substrate analog, but by the use of the allosteric inhibitor citrate. ATP is both a substrate and an allosteric inhibitor of this enzyme, and this inhibition is potentiated by citrate *(10)*.

1.4. Affinity Precipitation with Other Bifunctional Ligands

Examples of bis-ligand affinity precipitation in the literature are few, and the technique has not gained widespread popularity since it has only been found to work with a few proteins. In order to extend its use, other ligands have been synthesized, but these have been limited so far to triazine dyes and immobilized metal ions. Triazine dyes have been widely used as pseudo-affinity ligands (*see* Chapter 17 for a discussion of dye-ligand chromatography). A bis-derivative of Cibacron Blue F3GA was found to precipitate bovine serum albumin and lactate dehydrogenase *(11)*, and a monofunctional synthetic analog of Procion Blue H-B was found to precipitate rabbit muscle lactate dehydrogenase selectively *(13,14)*. The synthesis of this analog is reported in ref. *13*, and the method for large-scale LDH purification is described in ref. *14*. Only two examples of metal ion affinity precipitation have been reported to date, one of which consisted of the precipitation of recombinant galactose dehydrogenase containing a pentahistidine affinity tail by a $(Zn)_2$-EGTA chelate *(15)*. In the second case, myoglobin and hemoglobin were precipitated by an EGTA-Cu(II) chelate or, alternatively, by a bis-ligand consisting of Cu(II) cations chelated by molecules of iminodiacetic acid immobilized on each end of a molecule of polyethylene glycol *(17)*.

2. Materials
2.1. Synthesis of Coenzyme Derivatives
2.1.1. Synthesis of Bis-NAD$^+$

This compound is commercially obtainable, but is quite expensive ($90.00 for 5 mg, Sigma, St. Louis, MO, 1995). The starting material is

Affinity Precipitation Methods

N^6-carboxymethyl-NAD$^+$. This is also obtainable from Sigma, but may be readily synthesized. The following reagents are required.

2.1.1.1. N^6-Carboxymethyl-NAD$^+$

1. NAD$^+$ (98%, free acid).
2. Iodoacetic acid.
3. 2M LiOH.
4. 96% (w/v) Ethanol.
5. Sodium dithionite.
6. 0.24M NaHCO$_3$.
7. Yeast alcohol dehydrogenase (crystalline, Sigma or Boehringer Mannheim, Mannheim, Germany).
8. A sintered funnel (fairly large, at least 10 cm in diameter).
9. AG 1 × 2 anion-exchange resin (200–400 mesh, Cl$^-$ form, Bio-Rad, Hercules, CA).
10. CaCl$_2$.

2.1.1.2. Bis-NAD$^+$

1. Adipic acid dihydrazide dichloride (Sigma).
2. 2M NH$_4$OH.
3. NH$_4$HCO$_3$.
4. DEAE-cellulose (Whatman DE52, equilibrated with 10 column volumes of 1M NH$_4$HCO$_3$, and washed with water).

2.1.2. N_2,N_2'-Adipodihydrazido-Bis-(N^6-Carbonylmethyl-ATP)

As for N^6-carboxymethyl-NAD$^+$, except that NAD$^+$ is replaced by ATP and LiCl is used in place of CaCl$_2$. Neither this ATP derivative nor N^6-carboxymethyl-ATP is commercially available.

2.1.3. Other Bis-NAD$^+$ Derivatives

1. NAD$^+$ (Sigma, Boehringer; 98% free acid)
2. Ethyleneimine (Serva, Heidelberg, Germany). **Caution: This is toxic and carcinogenic.**
3. 70% HClO$_4$.
4. 96% Ethanol.
5. LiCl.
6. Bio-Rex 70 cation exchange resin (100–200 mesh, Na$^+$ form, Bio-Rad).
7. 1M LiOH.
8. 1M HCl.

Bio-Rex 70 can be converted to the H$^+$ form by washing it exhaustively with 0.5–1.0M HCl. The absence of Na$^+$ ions can be tested by flame photometry. The column is then washed with 1 mM HCl, pH 3.0, and equilibration is checked by pH and conductivity measurements.

For bis-NAD$^+$ synthesis: Glutaric, adipic, and pimelic acid (all available from Sigma) have been found to serve as satisfactory, water-soluble spacers. Also required: 1-ethyl-3-(3-dimethylaminopropyl)-carbodiimide hydrochloride (EDC), 5M NaOH, 1M HCl, DE52 cellulose, equilibrated with 0.5M ammonium acetate, pH 7.0, to convert it to the acetate form; this is then equilibrated for chromatography by washing it with 10 column volumes of 0.05M ammonium acetate, pH 7.0.

In addition to this, a rotary evaporator and chromatographic columns (dimensions variable) are needed for all described syntheses. Three different TLC solvent systems have been used for monitoring the products of these syntheses:

> System A: $(NH_4)_2SO_4$: 0.1M potassium phosphate, pH 6.8: 1-propanol (60:100:2 [w/v/v]).
> System B: isobutyric acid: 1M aqueous NH_3 (5:3 [v/v]), saturated with Na_2EDTA.
> System C: isobutyric acid: water: 25% aqueous NH_3 (66:33:1 [v/v/v]).

Plates: aluminum-backed, silica gel 60, fluorescent indicator F_{254}, layer thickness 0.2 mm (Merck). A source of fluorescent light (wavelength 254 nm) can be used to visualize the spots, which appear purple on a green background.

HPLC system: reverse-phase, µBondapak C-18 column (Waters; 3.9 × 300 mm), equilibrated with 0.1M potassium dihydrogen phosphate, pH 6.0, containing 10% (v/v) methanol for separating monosubstituted NAD$^+$ derivatives and 20% methanol for separation of bis-NAD$^+$ derivatives. All solutions used in HPLC must be degassed using a vacuum pump and filtered with a 0.22-µm filter to exclude particulate matter and avoid air bubbles. In addition, all samples must be centrifuged for 2 min in a minifuge to pellet particles before application to the column. The absorbance should be monitored at 254 nm.

2.2. Pilot Affinity Precipitation Studies

1. A source of the protein of interest, e.g., a crude tissue extract supernatant, or a commercially available preparation of the protein to ensure the reagent is effective.
2. The bis-ligand.
3. Stock solutions of a substrate analog, e.g., 560 mM oxalate for lactate dehydrogenase, 560 mM pyrazole for alcohol dehydrogenase, and 700 mM glutarate for GDH.

Affinity Precipitation Methods 225

4. 0.4M Potassium phosphate buffer (62.4 g/L NaH_2PO_4, titrated to pH 7.4 with 5M KOH), for some applications.
5. 1.5-mL Polypropylene minifuge tubes.
6. Assay reagents for enzymes (*see*, e.g., ref. *32*).

2.3. Enzyme Purification

Apart from the reagents and apparatus mentioned in Section 2.2., reagents and apparatus for electrophoresis, e.g., PAGE/SDS-PAGE are needed to check the purity of the affinity-precipitated protein *(33)*. In addition, reagents for a protein assay (e.g., by the Lowry or Bradford method) are required to determine the protein concentration.

For the separation of LDH isoenzymes (*see* Section 3.3.3.), materials for starch gel electrophoresis are appropriate. These include hydrolyzed potato starch (Sigma), electrophoresis-grade Trizma base to buffer the gel, and grade III NAD^+ (Sigma). The activity stain for LDH contains 1.21 g Tris, 7.72 mL 70% Na DL-lactate (8 g), 50 mg nitroblue tetrazolium, 50 mg grade III NAD^+, and 4 mg phenazine methosulfate. (**Caution: Nitroblue tetrazolium and phenazine methosulfate are toxic and the latter is light sensitive**). Make the volume up to 200 mL with water, and adjust the pH to 7.1 with 6M HCl.

3. Methods
3.1. Synthesis of Coenzyme Derivatives
3.1.1. Bis-NAD$^+$

The synthesis of N^6-carboxymethyl-NAD^+ is described in ref. *34*, but some modifications are given here. The synthesis of bis-NAD^+ is a modified and more detailed version of that described in ref. *4*.

1. Dissolve 9 g of fresh iodoacetic acid in approx 1 mL of water, neutralize the solution with 2M LiOH, and add 3 g of NAD^+. Adjust the pH to 6.5 with 2M HCl (the total volume should be about 30 mL), and leave the solution in darkness for 7 d at room temperature (approx 20°C) or, alternatively, at 37°C for 2 d. The pH should be checked daily (or every 4–6 h at the higher temperature), and readjusted to 6.5 with 2M HCl as required. The progress of the reaction should also be followed by TLC and/or HPLC.
2. When the reaction is complete, adjust the pH to 3.0 with 6M HCl, and add 2 vol of 96% (v/v) ethanol. This gives a milky, pink-tinged suspension, which precipitates on the addition of a further 10 vol of cold 96% ethanol. (The water/ethanol ratio is important; using a wet vessel can give rise to the formation of a brown, sticky substance that is water-soluble. This also applies to the corresponding ATP derivative.) Filter the crude $N(1)$-carboxymethyl-NAD^+

on a sintered funnel, wash with ethanol and diethyl ether, and dry under vacuum (average yield 2.9 g). Store the product at –20°C under vacuum.

3. Dissolve the crude $N(1)$-carboxymethyl-NAD$^+$ in $0.24M$ NaHCO$_3$ (90 mL), which gives a pale orange solution, and adjust the pH to 8.5 with $1M$ NaOH. Deoxygenate the solution by bubbling N$_2$ gas through it for 2 min, add 1.5 g of sodium dithionite, and leave the solution in the dark until maximum reduction is achieved. This depends on the dithionite—monitor the reaction by taking samples, diluting them 1:50 or 1:100, and measuring the increase in A_{340}.

4. Terminate the reaction by stirring vigorously for 10 min to oxygenate the solution, and then bubble N$_2$ gas through for 2 min. Adjust the pH to 11.5 with $1M$ NaOH, and leave in a water bath at 75°C to allow the Dimroth rearrangement from the $N(1)$ to the N^6-substituted derivative to occur. (*See* ref. 35 for a reaction mechanism.) Monitor the rearrangement by HPLC or TLC. Cool the reaction mixture to room temperature, and add 6 mL of $2M$ Tris and 1.5 mL of redistilled acetaldehyde. Adjust the pH to 7.5 with $1M$ HCl, and add 8–24 mg of yeast alcohol dehydrogenase (2500–7500 U). Monitor the reaction at 340 nm as in step 3. When a minimum A_{340} is reached, add 1 vol of 96% ethanol, and pour the milky flocculant precipitate into 10 vol of vigorously stirred 96% ethanol. Leave for 30 min (or overnight, if desired), and collect the precipitate by filtration. The crude N^6-carboxymethyl-NAD$^+$ can be stored at –20°C under vacuum for up to a month (yield approx 2.6 g).

5. The product from step 4 can be purified by dissolving the crude powder (2.6 g) in 30 mL water, adjusting the pH to 8.0 with $1M$ LiOH, and applying this solution to a column of Dowex AG 1X2 (200–400 mesh, Cl$^-$ form, 4 × 10 cm). Wash the column beforehand with 1 L of $3M$ HCl, followed by exhaustive washing with at least 20 L of water until neutral pH is reached. After applying the coenzyme solution, wash the column with 0.5 L of water, followed by 1 L of 5 mM CaCl$_2$, until the pH of the effluent is 2.8. Apply a linear gradient (2 × 1 L) from 5 mM CaCl$_2$, pH 2.7, to 50 mM CaCl$_2$, pH 2.0. Collect 10–20 mL fractions, and monitor the absorbance at 260 nm. The composition of the fractions can be monitored by TLC. R_f values: system A, NAD$^+$, 0.41, $N(1)$-carboxymethyl-NAD$^+$, 0.31, N^6-carboxymethyl NAD$^+$, 0.22; system B, NAD$^+$, 0.44, $N(1)$-carboxymethyl-NAD$^+$, 0.27, N^6-carboxymethyl NAD$^+$, 0.22. The values can vary slightly with changes in the solvent composition over time (*see* Note 2). Pool the fractions containing the N^6 derivative, adjust the pH to 7.0 with $2M$ Ca(OH)$_2$, and concentrate to 5–10 mL by rotary evaporation at 40°C. Precipitate with 96% ethanol as in step (4), and dry under vacuum. This gives a pale yellow compound (yield 0.82 g, 25%). $\varepsilon = 19,300/M/\text{cm}$ at 266 nm.

Affinity Precipitation Methods

6. To synthesize bis-NAD⁺, dissolve 0.82 g of purified N^6-carboxymethyl-NAD⁺ and 105 mg of adipic acid dihydrazide dihydrochloride in 20 mL water to give a brownish solution. Adjust the pH to 4.6 with $1M$ HCl, and then add $0.5M$ EDC at 0°C in 15 100-μL aliquots over a period of 35 min. Monitor the pH, and readjust it to 4.6 before each addition. Add water (2 L) to dilute the solution 100-fold, adjust the pH to 8.0 with $2M$ NH₄HCO₃, apply the solution to a DE52 column (2.5 × 30 cm) equilibrated with $1M$ NH₄HCO₃, pH 8.0, and then wash with water. Wash the column with water until the A_{260} is < 0.1, and apply a linear gradient (2 × 1 L) from 0–0.25M NH₄HCO₃, pH 8.0. Monitor the R_f values of the fractions by TLC (R_f of bis-NAD⁺: system A, 0.09; system B, 0.05; see Note 2). The yield from pure N^6-carboxymethyl-NAD⁺ is approx 14%. $\varepsilon = 42,800/M/cm$ at 266 nm.

3.1.2. Bis-ATP (36,37)

1. Dissolve 3.4 g of fresh iodoacetic acid in 1 mL of water, and adjust the pH to 7.0 with $2M$ LiOH. Dissolve 1.16 g of ATP in the solution, and leave it at pH 6.5 in the dark for 4 d at 30°C, adjusting the pH to 6.5 every day. Monitor the reaction by HPLC and/or TLC, or follow the decrease in absorbance at 640 nm. When the reaction is complete, as judged by TLC, HPLC, or absorbance change, add 1 vol of cold 96% ethanol to give a milky suspension, and add this to 6–8 vol of stirring absolute ethanol at 0°C. Filter, and wash as for the corresponding NAD⁺ derivative. (Typical yield 82–87%, about 1.4 g.)
2. Dissolve 1.4 g of product in 30 mL of water to give a reddish solution. Adjust the pH to 8.5 with $1M$ LiOH and incubate at 90°C for 100 min, with pH adjustment every 20 min. Monitor the reaction over this time by TLC or HPLC. The solution can be cooled on ice and stored overnight at 4°C, if required. HPLC retention times: ATP, 4.5 min, $N(1)$-carboxymethyl-ATP 3.1 min, N^6-carboxymethyl-ATP, 4.1 min.
3. Adjust the pH of the solution at 20°C to 2.75 with $6M$ HCl, and apply it to an AG 1 × 2 column (200–400 mesh, Cl⁻ form, 4 × 9 cm), equilibrated with water. Wash the column with 1 L of $0.3M$ LiCl, pH 2.75, until the A_{260} is < 1 0, and then apply a 2 × 1 L linear gradient, $0.3M$ LiCl, pH 2.75–$0.5M$ LiCl, pH 2.0. Monitor the R_f values of the fractions by TLC as previously. (R_f values: system A, ATP, 0.54, $N(1)$-carboxymethyl-ATP, 0.73, N^6-carboxymethyl-ATP, 0.54. system B, ATP, 0.35, $N(1)$-carboxymethyl-ATP, 0.27, N^6-carboxymethyl-ATP, 0.21; see Note 2.) Pool the fractions containing the N^6 derivative, and adjust the pH with $2M$ LiOH. Yield; 32–38% from ATP. $\varepsilon = 17,300/M/cm$ at 266 nm. This procedure should be carried out as quickly as possible, preferably within 1 d, since ATP is unstable under acidic conditions.

4. Bis-ATP: Dissolve 105 mg of adipic acid dihydrazide dihydrochloride in a solution containing 0.42g of N^6-carboxymethyl-ATP in 100 mL. Adjust the pH to 4.0 with 1M HCl, and add 3 1-mL aliquots of 1M EDC over a period of 45 min, while stirring continuously, and monitoring the pH. Dilute to 2 L with water to stop the reaction, and adjust the pH to 8.0 with 2M NH$_4$OH. This solution can be stored at –20°C for about a week.
5. The compound can be purified by applying the dilute solution to a DE-52 column (2.5 × 25 cm), equilibrated with 2–4 L 1M NH$_4$HCO$_3$, and then washed with 8 L water. Elute the bis-ATP with a 0–0.4M NH$_4$HCO$_3$ gradient, pH 8.0 (2 × 1 L). Collect 10–20 mL fractions and pool and freeze-dry the fractions containing bis-ATP, as judged by HPLC or by affinity precipitation of phosphofructokinase (*see* Section 3.3.4.) Ensure that the compound is completely dry before storage at –20°C, as it is very hygroscopic. Yield: 0.06 g from 0.42 g N^6-carboxymethyl-ATP (yield approx 2% from ATP). ε = 39,600/M/cm at 266 nm (*see* Note 3).

3.1.3. Other Bis-NAD⁺ Derivatives

1. The synthesis of N^6-(2-aminoethyl)-NAD⁺ is carried out essentially as described in ref. *38*. However, $N(1)$-(2-aminoethyl)NAD⁺ can be rearranged to the N^6-substituted derivative without removal of the unreacted NAD⁺ by ion-exchange chromatography *(39)*, and the following procedure is a modification of this. Dissolve crude $N(1)$-(2-aminoethyl)-NAD⁺ (contains approx 70% $N(1)$-(2-aminoethyl)NAD⁺; 3.05 g, 4.3 mmol) in 850 mL of water. Adjust the pH to 6.5 with 1M LiOH. Place the solution in a water bath at 50°C, and allow the rearrangement to proceed for 6–7 h. Monitor the reaction by HPLC/TLC/change in absorbance maximum over this time. The wavelength of maximum absorbance should change from 260 nm to at least 264 nm, since the λ_{max} for N^6-(2-aminoethyl)NAD⁺ is 266 nm. Terminate the reaction by cooling the solution to room temperature, and adjust the pH to 5.5 with 1M HCl. Concentrate the solution by rotary evaporation at 35°C to approx 20 mL. The compound is stable in concentrated solution for at least 1 wk at 4°C.
2. Apply the concentrated reaction mixture to a Bio-Rex 70 cation-exchange column (H⁺ form, 2.7 × 73 cm), pre-equilibrated with 1 mM HCl, pH 3.0. Wash the column with this solution, and collect fractions of 10–20 mL, in a total volume of approx 4 L. Monitor the A_{260} as an indicator of coenzyme concentration. The last peak to elute contains the N^6-(2-aminoethyl)NAD⁺. Some overlap may occur with the previous peak, containing the tricyclic NAD⁺ derivative 1, N^6-ethanoadenine-NAD⁺. Pool and rotary evaporate the fractions containing the desired compound to approx 5 mL. Lyophilization gives a fluffy pale yellow compound, stable for at least 12 mo at –20°C in an air-tight container. ε = 21,600/M/cm at 266 nm. HPLC retention times:

NAD$^+$, 6.3 min; N(1)-(2-aminoethyl)NAD$^+$, 4.4 min; N^6-(2-aminoethyl)NAD$^+$, 5.7 min; 1, N^6-ethanoadenine-NAD$^+$, 4.7 min. TLC R_f values: system C, NAD$^+$ 0.28, N(1)-(2-aminoethyl)NAD$^+$, 0.10, N^6-(2-aminoethyl)-NAD$^+$, 0.16, 1, N^6-ethanoadenine-NAD$^+$, 0.08 *(28)*.

3. Dissolve the spacer (glutaric, adipic, or pimelic acid) in water to give a 50-mM solution. Adjust the pH to 7.0 with 5M NaOH, and add 700 µL of this solution to 50 mg (67 µmol) of N^6-(2-aminoethyl)NAD$^+$ to give a 2:1 molar ratio of coenzyme to spacer arm. Adjust the pH from approx 3.0–5.4 with 5M NaOH. Add solid EDC (34 mg, 335 µmol) to the solution in six to seven portions over 20 min, while monitoring the reaction by TLC/HPLC. The pH tends to rise; readjust it to 5.5–6.0 when it reaches a value of 6.9–7.0. Stop the reaction after 40 min by freezing the reaction mixture at –20°C. TLC indicates that at least four side products are formed in addition to bis-NAD$^+$.

4. To purify the bis-NAD$^+$, dilute the sample twofold with 50 mM ammonium acetate, pH 7.0, and apply it to a Whatman DE-52 anion-exchange column (2.3 × 16 cm), equilibrated as described in Section 2.1.3. Wash the column with this buffer. The last major peak contains the bis-NAD$^+$ derivative. This can be checked by TLC, HPLC, or by the ability of the fractions to cause affinity precipitation of GDH. Pool and rotary evaporate the fractions containing bis-NAD$^+$ to 3–5 mL, and freeze-dry these (stable for at least 4 mo at 4°C or –20°C). The compound can also be stored for approx 1 mo in solution at 4°C, although bacterial contamination can result. TLC R_f values (system C) for bis-NAD$^+$ derivatives: N,N'-bis(N^6-ethylene-NAD$^+$)glutaramide, 0.09, N,N'-bis(N^6-ethylene-NAD$^+$) adipamide, 0.12, N,N'-bis(N^6-ethylene-NAD$^+$)pimelamide, 0.15. $\varepsilon = 44{,}200 \pm 1400/M/\text{cm}$. HPLC retention times (in 0.1M potassium phosphate buffer, pH 6.0, containing 20% methanol) are 7.8, 11.2, and 13 min for the compounds, respectively.

3.2. Pilot Precipitation Study

1. Take a source of the enzyme of interest (e.g., a crude supernatant), and assay it for enzyme activity (*see* ref. *32* for a wide range of assays) to determine the specific activity.
2. Calculate the approximate concentration of enzyme based on the specific activity.
3. Add 100–200 µL of the enzyme solution to a series of minifuge tubes. If the study is to be carried out with purified enzyme, dilute a stock solution of 400 mM potassium phosphate buffer, pH 7.4, to give a final concentration of 20 mM. Set up all samples in duplicate, including two control tubes to which water has been added in place of the coenzyme derivative. Set up duplicate controls in addition to these, which contain no bis-ligand or substrate analog. This enables the percent inhibition owing to the presence of the substrate analog to be calculated.

4. Calculate the concentration of a stock solution of bis-coenzyme spectrophotometrically, using the extinction coefficients given in Section 3.1.
5. Add bis-NAD$^+$ (or bis-ATP) to give a range of coenzyme Eq/enzyme subunit from approx 0.1–20, calculating the concentration from the extinction coefficient.
6. Add the appropriate substrate analog, and leave at 4°C for at least 30 min (or overnight, if required).
7. Centrifuge in a minifuge for 5 min, and assay the supernatant. Calculate the precipitation as a percentage of the activity remaining in the control sample. The tube giving the minimal residual activity shows the maximum affinity precipitation. This indicates the appropriate concentration of bis-coenzyme to be added to obtain maximum affinity precipitation.

3.3. Enzyme Purification Protocols

The following are a selection of protocols for purifying enzymes with bis-ligands.

3.3.1. Purification of Yeast Alcohol Dehydrogenase (YADH; EC 1.1.1.1)

1. The yeast lysis is carried out according to a modification of the procedure in ref. 40. Crumble 40 g of fresh baker's yeast into 21 mL of toluene, preheated to 45°C in a water bath. **(Caution: Keep the vessel covered, or incubate in a fume cupboard.)** Allow the mixture to liquefy over a period of 90 min with occasional stirring. (*see* Note 4).
2. Leave the mixture at room temperature for 3 h, and add 42 mL of 1 mM Na$_2$EDTA as a protease inhibitor. Stir for 2 h at 4°C, and leave the mixture at that temperature overnight.
3. Centrifuge the lysate at 47,800g for 10 min, remove the top fatty layer by aspiration, and retain the supernatant. Add solid (NH$_4$)$_2$SO$_4$ slowly, until 60% saturation is reached (7.22 g to 20 mL), and centrifuge for 10 min at 47,800g. The ammonium precipitation step removes some proteins that would otherwise precipitate on addition of bis-NAD$^+$ and gives a purification of approx three- to fourfold. Dissolve the pellet in approx 3 mL of 20 mM potassium phosphate buffer, pH 7.4, containing 1 mM Na$_2$EDTA and dialyze against two changes of 2 L of this buffer to remove the salt (alternatively, gel filter on a column of Sephadex G-25, at least 1 × 50-cm long).
4. Carry out a pilot precipitation of this supernatant. This can be performed before or after removing the (NH$_4$)$_2$SO$_4$. However, if the (NH$_4$)$_2$SO$_4$ has been removed, it must be added back to a concentration of at least 0.5M, since it is required to promote precipitation. Dilute the supernatant 1:4 with buffer, and add pyrazole to a final concentration of 28 mM. A suitable range of concentrations of NAD$^+$ equivalents is 0–20 µM. Allow precipitation to occur for at least 3 h (preferably overnight).

5. Assay the supernatants as described in ref. *32,* diluting the enzyme solution if necessary, and determine which concentration of bis-NAD$^+$ gives maximum precipitation. Add the appropriate concentration of bis-NAD$^+$ to whatever vol of supernatant is required, along with $(NH_4)_2SO_4$ and pyrazole. Allow precipitationto occur as described above, and centrifuge the precipitate for 10 min at 12,000g to pellet the crosslinked enzyme aggregate.
6. Resolubilize the pellet by adding 0.6 mM NADH in 20 mM potassium phosphate buffer, pH 7.4 (the volume can be varied, depending on what final enzyme concentration is required). Allow resolubilization to proceed for at least 3 h, preferably 12 h, at 4°C on a rocking tray or with occasional stirring (*see* Note 4). The enzyme should be essentially homogeneous, as shown by SDS-PAGE on a 12.5% gel.

3.3.2. Purification of GDH (EC 1.4.1.3) from Beef and Rat Liver (9)

1. If beef liver is used, transport this from the abattoir on ice.
2. Homogenize the tissue, and carry out ammonium sulfate precipitation and DEAE-cellulose chromatography as described in ref. *41.*
3. Pool the fractions from the DE-52 column that contain GDH activity *(32),* and concentrate them to a volume of approx 10 mL by ultrafiltration through an Amicon XM-50 membrane. Dialyze the solution overnight at 4°C against 200 mL of sodium or potassium phosphate buffer, pH 7.4, with at least one change of buffer.
4. Carry out a pilot precipitation as described in Section 3.2. Incubate 100-µL samples of the dialyzed solution in the presence of 12.7 µL of 0.7M glutarate, pH 7.0, and bis-NAD$^+$ for 15 min. Assume a subunit M_r for GDH of 56,700. The optimum ratio of NAD$^+$ Eq/enzyme subunit may vary with the preparation used; for example, with a preparation of beef liver enzyme that had a specific activity of 2.7 U/mg, about half the activity was precipitated at a ratio of 2 NAD$^+$ Eq/GDH subunit. However, a preparation with the lower specific activity of 0.8 U/mg required approx 8 NAD$^+$ Eq/GDH subunit to precipitate half the enzyme activity. The precipitation yield also depends on the protein concentration.
5. Take the dialyzed solution, and add glutarate to a final concentration of 79 mM. Add the appropriate amount of bis-NAD$^+$ to achieve maximum precipitation. Keep the mixture on ice overnight, and centrifuge at 10,000g for 15 min. Redissolve the pellet by stirring for 6 h in 1 mL of 20 mM sodium or potassium phosphate buffer, pH 7.4, containing 0.6 mM NADH. Dialyze the redissolved pellet against 200 mL of the same buffer as in step 3. A purification summary is given in Table 2. This method appears to avoid some of the proteolysis encountered in more conventional methods of purifying this enzyme. Although affinity precipitation was obtained with

Table 2
Purification of Beef and Rat Liver GDH[a]

Step	Volume, mL		Total protein, mg		Total activity, U		Specific activity, U/mg		Purification factor		Yield, %	
	Beef	Rat	Beef	Rat	Beef	Rat	Beef	Rat	Beef	Rat	Beef	Rat
Homogenate	370	90	8580	2720	1920	760	0.2	0.3	—	—	100	100
Resuspended (NH$_4$)$_2$SO$_4$ precipitate	75	35	2010	980	1640	505	0.3	0.5	4	2	86	67
DEAE-cellulose chromatography	8.4	9.8	139	130	408	195	2.9	1.5	13	5	21	26
Affinity precipitation	1.2	1.0	9.4	5.1	376	190	40	37	180	140	20	25

[a]Reproduced from Graham et al. (9). The first three steps were carried out according to the method of Mc Carthy et al. (41). The purification of the enzyme from beef liver used 46 g and that from rat used 10 g liver.

preparations that had not been subjected to chromatography on DEAE-cellulose, the yield was much lower and the precipitated material was not completely pure.

3.3.3. Separation of Lactate Dehydrogenase (LDH; EC 1.1.1.27) Isoenzymes

This technique is based on the principle that both M and H isoenzyme subunits of LDH form abortive complexes with NAD^+ and oxalate, whereas the H form gives rise to abortive ternary complexes with NAD^+ and oxamate. In a mixture of H and M subunits, ternary complexes and, thus, crosslinks will form only with the H subunits, and tetramers with predominantly M subunits will tend not to be precipitated.

1. Select a source of LDH isoenzymes (e.g., a crude supernatant or Sigma Type X bovine muscle LDH, which contains varying proportions of all five isoenzymes). Determine the specific activity, and hence, estimate the concentration of the enzyme *(32)*.
2. Carry out a pilot precipitation as described in Section 3.2. The precipitation solution should contain 20 mM potassium phosphate buffer, pH 7.4, and potassium oxamate to a final concentration of 20 mM. Optimum affinity precipitation should be obtained with a ratio of approx 1 NAD^+ Eq/LDH subunit. Allow precipitation to occur overnight at 4°C. Centrifuge for 10 min and retain the supernatant for further analysis.
3. Resolubilize the precipitated enzyme by adding 50 µL of 0.6 mM NADH in the above buffer and incubate overnight. Centrifuge the samples for 10 min at 12,000g to pellet any remaining crosslinked enzyme.
4. The isoenzymic composition of the pellet and supernatant can be analyzed by starch gel electrophoresis. This is carried out by the method described in ref. *42*. Make an 11% starch gel. Apply 10-µL samples of the enzyme solutionto the gel on small squares of filter paper (about 0.5 × 0.5 cm), and carefully insert these, using a forceps, in wells cut across the middle of the gel. Place a piece of polyethylene wrap over the gel, and put a cooling plate on top. Run the gel for 4 h at 550 V and 60 mA in a cold room at 4°C. After electrophoresis, stain the gel with an activity stain for LDH (*see* Section 2.3.) until the bands appear. Wash the gel exhaustively with water, otherwise overstaining can easily occur, and treat the gel gently, since starch gels are fragile. Figure 3 shows a starch gel on which the separation pattern can be seen.

3.3.4. Purification of Beef Heart Phosphofructokinase (PFK; EC 2.7.1.11)

1. Obtain a fresh portion of beef heart from a freshly slaughtered animal, and keep it on ice. Remove the ventricular muscles and adipose tissue, dice the

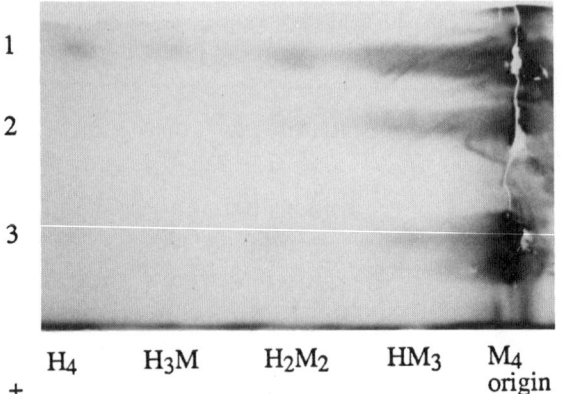

Fig. 3. Starch gel electrophoresis of LDH isoenzymes. The gel was stained with an activity stain for LDH, as described in the text. Lane 1 consists of Sigma Type X bovine muscle LDH (1.4 U) containing all isoenzymes of LDH with M and H subunits. Lane 2 contains the supernatant after affinity precipitation in the presence of bis-NAD^+ (1 NAD^+ Eq/LDH subunit) and 20 mM oxamate. Lane 3 contains the resolubilized pellet from the same sample as lane 2. The pellet was resolubilized in 20 mM potassium phosphate, pH 7.4, containing 0.6 mM NADH. Affinity precipitation and resolubilization were carried out in a final vol of 100 µL.

remaining tissue, and wash it in distilled water. Homogenize it in a precooled blender for 10 s, take 500 g of mince, and add 10 mM Tris-HCl buffer containing 2mM EDTA, pH 8.2, to a total vol of 1.25 L. Homogenize this for 90 s at high speed, and then centrifuge the homogenate at 2,600g for 10 min. Discard the supernatant.

2. Resuspend the pellet in 10 mM Tris-HCl containing 0.5 mM ATP, 50 mM $MgSO_4$, and 5 mM 2-mercaptoethanol, pH 8.0, to a total vol of 1 L. Prewarm the buffer to 37°C. Stir the suspension at 37°C for 30 min. This extraction procedure gives rise to a decrease in pH of 0.6–0.8 pH units. Remove solid material by centrifugation for 15 min at 8000g, followed by a further centrifugation at 12,000g. The supernatant contains about 90% of the PFK activity.

3. Set up a pilot precipitation (*see* Section 3.2.) to determine the concentration of bis-ATP giving optimum precipitation. The following volumes are appropriate: 100 µL enzyme solution, 30 µL of 30 mM citrate, pH 7.0, 60 µL 0.16M $MgSO_4$. Add bis-ATP to give a range of ATP Eq/PFK subunit ratios (0–8 is suitable, at an enzyme concentration of 0.38 mg/mL PFK). Make the volume up to 500 µL with extraction buffer. A ratio of 4–6 is usually found

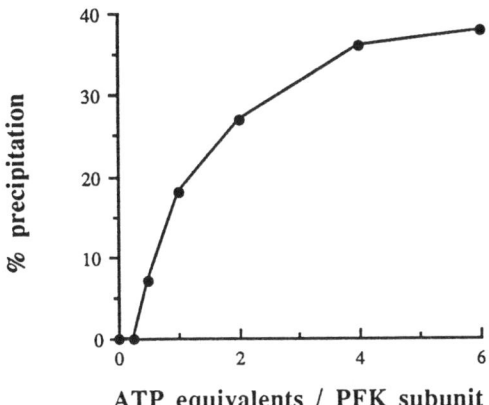

Fig. 4. Affinity precipitation of PFK. A pilot precipitation was set up as described in the text. The precipitation mixture contained 0.1 mL enzyme supernatant, 20 µL 0.5M citrate, pH 7.0, bis-ATP (0.25–6 ATP Eq/ATP subunit), and 20 µL 0.5M MgSO$_4$. The tubes were left overnight at 0°C, and centrifuged for 10 min to pellet the enzyme.

Table 3
Purification of PFK from Bovine Heart (200 g Tissue) (36)[a]

Step	Volume, mL	Total protein, mg	Total activity, U	Specific activity, U/mg	Purification, factor	Yield, %
Crude extract Resuspended Precipitate	490	3675	970	0.26	1	100
Supernatant After extraction	548	376	883	3.2	12.3	91
Affinity precipitation	3	7.1	214	30.1	115.7	22

[a]The method used is that given in Section 3.3.4.

to be optimum for precipitation (see Fig. 4). A similar pilot experiment varying the enzyme concentration can also be set up at a fixed concentration of bis-coenzyme. A sample purification table is given (Table 3).

4. The volume for precipitation can be scaled up as required. The enzyme can be recovered by dialysis, which causes resolubilization of the pellet.

4. Notes

1. Bis-NAD$^+$ has been used as a ligand with several NAD$^+$-linked dehydrogenases, but only for three enzymes (YADH, LDH, GDH) has it successfully

been applied for purification purposes. A partial purification of isocitrate dehydrogenase giving rise to a yield of 18% has been described *(7)*. Horse liver alcohol dehydrogenase has also been observed to form aggregates with bis-NAD$^+$ and pyrazole *(5,8)*, but it does not affinity precipitate, despite the fact that it may form larger polymers. Other enzymes that do not successfully affinity precipitate include malate, alanine, and glyceraldehyde-3-phosphate dehydrogenases *(28)*. Malate dehydrogenase is a dimer, and the other two enzymes, although tetrameric, may fail to form crosslinked ternary complexes, possibly owing to steric or kinetic factors.

2. The R_f values given for TLC of coenzyme derivatives are not exact. They can vary by up to ±0.05. This may be owing to slight changes in the solvent composition over time, possibly because of evaporation.
3. To date, bis-ATP has only been used in the purification of PFK. The yields obtained for this compound are low, and it is unstable, possibly owing to hydrolysis of the spacer arm, catalyzed intramolecularly by a phosphate group. It should be used within 4 d of synthesis. Because of this, the reagent must be regarded as being of limited practical use, and further studies on the synthesis of a more stable bis-ATP derivative are necessary.
4. YADH has essential thiol groups, and if these are oxidized, the enzyme loses activity. They should be kept reduced by the addition of 1 mM dithiothreitol or 1 mM 2-mercaptoethanol. The breaking of the yeast cells may also be carried out using glass beads.

Acknowledgments

Support from Trinity College, Dublin and EOLAS (Forbairt; The Irish Science and Technology Agency) is gratefully acknowledged.

References

1. Chen, J.-P. (1990) Novel affinity based processes for protein purification. *J. Ferment. Bioeng.* **70**, 199–209.
2. Luong, J. H. T. and Nguyen, A.-L. (1992) Novel separations based on affinity interactions. *Adv. Biochem. Eng. Biotechnol.* **47**, 138–158.
3. Irwin, J. A. and Tipton, K. F. (1995) Affinity precipitation: a novel approach to protein purification. *Essays Biochem.* **29**, 137–156.
4. Larsson, P.-O. and Mosbach, K. (1979) Affinity precipitation of enzymes. *FEBS Lett.* **98**, 333–338.
5. Flygare, S., Griffin, T., Larsson, P.-O., and Mosbach, K. (1983) Affinity precipitation of dehydrogenases. *Anal. Biochem.* **133**, 409–416.
6. Larsson, P.-O., Flygare, S., and Mosbach, K. (1984) Affinity precipitation of dehydrogenases. *Methods Enzymol.* **104**, 364–369.
7. Beattie, R. E., Graham, L. D., Griffin, T. O., and Tipton, K. F. (1985) Purification of NAD$^+$-dependent dehydrogenases by affinity precipitation with adipo-N_2,N_2'-dihydrazido bis-(N^6-carboxymethyl-NAD$^+$) (bis-NAD$^+$) *Biochem. Soc. Trans.* **12**, 433.

8. Buchanan, M., O'Dea, C. D., Griffin, T. O., and Tipton, K. F. (1989) Reversible cross-linking of alcohol and lactate dehydrogenases with the bifunctional reagent N_2, N_2'-adipodihydrazido-bis-(N^6-carboxymethyl-NAD$^+$). *Biochem. Soc. Trans.* **17**, 422.
9. Graham, L. D., Griffin, T. O., Beatty, R. E., Mc Carthy, A. D., and Tipton, K. F. (1985) Purification of liver GDH by affinity precipitation and studies on its denaturation. *Biochim. Biophys. Acta* **828**, 266–269.
10. Beattie, R. E., Buchanan, M., and Tipton, K. F. (1987) The synthesis of N_2,N_2'-adipodihydrazido-bis-(N^6-carboxymethyl-ATP) and its use in the purification of phosphofructokinase. *Biochem. Soc. Trans.* **15**, 1043,1044.
11. Hayet, M. and Vijayalakshmi, M. A. (1986) Affinity precipitation of proteins using bis-dyes. *J. Chromatogr.* **376**, 157–161.
12. Lowe, C. R. and Pearson, J. C. (1983) Bio-mimetic dyes, in *Affinity Chromatography and Biological Recognition* (Chaiken, I. M., Wilchek, M., and Parikh, I., eds.), Academic, London, pp. 421–432.
13. Pearson, J. C., Burton, S. J., and Lowe, C. R. (1986) Affinity precipitation of lactate dehydrogenase with a triazine dye derivative: selective precipitation of rabbit muscle lactate dehydrogenase with a Procion Blue H-B analog. *Anal. Biochem.* **158**, 382–389.
14. Pearson, J. C., Clonis, Y. D., and Lowe, C. R. (1989) Preparative affinity preparation of L-lactate dehydrogenase. *J. Biotechnol.* **11**, 267–274.
15. Lilius, G., Persson, M., Bülow, L., and Mosbach, K. (1991) Metal affinity precipitation of proteins carrying genetically attached polyhistidine affinity tails. *Eur. J. Biochem.* **198**, 499–504.
16. Van Dam, M. E., Wuenschell, G. E., and Arnold, F. H. (1989) Metal affinity precipitation of proteins. *Biotechnol. Appl. Biochem.* **11**, 492–502.
17. Schneider, M., Guillot, C., and Lamy, B. (1981) The affinity precipitation technique. Application to the isolation and purification of trypsin from bovine pancreas. *Ann. N.Y. Acad. Sci.* **369**, 257–263.
18. Senstad, C. and Mattiasson, B. (1989) Affinity-precipitation using chitosan as ligand carrier. *Biotechnol. Bioeng.* **33**, 216–220.
19. Senstad, C. and Mattiasson, B. (1989) Purification of wheat germ agglutinin using affinity flocculation with chitosan and a subsequent centrifugation or flotation step. *Biotechnol. Bioeng.* **34**, 387–393.
20. Taniguchi, M., Kobayashi, M., Natsui, K., and Fujii, M. (1989) Purification of staphylococcal protein A by affinity precipitation using a reversibly soluble–insoluble polymer with human IgG as a ligand. *J. Ferment. Bioeng.* **68**, 32–36.
21. Kamihira, M., Kaul, R., and Mattiasson, B. (1992) Purification of recombinant protein A by aqueous two-phase extraction integrated with affinity precipitation. *Biotechnol. Bioeng.* **40**, 1381–1387.
22. Bradshaw, A. P. and Sturgeon, R. J. (1990) The synthesis of soluble polymer-ligand complexes for affinity precipitation studies. *Biotechnol. Techniques* **4**, 67–71.
23. Senstad, C. and Mattiasson, B. (1989) Preparation of soluble affinity complexes by a second affinity interaction: a model study. *Biotechnol. Appl. Biochem.* **11**, 41–48.
24. Linné, E., Garg, N., Kaul, R., and Mattiasson, B. (1992) Evaluation of alginate as a ligand carrier in affinity precipitation. *Biotechnol. Appl. Biochem.* **16**, 48–56.

25. Gupta, M. N., Dong, G. Q., and Matiasson, B. (1993) Purification of endopolygalacturonase by affinity precipitation using alginate. *Biotechnol. Appl. Biochem.* **18**, 321–328.
26. Chen, J. P. and Hoffman, A. S. (1990) Polymer–protein conjugates. II. Affinity precipitation separation of human immunogammaglobulin by a poly (*N*-isopropylacrylamide)-protein A conjugate. *Biomaterials* **11**, 631–634.
27. Nguyen, A. L. and Luong, J. H. T. (1989) Syntheses and applications of water-soluble reactive polymers for purification and immobilization of biomolecules. *Biotechnol. Bioeng.* **34**, 1186–1190.
28. Irwin, J. A. and Tipton, K. F. (1995) Resolution of lactate dehydrogenase isoforms by affinity precipitation. *Biochem. Soc. Trans.* **23**, 3655.
29. Larsson, P.-O. and Mosbach, K. (1981) Novel affinity techniques. *Biochem. Soc. Trans.* **9**, 285–287.
30. Feinstein, A. and Rowe, A. J. (1965) Molecular mechanism of formation of an antigen–antibody complex. *Nature* **205**, 147–149.
31. O'Carra, P. (1978) Theory and practice of affinity chromatography, in *Chromatography of Synthetic and Biological Polymers,* 2nd ed. (Epton, R., ed.), Ellis Horwood for the Chemical Society, London, Chapter 11, pp. 131–158.
32. Bergmeyer, H. U., Grassl, M., and Walter, H.-E. (1983) Reagents for enzymatic analysis, in *Methods of Enzymatic Analysis*, vol. 2, 3rd ed. (Bergmeyer, H. U., Bergmeyer, J., and Grassl, M., eds.), Verlag Chemie, Weinheim, pp. 126–328.
33. Laemmli, U. K (1970) Cleavage of structural proteins during the assembly of the head of bacteriophage T4. *Nature* **227**, 680–685.
34. Mosbach, K., Larsson, P.-O., and Lowe, C. (1976) Immobilised coenzymes. *Methods Enzymol.* **44**, 859–887.
35. Engel, J. D. (1975) Mechanism of the Dimroth rearrangement in adenine. *Biochem. Biophys. Res. Comm.* **64**, 581–585.
36. Buchanan, M. (1988) The synthesis of N_2,N_2'-Adipodihydrazido-Bis-(N^6-Carboxymethyl-NAD) and N_2,N_2'-Adipodihydrazido-Bis-(N^6-Carboxymethyl-ATP) and Subsequent Affinity Precipitation of Enzymes. MSc Thesis, University of Dublin.
37. Beattie, R. E. (1984) The Synthesis of N_2,N_2'-Adipodihydrazido-Bis-(N^6-Carboxymethyl-NAD$^+$) and its use in the Purification of Dehydrogenases. MSc Thesis, University of Dublin.
38. Bückmann, A. F. (1987) A new synthesis of coenzymically active water-soluble macromolecular NAD and NADP derivatives. *Biocatalysis* **1**, 173–186.
39. Bückmann, A. F. and Wray, V. (1992) A simplified procedure for the synthesis and purification of N^6-(2-aminoethyl)-NAD and tricyclic 1,N^6-ethanoadenine NAD. *Biotechnol. Appl. Biochem.* **15**, 303–310.
40. Butler, P. J. G. and Thelwall Jones, G. M. (1970) The preparation of alcohol dehydrogenase and glyceraldehyde-3-phosphate dehydrogenase from baker's yeast. *Biochem. J.* **118**, 375–378.
41. Mc Carthy, A. D., Walker, J. M., and Tipton, K. F. (1980) Purification of glutamate dehydrogenase from ox brain and liver. *Biochem J.* **191**, 605–611.
42. Phelps, C. (1984) in *Techniques in the Life Sciences*: vol. B1/1, suppl. BS 104, *Protein and Enzyme Biochemistry* (Tipton, K. F., ed.), Elsevier, Shannon Industrial Estate, Ireland, pp. 1–16.

CHAPTER 23

Isoelectric Focusing

Reiner Westermeier

1. Introduction

Isoelectric focusing (IEF) is performed in a pH gradient in an electric field. The charged proteins migrate toward the anode or the cathode—according to the sign of their net charge—until they reach the position in the pH gradient where their net charges are zero. This pH value is the isoelectric point (pI) of the substance, an exactly defined physicochemical constant. Since the molecule is no longer charged, it stays there; the electric field does not have any influence on it. Should the protein diffuse away, it will gain a net charge and the applied electric field will cause it to migrate back to its pI. This concentrating effect leads to the name "focusing" and makes the method very useful for purification purposes.

The principle of isoelectric focusing is employed in varying technical approaches. The most complete information about these methods is found in the works of Righetti *(1,2)* and Andrews *(3)*.

The procedure described here is performed in a horizontal flat bed of a granulated dextran gel. Large sample volumes can be mixed with the original gel slurry from which the gel bed is prepared. Labile samples are applied at a defined zone of the gradient. The separation is run at a constant power of 8 W for 14–16 h at controlled temperature. pH measurements and prints on filter paper can be made directly on the gel surface. The sample fractions are collected by sectioning the gel, and then eluting the proteins from these sections. The technique, introduced by Radola in 1969 *(4)*, has been selected for several reasons:

1. It can be performed with standard horizontal electrophoresis equipment, which is also used for analytical methods.

2. The method has a high loading capacity (up to gram quantities).
3. A large number of different applications and references are available for this technique; some examples are listed in refs. *5–15*.
4. The technique is much less sensitive to precipitation of proteins at their isoelectric points compared to preparative isoelectric focusing methods in a free liquid because the precipitate is trapped within the gel bed.
5. The recovery of the protein fractions from a granulated dextran gel is much easier and gives a higher yield compared to compact gel media like agarose and polyacrylamide.

The original method has been refined and optimized by Winter et al. *(16)*.

2. Materials

2.1. Equipment

In addition to standard laboratory equipment, the following are needed:

1. Horizontal electrophoresis chamber (Multiphor II) and Preparative IEF kit (Pharmacia Biotech, Uppsala Sweden). The Preparative IEF Kit contains a tray with a 5-mm silicone rim, a fractionating grid frame with 20 fractionation blades, a sample applicator, sample elution columns, IEF electrode strips, and print papers.
2. Thermostatic circulator at 10°C.
3. Constant power supply (>1 kV).

2.2. Consumables, Chemicals, and Solutions

1. Desalting columns for sample preparation: PD-10 columns prepacked with Sephadex G-25 (Pharmacia Biotech, product code 17-0851-01).
2. IEF electrode strips (clean filter paper strips approx 5 × 2 mm).
3. Print paper (clean filter paper approx 110 × 250 cm).
4. Granulated gel media with a very low level of charged contaminants: Ultrodex® (Pharmacia Biotech, product code 80-1130-01) (*see* Note 1).
5. Carrier ampholytes of wide pH range (e.g., pH 3.5–9.5), and narrow pH ranges (2–3 pH units) depending on the pIs of the proteins to be purified: Ampholine® and Pharmalytes® (Pharmacia Biotech).
6. Detergent solution (0.1% [v/v] Triton X-100): 1 mL Triton X-100 made up to 1 L with distilled water.
7. Anode solution ($1M$ phosphoric acid) 2.8 mL concentrated H_3PO_4 made up to 50 mL with distilled water.
8. Cathode solution ($1M$ sodium hydroxide): 2 g NaOH made up to 50 mL with distilled water.
9. Fixing solution (10% [w/v] trichloroacetic acid): 100 g trichloroacetic acid made up to 1 L with distilled water.

Isoelectric Focusing

Table 1
Volumes for the Carrier Ampholyte Solution

Ampholytes	pH	pH range, mL				
		2.5–5.0	3.5–5.5	5.0–7.5	6.0–8.5	7.0–10.0
Pharmalyte	2.5–5.0	5.5				
Ampholine	3.5–5.0		2.75			
Ampholine	4.0–6.0		2.75			
Ampholine	5.0–8.0			5.5		
Ampholine	6.0–8.0				2.75	
Ampholine	7.0–9.0				2.75	2.75
Ampholine	9.0–11.0					2.75
Distilled water (+ sample)		104.5	104.5	104.5	104.5	104.5
Total volume		110	110	110	110	110

10. Destaining solution (10% [v/v] methanol/20% [v/v] acetic acid): Mix 100 mL methanol with 200 mL acetic acid and make up to 1 L with distilled water.
11. Staining solution (0.2% [w/v] Coomassie blue): Dissolve 0.6 g Coomassie blue R 250 in 300 mL destaining solution.

3. Methods

3.1. Gel Preparation

When large volumes of stable sample solution have to be separated, the sample is included in the initial gel slurry. Labile samples are applied in a certain zone of the gel bed with a sample applicator.

3.1.1. Preparing the Carrier Ampholyte Solution

Prepare a total of 110 mL of a 2% (w/v) carrier ampholyte solution (100 mL for the gel bed and 10 mL for the electrode strips) (*see* Note 2). The ampholytes are diluted in distilled water or with sample solution if large volumes of sample are to be separated (*see* Section 3.1.). The pH range of the carrier ampholytes is selected so that the pI(s) of the protein(s) to be purified fall(s) approximately in the middle of the range. This pH value can be identified by a previous analytical experiment *(17)* or from information in the literature. A large number of proteins are listed with their isoelectric points in two tables that have been published by Righetti et al. *(18,19)*. Some carrier ampholyte solution recipes are listed in Table 1.

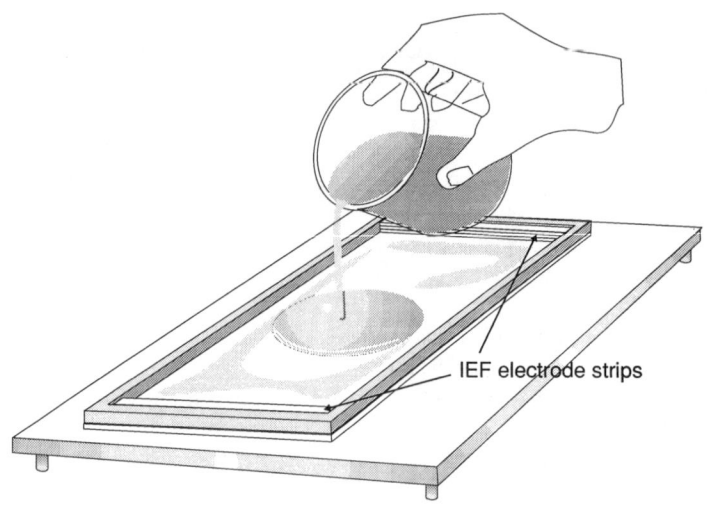

Fig. 1. Pouring the gel slurry into the tray.

3.1.2. Preparing the Gel Bed *(see Note 3)*

1. Clean the glass plate of the tray thoroughly before use in order to avoid uneven spreading of the gel layer.
2. Cut six IEF electrode strips to 10.5 cm, and soak them in 10 mL of the carrier ampholyte solution. Place three layers of strips at each end of the tray, weigh the tray, and place it on a horizontal table.
3. Weigh out 4 g of Ultrodex, and sprinkle it in small portions over 100 mL of carrier ampholyte solution in a 200-mL beaker. Let the gel sink down into the liquid before adding the next portion.
4. Weigh the beaker and its contents. Homogenize the suspension by gently stirring with a spatula (do not use a magnetic stirrer). When a basic gradient range is selected, degas the slurry with a water jet pump or comparable equipment, in order to remove the atmospheric carbon dioxide.
5. Pour the suspension into the tray (Fig. 1). Gently tap against the ends of the tray, so that the suspension spreads evenly.
6. Weigh the beaker and the remainder of the slurry.

3.1.3. Evaporating the Gel Bed to the Correct Water Content

1. Mount a small fan 70 cm above the tray, and evaporate excess water with a light stream of air. The speed and distance of the fan must be adjusted so that no ripples are formed on the surface.
2. Control the weight of the tray and its content from time to time. After 1.5–2 h, it should have reached the final weight, which was calculated from the

weight difference of the beaker, the weight of the tray with the strips, and the evaporation limit (*see* Notes 4 and 5).

3.1.4. Arranging the Tray on the Cooling Plate

1. Apply 2 mL of a 0.1% (v/v) Triton X-100 solution on the cooling plate of the Multiphor electrophoresis chamber, which has already been cooled to 10°C.
2. Transfer the tray onto the cooling plate avoiding air bubbles.
3. Place an electrode strip soaked in anode solution at the anodal side and another strip soaked in cathode solution at the cathodal side, each on top of the strips already in the tray. Cathodal and anodal sides are marked on the cooling plate.
4. Cut off the protruding parts so that the electrode strips fit exactly into the tray. No gaps must exist between the ends of the strips and the silicone rubber of the tray.

3.2. Sample Application

3.2.1. Preparing the Sample

1. The sample should not contain any particles or grease. For isoelectric focusing, salt and buffer concentrations over 50 mM should be avoided. Ideally, it should not be higher than 10 mM. The easiest and quickest way to desalt a sample is to perform gel filtration with a PD-10 column or another column packed with Sephadex G-25.
2. The maximum loading capacity is 5–10 mg of protein mixture/mL gel bed (2–4 mg/mL of one single protein) when a narrow pH range is employed, e.g., pH 4.0–6.0. The wider the pH range, the lower the loading capacity.

3.2.2. Loading the Sample

The sample can be already mixed with the initial slurry as described in Section 3.1. Labile samples are applied with the sample applicator as a narrow zone in the gel bed as follows (*see* Note 6):

1. The sample solution should have a volume of 3 mL. Otherwise dilute it with the remainder of the carrier ampholyte solution, which was used for soaking the electrode strips.
2. Press the sample applicator through the gel bed at the desired position (Fig. 2). Scrape off the gel in the applicator, and mix it with the sample in a small beaker.
3. Pour the sample into the applicator, remove the applicator, and allow the the bed to equilibrate for a few minutes.

3.3. Isoelectric Focusing

The resulting separation pattern is dependent on the temperature; conventionally 10°C is used (*see* Notes 7 and 8).

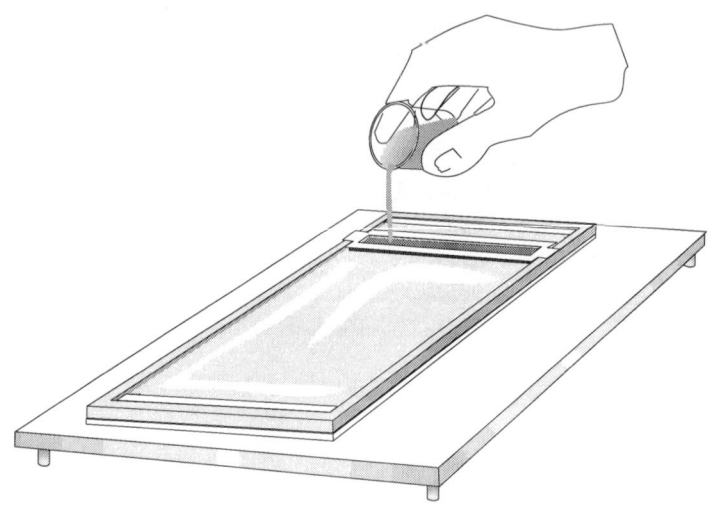

Fig. 2. Pouring the sample into the applicator.

1. Place the platinum electrodes on the respective electrode strips.
2. Perform electrofocusing at a constant current of 8 W for 14–16 h (overnight). The maximal current is 25 mA, and the maximal voltage 1.5 kV.

3.4. Detection of Protein Zones

1. Mark the anode (+) on a sheet of print paper.
2. Carefully roll the paper onto the gel surface avoiding air bubbles. Leave it for 1 min.
3. Carefully remove the print. Apply the electric field again to the gel during the staining of the print in order to avoid diffusion of the zones.
4. Treat the print as follows:
    a. Three times washing out of the carrier ampholytes in fixing solution for 15 min.
    b. Staining for 10 min in staining solution (*see* Note 9).
    c. Destaining for 1 h in destaining solution (*see* Notes 10–12).
    d. Dry the print with hot air.

    If colored protein fractions are separated, no staining is necessary. Figure 3 shows the stained paper print of a hemoglobin separation from a pH gradient 5.0–7.5.

3.5. Measurement of the pH Gradient

1. Press the fractionating grid onto the gel, which is still on the cooling plate at 10°C (Fig. 4).

Isoelectric Focusing

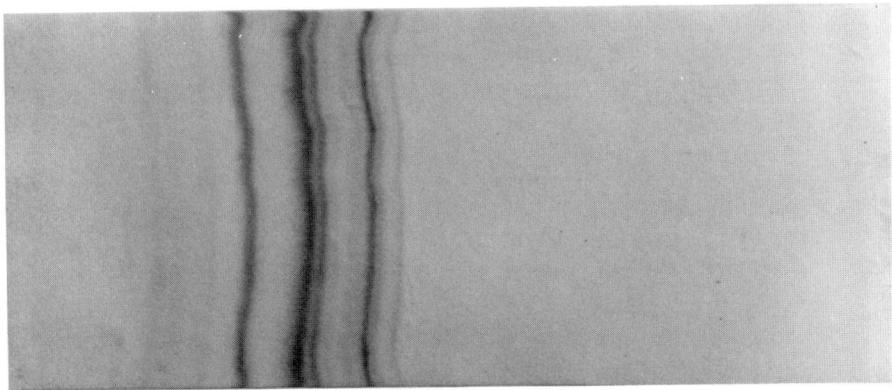

Fig. 3. Paper print of focused hemoglobin bands from a preparative bed of granulated gel with a pH gradient of 6.9–8.5. Reproduced with permission from Pharmacia Biotech AB.

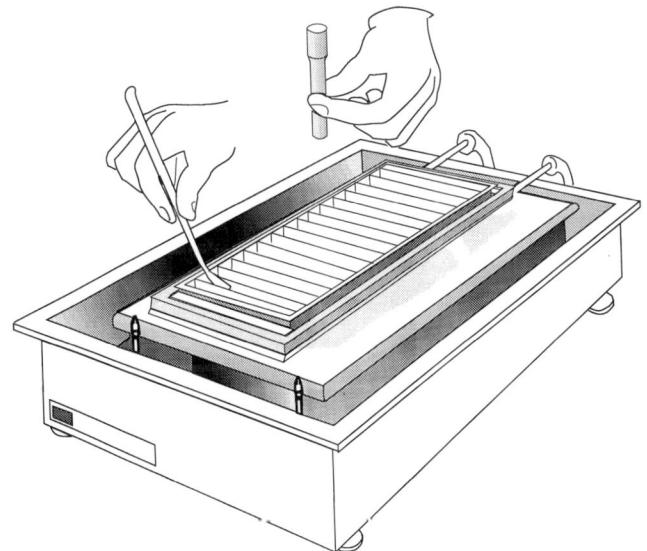

Fig. 4. Scraping the protein sections from the gel using the fractionation grid.

2. Measure the pH gradient by inserting a surface electrode into every second compartment of the 20 fractions of the grid. Wait until the reading has stabilized. For exact results, the electrode has to be calibrated at 10°C. Always start at the cathodal side of the gradient because carbon dioxide

from the air will slowly diffuse into the gel surface and the HCO_3^- ions will shift the measured pH to a lower value.

3.6. Collection of the Zones

1. Place the print near the tray to locate the zones of interest.
2. Scrape out relevant bands, and transfer to an elution column with a spatula.
3. Resuspend the gel in the column with a suitable buffer, and leave to settle until all buffer has entered the gel.
4. Add one gel volume of buffer, and elute the protein from the gel. Alternatively, all fractions can be collected with the help of the fractionating grid (Fig. 4) and then transferred to the elution columns.

3.7. Separation of the Protein from the Carrier Ampholytes

In most cases, it is not necessary to remove the carrier ampholytes from the protein fractions since:

1. They do not interfere with activity measurements and binding studies.
2. They are not toxic.
3. The fraction can be directly injected for antibody production.

If necessary, several methods can be employed for separating the protein from the carrier ampholytes, such as gel filtration (Chapter 11), ultrafiltration (Chapter 12), dialysis (Chapter 11), ammonium sulphate precipitation (Chapter 10), and electrophoresis (Chapters 33 and 34).

4. Notes

1. Ultrodex is a specially prepared dextran gel with very weak electroendosmotic effects. Untreated Sephadex G-75 or G-100 superfine gels may contain varying amounts of charged, water-soluble contaminants that interfere with the formation of the pH gradient. These gels have to be washed with 10 vol of distilled water and filtered under gentle vacuum.
2. Do not use < 2% (w/v) carrier ampholytes. Otherwise distortions and uneven bands appear in the gel during focusing. Using a higher concentration of carrier ampholytes generally does not improve the separation or the loading capacity.
3. The water content of the final gel bed is very important. If it is too low, the gel cracks. If it is too high, the focused proteins sediment to the lower part of the gel bed. The optimal consistency of the gel bed is controlled by weighing the initial slurry and repeated weighing during an evaporation until a certain percentage of the original weight is reached. This percentage is calculated using the defined evaporation limit, which is indicated on each Ultrodex bottle. For example, a typical evaporation limit might be

Isoelectric Focusing

34%. Suppose that the weights of the beaker plus slurry before and after filling the tray were 162.6 and 60.3 g, respectively, i.e., 102.3 g of slurry were added to the tray. The evaporation limit is 34% of this, i.e., 34.8 g, and the target weight of slurry is 67.5 g. If the initial weight of the tray and strips was 145.1 g, then the desired final weight after evaporation would be 145.1 + 67.5 = 212.6 g.

4. Alternative to the weighing procedure, the water content of the gel bed can be adjusted in the following way: Cover the mouth of the bottle of Ultrodex with a fine gauze net, and fix it with an elastic rubber ring. Sprinkle the dry material onto the surface until the correct consistency has been achieved. This is controlled by tilting the tray to an angle of 45°; when the gel surface does not move, the water content is correct.
5. If the gel cracks during the run in spite of all the precautions described, check the cooling efficiency of the system; the water flow should be 4–12 L/min at 4–10°C.
6. In the case of very labile proteins, sample application should be performed after a prefocusing step of 30 min at a constant 8 W. In this way, the sample can be loaded at a defined optimal pH value that can be predetermined from a previous analytical experiment. Often it is very helpful to check the literature to find out the optimal place of sample application.
7. When oxygen-sensitive enzymes or very basic proteins are being separated, flush the chamber with a light stream of nitrogen during the focusing experiment.
8. Some enzymes have to be separated at a lower temperature to maintain their biological activity, e.g., at 4°C. pH measurement is then almost impossible, because the electrode would respond extremely slowly at this low temperature. The pH values must always be measured at the focusing temperature because the pK_a values of the carrier ampholytes and of the proteins are highly dependent on temperature.
9. Alternatively or additionally to general protein staining of the paper print, specific staining for glycoproteins or zymogram techniques can be employed. The print can be cut into several strips and stained using different methods.
10. If skew bands are produced, then either the salt content of the sample could be too high or there could be too much electrolyte in the IEF electrode strips. Desalt the sample and/or blot the IEF electrode strips on a filter paper before placing them into the tray.
11. Irregular bands can also appear owing to overloading in a zone; reduce sample concentration, or choose a narrower pH range.
12. In some cases, band distortions can be cured by adding nonionic detergents (e.g., 0.5% [v/v] Triton X-100), 10% [v/v] glycerol, 7% [v/v] monoethyleneglycol, or 3–7M urea to the the gel and the sample.

References

1. Righetti, P. G. (1983) *Isoelectric Focusing: Theory, Methodology and Applications.* Elsevier Biomedical, Amsterdam.
2. Righetti, P. G. (1990) *Immobilized pH Gradients: Theory and Methodology.* Elsevier, Amsterdam.
3. Andrews, A. T. (1986) E*lectrophoresis: Theory, Techniques and Biochemical and Clinical Applications.* 2nd ed. Clarendon, Oxford.
4. Radola, B. J. (1969) Thin-layer isoelectric focusing of proteins. *Biochim. Biophys. Acta* **194,** 335–338.
5. Delincee, H. and Radola, B. J. (1970) Thin-layer isoelectric focusing on Sephadex layers of horseradish peroxidase. *Biochim. Biophys. Acta* **200,** 404–407.
6. Delincee, H. and Radola, B. J. (1970) Some size and charge properties of tomato pectin methylesterase. *Biochim. Biophys. Acta* **214,** 178–189.
7. Radola, B. J. (1970) Thin-layer isoelectric focusing of sarcoplasmic proteins from irradiated meat. *Report Eur.* **4695,** 73.
8. Gordon, B. J. and Dykes, P. J. (1972) Alpha$_1$-acute-phase globulins of rats: microheterogeneity after isoelectric focusing. *Biochem. J.* **130,** 95.
9. La Gow, J. and Parkhurst, L. J. (1972) Kinetics of carbon monoxide and oxygen binding for eight electrophoretic components of sperm-whale myoglobin. *Biochemistry* **24,** 4520–4525.
10. Coffer, A. J. and King, J. B. (1974) Isoelectric focusing of ^{3}H Oestradiol-17s receptor in flat beds of Sephadex. *Bioch. Soc. Trans.* **2,** 1269–1272.
11. Thorpe, R. and Robinson, D. (1975) Isoelectric focusing of isoenzymes of human liver alpha-L-fucosidase. *FEBS Lett.* **54,** 89–92.
12. Parkhurst, L. J. and La Gow, J. (1975) Kinetic and equilibrium studies of the ligand binding reactions of eight electrophoretic components of sperm-whale ferrimyoglobin. *Biochemistry* **14,** 1200–1205.
13. Steinmeier, R. C. and Parkhurst, L. J. (1975) Kinetic studies on the five principal components of normal adult human hemoglobin. *Biochemistry* **14,** 1564–1572.
14. Vacca, C. V., Hall, P. W. III, and Crowley, A. Q. (1979) Preparative electrofocusing in the isolation and purification of human betas microglobulin, in *Electrofocus 78* (Haglund, H., Westerfeld, J., and Ball, J., Jr., eds.), Elsevier, North Holland, pp. 49–51.
15. Bours, J., Garbers, M., and Hockwin, O. (1980) Preparative flat-bed isoelectric focusing of LDH, MDH, SDH, and G-6-PDH from rabbit lens, in *Electrophoresis 79* (Radola, B. J., ed.), Walter de Gruyter, Berlin, pp. 539–544.
16. Winter, A., Perlmutter, H., and Davies, H. (1980) Preparative flat-bed electrofocusing in a granulated gel with the LKB 2117 Multiphor. *LKB Application Note* 198.
17. Westermeier, R. (1993) *Electrophoresis in Practice.* VCH, Weinheim, Germany.
18. Righetti, P. G. and Caravaggio, T. (1976) Isoelectric points and molecular weights of proteins. A table. *J. Chromatogr.* **127,** 1–28.
19. Righetti, P. G., Tudor, G., and Ek, K. (1981) Isoelectric points and molecular weights of proteins. A new table. *J. Chromatogr.* **220,** 115–194.

CHAPTER 24

Chromatofocusing

Timothy J. Mantle and Patricia Noone

1. Introduction

Chromatofocusing *(1)* is essentially ion-exchange chromatography (conventionally using an anion exchanger) where the elution conditions are obtained by dropping the pH so that the proteins elute in order of their isoelectric points (pI). Although in principle this can be achieved by equilibrating an ion exchanger at a relatively alkaline pH and then titrating the resin with a buffer at a lower, more acidic pH, in practice it is more usual to utilize amphoteric buffers titrated to the lower pH to generate a more linear pH gradient. Since the pH gradient is generated gradually as the eluting buffer moves down the column, it is possible to add large volumes to the column or to add a second batch during the run and, in either case, single identical components with a particular pI will focus and elute as a single peak.

In practice this method is extremely straightforward since no gradient makers are required and the separation is normally superior to ion-exchange chromatography.

The protocol described here allows the purification of five major forms of pig liver glutathione *S*-transferase (GST) by chromatofocussing on polybuffer exchanger PBE 94 (*see* Note 1) following the preparation of a GST affinity pool using *S*-hexylglutathione-Sepharose. It is typical of the method used for proteins with pI values in the range 6.5–8.5 (*see* Note 2 for variations in the protocol for proteins with pI values outside this range).

2. Materials

2.1. Buffers

Chromatofocusing buffers are available commercially as polybuffers (Pharmacia LKB, Uppsala, Sweden). This protocol uses Polybuffer 96 (*see* Note 1) which has a characteristic low absorbance at 280 nm, but can be detected at 254 nm. For this reason eluent from a chromatofocusing column must be measured at 280 nm. All other reagents are made up in distilled deionized water at 20°C, unless otherwise indicated.

1. Buffer A: 10% (v/v) Polybuffer 96, pH 6.0 (adjusted with glacial acetic acid).
2. Buffer B: $0.025M$ ethanolamine, pH 9.4 (adjusted with glacial acetic acid).
3. Buffer C: $0.1M$ sodium phosphate, pH 6.5.
4. Buffer D: $0.01M$ sodium phosphate, pH 7.2.
5. Buffer E: 0.5% (w/v) SDS, 0.006% (w/v) bromophenol blue, 50% (v/v) glycerol, 63 mM Tris-HCl, pH 6.8, 0.5% (v/v) 2-mercaptoethanol added immediately prior to use.

2.2. Resins

Pack a 25 × 1.5-cm diameter column (total bed volume 44.3 mL, total capacity 3.5 meq/100 mL of resin) with PBE 94 (*see* Note 2).

2.3. Electrophoresis

For SDS-PAGE, Analar-grade chemicals are used with the buffer system of Laemmli *(2)*. A 12% resolving gel was used in the present case.

3. Method

1. Pre-equilibrate the gel to be used (PBE 94) with start buffer (buffer B) and pack into the column.
2. Wash the column with buffer B until the pH of the eluent is the same as that of buffer B (approx 10 column volumes).
3. Precede the application of the sample by the application of 5 mL of the eluent buffer; this ensures the protein sample is never exposed to an extreme of pH.
4. Apply the sample at a flow rate of 1 mL/min (*see* Note 3).
5. Elute by the application of the eluent buffer to the column. The volume of the elution buffer depends on the strength of the eluent solution. For pH 6.0–9.0, an elution buffer volume of 10.5 column volumes is recommended (*see* Note 4).
6. Assay the fractions for the protein of interest (in this case GST; *see* Note 5).
7. Monitor the protein concentration by measuring the absorbance at 280 nm. It is assumed that an absorbance reading of 1 is equivalent to a protein concentration of 1 mg/mL *(4)*. A typical elution profile is shown in Fig. 1.

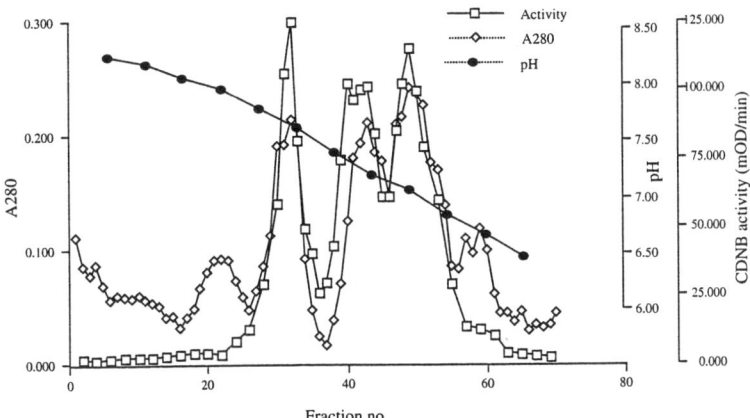

Fig. 1. Chromatofocusing of hepatic porcine GSTs, pH 6.0–9.0.

8. Prepare the samples for electrophoresis by adding buffer D to give a final protein concentration of 0.1 mg/mL (*see* Note 6).
9. Add buffer E (5 µL) to 20 µL of each sample, mix, and boil on a heating block for 5 min to ensure complete inactivation of proteases and total protein denaturation by SDS.
10. Load the samples (10 µL) on the gel using a Hamilton syringe. Run the gel at 30 mA until the dye front has run off the gel. An example is shown in Fig. 2.
11. Remove polybuffer, if required, by extensive dialysis against buffer D or by gel filtration on Sephadex G-25 (*see* Note 7 and Chapter 11).

4. Notes

1. For most work all that is required is the PBE 94 and the two amphoteric buffers, Polybuffer 96 and Polybuffer 74, which cover the pH ranges 9.0–6.0 and 7.0–4.0, respectively. If the protein of interest has a more basic pI, then it will be necesssary to invest additionally in PBE 118 and Pharmalyte pH 8.0–10.5. All of these materials can be obtained from Pharmacia. Sophisticated columns are available commercially together with a range of peristaltic pumps and on-line monitors for A_{280} and pH. However we have always used successfully a long straight glass column with a sinter that allows a good flow rate and gravity.
2. For most separations a bed volume of approx 20 mL is sufficient for samples containing up to 200 mg of protein; however this is only a guide and may require adjusting. If the pI of the target protein is known (from isoelectric focusing) then Table 1 should be consulted for a suitable start buffer. If there is no information on the pI of the target protein, then it is probably sensible to start with PBE 94 and use the pH range 7.0–4.0

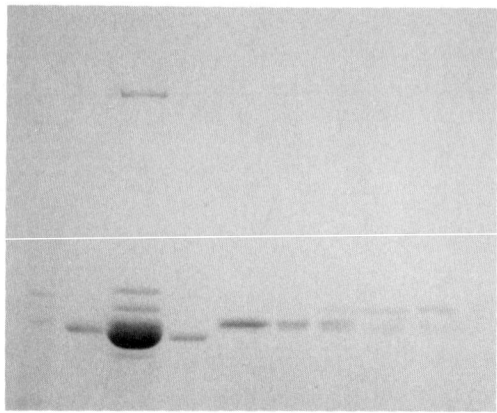

Fig. 2. Chromatofocusing of hepatic porcine GSTs, pH 6.0–9.0 analyzed by SDS-PAGE. Cytosolic GSTs were applied to a chromatofocusing column as described in Section 3. The peaks from the column were applied to a 12% SDS-PAGE gel and run as described. Lane 1, molecular weight markers; Lane 2, purified pig GST pi; Lane 3, mixture of cytosolic GSTs as purified on S-hexyl-glutathione-Sepharose; Lane 4, flow-through from the chromatofocusing column; Lanes 5–9, protein and activity peaks from the chromatofocusing column.

Table 1
Buffers Used in Chromatofocusing[a]

pH range	Start buffer	Eluent	Dilution factor
10.5–8.0[b]	pH 11.0, 0.25M tri-ethylamine/HCl	pH 8.0, Pharmalyte pH 8.0–10.5/HCl	1:45
9.0–6.0	pH 9.4, 0.025M ethanolamine/HCl	pH 6.0, Polybuffer 96/ acetic acid	1:10
7.0–4.0	pH 7.4, 0.025M imadazole/HCl	pH 4.0, Polybuffer 74/ HCl	1:8

[a]The gradient volume should be 10 times the column volume in all cases.
[b]The exchanger used in this case is PBE 118; in the other cases it is PBE 94 (see Note 1).

(Polybuffer 74). If the target protein does not bind then try using the pH range 9.0–7.0 (Polybuffer 96). By using these two sets of conditions (pH range 9.0–4.0), the pI of at least 85% of all known proteins (3) will have been covered. If the target protein is still not binding to the resin, then it is a very basic protein and PBE 118 should be used. When the choice of

exchanger and start buffer has been made, equilibration with start buffer should be commenced. This can take place in the column or more rapidly in a sintered glass funnel. In either case equilibration usually requires 10–15 bed volumes of start buffer. Equilibration should always be checked so that the pH and conductivity of the eluent is identical with that of the start buffer.
3. The sample volume should normally be within 0.5 bed volumes, although the sample volume is unimportant as long as all of the sample is applied to the column before the component of interest has eluted. The sample should not contain large amounts of salt and the ionic strength should be <0.05M. If the buffer concentration of the sample is low then the pH of the sample is unimportant, however it is normal to load the sample in either start buffer or eluent buffer, whichever maintains the activity of the component of interest. The loading and elution sequence is:
   a. 5 mL of eluent buffer;
   b. Sample (in eluent or start buffer);
   c. Eluent buffer.
4. Elution is achieved simply by running 10 column volumes of eluent buffer through the column. The flow rate can be variable and may be regulated by a peristaltic pump or by gravity. Protein should be monitored at 280 nm and the pH of each fraction measured immediately. There is a dead volume of 1.5–2 column volumes so in practice it is usual to use 12 column volumes.
5. The assay conditions for the activity of GST have been described previously *(4)*. Reactions are monitored by following a change in absorbance when substrate is conjugated to glutathione.

    Assays are carried out in a final volume of 2 mL in a quartz cuvet at 30°C. The absorbance change at 340 nm is recorded with a Philips (Eindhoven, Holland) PU8625 UV/VIS spectrophotometer attached to a Philips PM 8261 Xt recorder. Glutathione (GSH; 40 mM stock solution) is prepared freshly in buffer C. 1-Chloro-2,4-dinitrobenzene (CDNB; 40 mM), the electrophilic substrate, is dissolved in ethanol. Buffer C (1.8 mL) incubated at 30°C is placed in the cuvet, glutathione (50 µL) and CDNB (50 µL) are then added. A blank rate is observed before eluent fraction (100 µL) is added.

    Alternatively the assays are carried out in a microwell plate (Nunclon, Kamstrup, Denmark) with a final volume of 200 µL. The absorbance change is monitored at 340 nm (30°C) using a Molecular Devices Thermomax plate reader attached to a Macintosh SE 1/40. The results are analyzed using the Softmax® software. Buffer C (140 µL) is applied to each well followed by protein (10 µL) except for the blanks where buffer C (10 µL) is added. CDNB (0.5 mL) and GSH (0.5 mL) are added to buffer C (9 mL). The reaction mix (50 µL) is added to each well and the absorbance change

is observed for 2 min at 6-s intervals. This second method is particularly useful if only small quantities of protein are obtained.
6. If the protein concentration is <0.1 mg/mL, the sample can be concentrated using trichloroacetic acid (TCA). The desired amount of protein solution is mixed with an equal volume of ice-cold TCA (25% [w/v]). The samples are incubated on ice for 20 min and spun in a microcentrifuge at 12,000g for 3 min. The supernatant is discarded and the pellet is washed twice in ice-cold acetone (1 mL) and allowed to dry. The pellet is resuspended in buffer D (20 µL).
7. There are a number of ways that removal of polybuffer from the sample can be achieved:
    a. Precipitation with ammonium sulphate. Add sufficient solid ammonium sulphate to obtain 85% saturation (608 g/L). After 1 h precipitate the suspension using centrifugation, wash twice with saturated ammonium sulphate, and then store as an ammonium sulphate suspension or dialyse/gel filter to remove the ammonium sulphate. All of these steps should be conducted at 4°C.
    b. Polybuffer can be removed from most proteins by gel filtration providing that the protein of interest has an M_r >15,000.
    c. If the protein of interest binds to an affinity matrix then this is the most ideal way of obtaining rapid andcomplete separation.
8. The following cautions should be noted:
    a. In the absence of any data on the pI of the protein of interest, it is well worth a number of preliminary experiments to optimize the binding and eluting conditions.
    b. Care should be taken with regard to the buffer counterion which should have a pK_a at least 2.0 pH units below the lowest point of the chosen gradient. For this reason chloride is the counterion of choice.
    c. Atmospheric carbon dioxide may casuse a plateau in the region pH 5.5–6.5 and if this is a problem all buffers should be degassed.

References

1. Sluyterman, L. A. and Widnes, J. (1977) Chromatofocussing: isoelectric focussing on ion exchangers in the absence of an externally applied potential, in *Proceedings of the International Symposium Electrofocusing and Isotachophoresis* (Radola, B. J. and Graesslin, D., eds.), deGruyter, Berlin, pp. 463–466.
2. Laemmli, U. K. (1970) Cleavage of structural proteins during the assembly of the head of bacteriophage T4. *Nature* **227,** 680–685.
3. Gianazza, E. and Righetti, P. G. (1980) Size and charge distribution of macromolecules in living systems. *J. Chromatogr.* **193,** 1–8.
4. Habig W. H., Pabst, M. J., and Jakoby, W. B. (1974) Glutathione S-transferase: the first enzymatic step in mercapturic acid formation. *J. Biol. Chem.* **249,** 7130–7139.

CHAPTER 25

Size-Exclusion Chromatography

Paul Cutler

1. Introduction

Size-exclusion chromatography (also known as gel-filtration chromatography) is a technique for separating proteins and other biological macromolecules on the basis of molecular size. Size-exclusion chromatography is a commonly used technique owing to the diversity of the molecular weights of proteins in biological tissues and extracts. It also has the important advantage of being compatible with physiological conditions *(1,2)*.

The solid-phase matrix consists of porous beads (100–250 µm) that are packed into a column with a mobile liquid phase flowing through the column (Fig. 1). The mobile phase has access to both the volume inside the pores and the volume external to the beads. The high porosity typically leads to a total liquid volume of >95% of the packed column.

Separation can be visualized as reversible partitioning into the two liquid volumes. Large molecules remain in the volume external to the beads, since they are unable to enter the pores. The resulting shorter flow path means that they pass through the column relatively rapidly, emerging early. Proteins that are excluded from the pores completely elute in what is designated the void volume, V_o. This is often determined experimentally by the use of a high-mol-wt component, such as Blue Dextran or calf thymus DNA. Small molecules that can access the liquid within the pores of the beads are retained longer and, therefore, pass more slowly through the column. The elution volume for material included in the pores is designated the total volume, V_t. This represents the total liquid volume of the column and is often determined by small molecules, such as vitamin B12.

From: *Methods in Molecular Biology, Vol. 59: Protein Purification Protocols*
Edited by: S. Doonan Humana Press Inc., Totowa, NJ

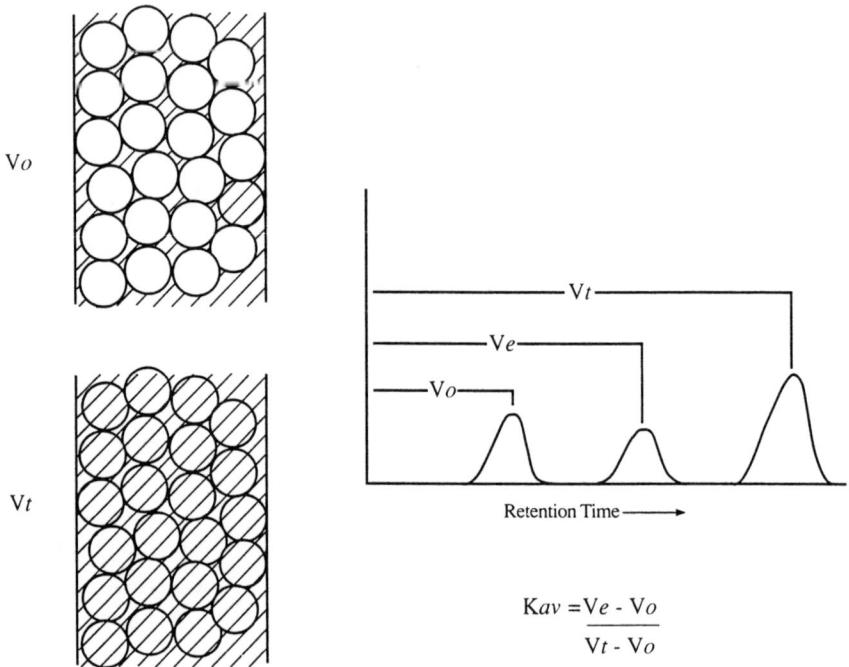

Fig. 1. The basic principle of size exclusion. Solutes are separated according to their molecular size. Large molecules are eluted in the void volume (V_o), and small molecules are eluted in the total volume (V_t). Solutes within the separation range of the matrix are fractionally excluded with a characteristic elution volume (V_e).

The elution volume for a given protein will lie between V_o and V_t and is designated the elution volume, V_e. Intermediate-sized proteins will be fractionally excluded with a characteristic value for V_e. A partition coefficient can be determined for each protein as K_{av}. In size exclusion, the macromolecules are not physically retained, unlike adsorption techniques. Therefore, the protein will elute in a defined volume between V_o and V_t (see Note 1). Size exclusion tends to be used at the end of a purification scheme when impurities are low in number, and the target protein has been purified and concentrated by earlier chromatography steps. An exception to this is membrane proteins, where gel filtration may be used first since concentration techniques are not readily used and the material will be progressively diluted during the purification scheme (3).

A range of different preparative and analytical matrices are available commercially. High-performance columns are used analytically for studying protein purity, protein folding, protein–protein interactions, and so forth. *(4)*. Preparative separations performed on low-pressure matrices are used to resolve proteins from proteins of different molecular weights, proteins from other biological macromolecules, and for the separation of aggregated proteins from monomers (*see* ref. *5* and Note 2).

Several parameters are important in size-exclusion chromatography. The pore diameter controlling the separation is selected for the relative size of proteins to be separated. Many types of matrices are available. Some are used for desalting techniques where proteins are separated from buffer salts (*see* ref. *6* and Note 3). Some matrices offer a wide range of mol-wt separations, and others are high resolution matrices with a narrow range of operation (Table 1).

2. Materials
2.1. Equipment

The preparative separation of proteins by size exclusion is suited to commercially available standard low-pressure chromatography systems. Systems require a column packed with a matrix offering a suitable fractionation range, a method for mobile-phase delivery, a detector to monitor the eluting proteins, a chart recorder for viewing the detector response, and a fraction collector for recovery of eluted proteins (Fig. 2). The system should be plumbed with capillary tubing with a minimum hold-up volume.

Early systems were less sophisticated with a gravimetric feed of the mobile phase from a suspended reservoir, whereas the most modern systems now have computers to control operating parameters and to collect and store data. The principle of separation, however, remains the same, and high resolution is attainable with relatively simple equipment.

1. Pumps: An important factor in size exclusion is a reproducible and accurate flow rate. The most commonly used pumps are peristaltic pumps, which are relatively effective at low flow rates, inexpensive, and sanitizable. Peristaltic pumps do, however, create a pulsed flow, and often a bubble trap is incorporated both to prevent air from entering the system and to dampen the pulsing effect. More expensive, yet more accurate pumps are syringe pumps, such as those seen on the FPLC® system (Pharmacia, Uppsala, Sweden).
2. Column: Size exclusion, unlike some commonly used adsorption methods of protein separation, is a true chromatography method based on continuous partitioning. Hence, resolution is dependent on column length. Col-

Table 1
Commonly Used Preparative Size-Exclusion Matrices

Supplier/matrix	Material type, pH stability[a]	Separation range[b], kDa
Sephadex		
G25	Dextran, 2.0–10.0[d]	1–5
G50		1.5–30
G75		3–80[c]
G100		4–100[c]
G150		5–150[c]
G200		5–600[c]
Sepharose		
6B	Agarose, 3.0–13.0	10–4000
4B		60–20,000
2B		70–40,000
Superdex		
30	Agarose/dextran, 3.0–12.0	0–10
75		3–70
200		10–600
Sephacryl		
S100HR	Dextran/bisacrylamide, 3.0–11.0	1–100
S200HR		5–250
S300HR		10–1500
Biogel		
P-2 gel	Polyacrylamide, 2.0–10.0	0.1–1.8
P-4 gel		0.8–4
P-10 gel		1.5–20
P-60 gel		3–60
P-100 gel		5–100

[a]In aqueous buffers.
[b]For globular protein.
[c]Different grades are available, which affects performance.
[d]pH stability may vary depending on grade and exclusion limit.

umns tend toward being long and thin, typically 70–100 cm long (*see* Note 4). The column must be able to withstand the moderate pressures generated during operation and be resistant to the mobile phase. The use of columns with flow adapters is recommended to allow the packing volume to be varied and provide a finished support with the required minimum of dead space.

3. Detectors: Protein elution is most often monitored by absorbance in the UV range, either at 280 nm, which is suitable for proteins with aromatic amino acids, or at 206 nm, which detects the peptide bond. Detection at lower wavelengths may be complicated by the absorbance characteristics

Size-Exclusion Chromatography

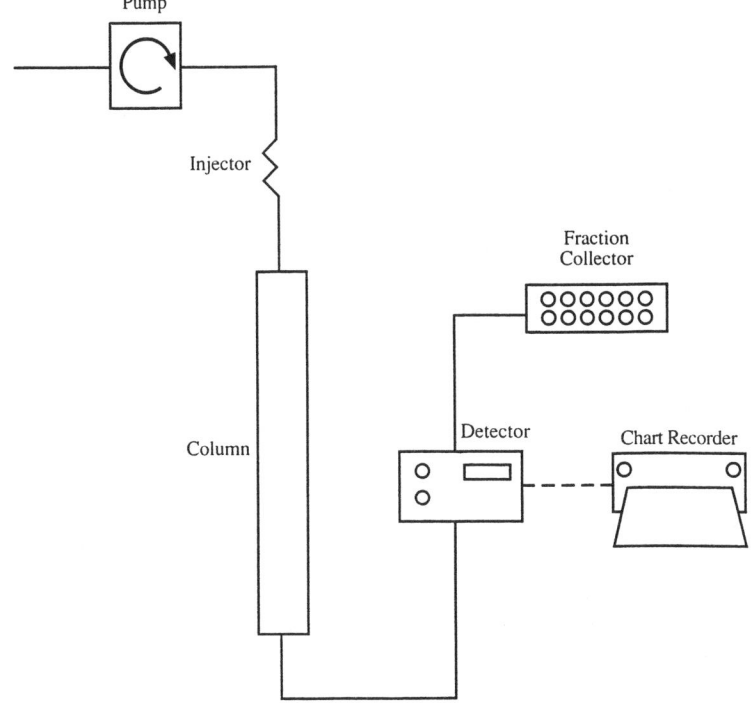

Fig. 2. Schematic diagram of the equipment for preparative size-exclusion chromatography.

of certain mobile phases. The advent of diode array detectors has enabled continuous detection at multiple wavelengths, enabling characterization of the elutes via analysis of spectral data.

Fluorescence detection either by direct detection of fluorescent tryptophan and tyrosine residues or after chemical derivitization (e.g., by fluorescein) has been used, as have refractive index, radiochemical and electrochemical methods, and molecular size (by laser light scattering). In addition to these nonspecific on-line monitoring systems, it is quite common, particularly when purifying enzymes to make use of specific assays for individual target molecules.

4. Fraction collectors: A key factor in preparative protein purification is the ability to collect accurate fractions. No matter how efficiently the column may have separated the proteins, the accurate collection of fractions is critical. For the detector to reflect as near as possible in real time the fraction collector, the volume between the detector and the fraction collector should be minimal.

2.2. Buffers

Size-exclusion matrices tend to be compatible with most aqueous buffer systems even in the presence of surfactants, reducing agents, or denaturing agents. Size-exclusion matrices are extremely stable with effective pH ranges of approx 2.0–12.0 (Table 1). An important exception to this are the silica-based matrices, which offer good mechanical rigidity, but low chemical stability at alkaline pHs. Some silica matrices have been coated with dextran, and so forth, to increase the chemical stability and increase hydrophilicity.

The choice of mobile phase is most often dependent on protein stability necessitating considerations of appropriate pH, solvent composition, as well as the presence or absence of cofactors, protease inhibitors, and so on, which may be essential to maintain the structural and functional integrity of the target molecule. All buffers used in size exclusion should ideally be filtered through a 0.2-μm filter, and degassed by low vacuum or sparging with an inert gas, such as helium (*see* Note 5).

Many matrices retain a residual charge owing, for example, to sulfate groups in agarose or carboxyl residues in dextran. The ionic strength of the buffer should be kept at $0.15–2.0M$ to avoid electrostatic or van der Waals interactions, that can lead to nonideal size exclusion *(7,8)*. Crosslinking agents, such as those used in polyacrylamide, may reduce the hydrophilicity of the matrix leading to the retention of some small proteins, particularly those rich in aromatic amino acid residues. These interactions have been exploited effectively to enhance purification in some cases, but are generally best avoided (*see* Note 6).

2.3. Selection of Matrix

The beads used for size exclusion have a closely controlled pore size, with a high chemical and physical stability. They are hydrophilic and inert to minimize chemical interactions between the solutes (proteins) and the matrix itself.

The performance and resolution of the technique have been enhanced by the development of newer matrices with improved properties. Historically gels were based on starch, although these were superseded by the crosslinked dextran gels (e.g., Sephadex). In addition, polystyrene-based matrices were developed for the use of size exclusion in nonaqueous solutions. Polyacrylamide gels (e.g., Biogel P series) are particularly suited to separation at the lower mol wt range owing to their microreticular structure. More recently, composite matrices have been developed, such as the Superdex® gel (Pharmacia), where dextran chains have been chemically bonded to a highly crosslinked agarose for high-speed size exclusion (Table 1).

3. Methods

3.1. Flow Rates

In chromatography, flow rates should be standardized for columns of different dimensions by quoting linear flow rate (cm/h). This is defined as the volumetric flow rate (cm^3/h)/unit cross-sectional area (cm^2) of a given column. Since the principle of size exclusion is based on partitioning, success of the technique is particularly susceptible to variations in flow rates. Conventional low-pressure size-exclusion matrices tend to operate at linear flow rates of 5–15 cm/h. Too high a flow rate leads to incomplete partitioning and band spreading. Conversely, very low flow rates may lead to diffusion and band spreading.

3.2. Preparation and Packing

3.2.1. Preparation

Gel matrices are supplied as either preswollen gels or as dry powder. If the gel is supplied as a dry powder, it should be swollen in excess mobile phase as directed by the manufacturers. Swollen gels must be transferred to the appropriate mobile phase. This can be achieved by washing in a scintered glass funnel under low vacuum. During preparation, the gel should not be allowed to dry. The equilibrated gel should be decanted in to a Buchner flask, allowed to settle, and then fines removed from the top by decanting. The equilibrated gel (approx 75% slurry) is then degassed under low vacuum.

3.2.2. Packing

Good column packing is an essential prerequisite for efficient resolution in size-exclusion chromatography. The column should be held vertically in a retort stand avoiding adverse drafts, direct sunlight, or changes in temperature. The gel should be equilibrated and packed at the final operating temperature. With the bottom frit or flow adaptor in place, degassed buffer (5–10% of the bed volume) is poured down the column side to remove any air from the system. In one manipulation, the degassed gel slurry is poured in to the column using a glass rod to direct the gel down the side of the column, avoiding air entrapment. If available, a packing reservoir should be fitted to columns to facilitate easier packing. The column can be packed under gravity, although a more efficient method is to use a pump to push buffer through the packing matrix. The flow rate during packing should be approx 50% higher than the operating flow rate (e.g., 15 cm/h for a 10 cm/h final flow rate). The bed height can be monitored by careful inspection of the column as it is packing. Once packed, a clear layer of buffer will appear above the bed, and the level of the gel will remain constant. The flow should then be stopped and excess buffer removed from the top of the column, leaving approx 2 cm buffer above the gel. The outlet from the column should be closed, and the top flow adaptor carefully placed on top of the gel avoiding trapped air or disturbing the gel bed.

3.3. Equilibration

The packed column should be equilibrated by passing the final buffer through the column at the packing flow rate for at least one column volume. The pump should always be connected so as to pump the eluent onto the column under positive pressure. Drawing buffer through the column under negative pressure may lead to bubbles forming as a result of the suction. The effluent of the column should be sampled and tested for pH and conductivity in order to establish equilibration in the desired buffer.

3.4. Sample Application

Several methods exist for sample application. It is critical to deliver the sample to the top of the column as a narrow sample zone. This can be achieved by manually loading via a syringe directly onto the column, although this requires skill and practice. The material may be applied through a peristaltic pump, although this will inevitably lead to band spreading owing to sample dilution. The sample should never be loaded through a pump with a large hold-up volume, such as a syringe pump or upstream of a bubble trap. Arguably the best method of applying the sample is via a sample loop in conjunction with a switching valve, allowing the sample to be diverted manually or electronically through the loop and directly onto the top of the column.

3.5. Evaluation of Column Packing

The partitioning process occurs as the bulk flow of liquid moves down the column. As the sample is loaded, it forms a sample band on the column. In considering the efficiency of the column, partitioning can be

Size-Exclusion Chromatography

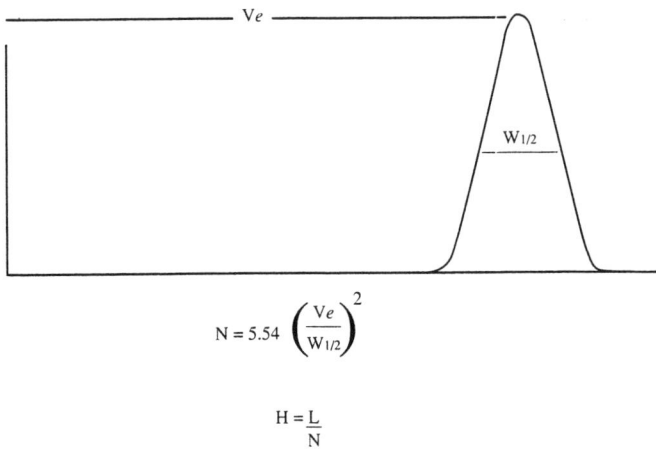

Fig. 3. Calculation of the theoretical plate height as a measure of column performance. The number of theoritical plates *(N)* is related to the peak width at half height ($W_{1/2}$) and the elution volume *(V_e)*. The height of the theoretical plate *(H)* is related to the column length *(L)*.

perceived as occurring in discrete zones along the axis of the column's length. Each zone is referred to as a theoretical plate, and the length of the zone termed the theoretical plate height *(H)*. The value of *H* is a function of the physical properties of the column, the exclusion limit, and the operating conditions, such as flow rate, and so forth. Column efficiency is defined by the number of theoretical plates *(N)*, which can be measured experimentally using a suitable sample, e.g., 1% (v/v) acetone (Fig. 3). Resolution, and hence, the number of theoretical plates are enhanced by increasing column length (*see* Note 7).

3.6. Standards and Calibration

Calibration is obtained by use of standard proteins and plotting retention time (V_e) against log molecular weight. Successful calibration requires accurate flow rates. The resultant plot gives a sigmoidal curve approaching linearity in the effective separation range of the gel.

It should be noted that it is not only the molecular weight that is important in size exclusion, but the hydrodynamic volume or the Stokes radius of the molecule. Globular proteins appear to have a lower molecular size than proteins with a similar molecular weight that are in an α helical form. These, in turn, appear smaller than proteins in random coil form. It

Table 2
Molecular-Weight Standards
for Size-Exclusion Chromatography

Vitamin B_{12}	1350
Ribonucelase A	13,700
Myoglobin	17,000
Chymotrypsinogen A	25,000
Ovalbumin	43,000
Bovine serum albumin	67,000
Bovine γ globulin	158,000
Blue dextran 2000	~2,000,000

is critical to use the appropriate standards for size exclusion where proteins have a similar shape. This has been useful in some cases for studying protein folding and unfolding. Commonly used mol-wt standards are given in Table 2.

3.7. Separation of Proteins

The optimum load of size-exclusion columns is restricted to <5% (typically 2%) of the column volume in order to maximize resolution. Gel-filtration columns are often loaded at relatively high concentrations of protein, such as 2–20 mg/mL. The concentration is limited by solubility of the protein and the potential for increased viscosity, which begins to have a detrimental effect on resolution. This becomes evident around 50 mg/mL. It is important to remove any insoluble matter prior to loading by either centrifugation or filtration. Owing to the limitation on loading, it is often wise to consider the ability of the method for scale-up when optimizing the operating parameters (ref. 9 and Fig. 4).

3.8. Column Cleaning and Storage

Size-exclusion matrices can be cleaned in situ or as loose gel in a sintered glass funnel. Suppliers usually offer specific guidelines for cleaning gels. Common general cleaning agents include nonionic detergents (e.g., 1% [v/v] Triton X-100) for lipids, and 0.2–0.5M NaOH for proteins and pyrogens (not recommended for silica-based matrices). In extreme circumstances, contaminating protein can be removed by use of enzymic digestion (pepsin for proteins and nucleases for RNA and DNA). The gel should be stored in a buffer with antimicrobial activity, such as 20% (v/v) ethanol or 0.02–0.05% (w/v) sodium azide. NaOH is a good storage

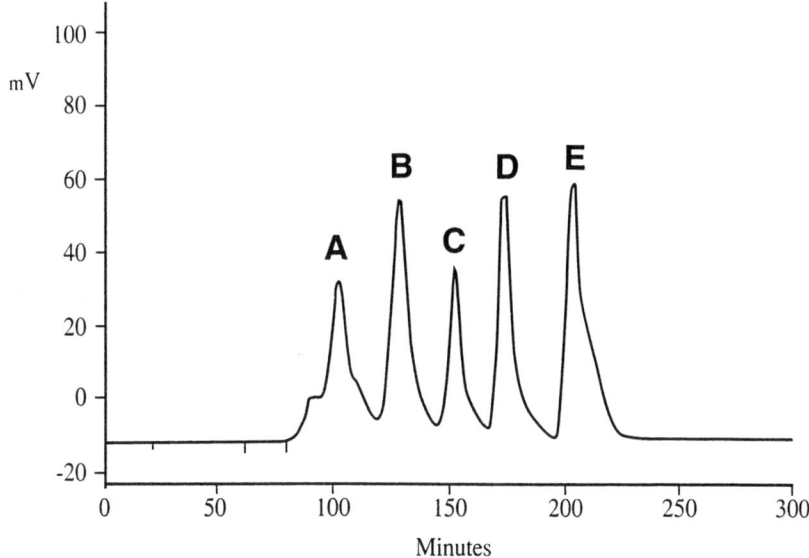

Fig. 4. Size exclusion of five mol-wt markers on Superdex 200 (Pharmacia) column (1.6 × 75 cm). Thyroglubin, 670,000 (A); γ globulin, 158,000 (B); ovalbumin, 44,000 (C); myoglobin, 17,000 (D), and vitamin B_{12}, 1350 (D).

agent, which combines good solubilizing activity with prevention of endotoxin formation. It may, however, lead to chemical breakdown of certain matrices.

4. Notes

1. If the protein elutes before the void volume ($V_e < V_o$), this suggests chanelling through the column owing to improper packing or operation of the column. If the protein elutes after the total volume ($V_e > V_t$), then some interaction must have occured between the matrix and the protein of interest.
2. Size exclusion is particularly suited for the resolution of protein aggregates from monomers. Aggregates are often formed as a result of the purification procedures used. Size exclusion is often incorporated as a final polishing step to remove aggregates and act as a buffer exchange mechanism into the final solution.
3. Desalting gels are used to remove low-mol-wt material rapidly, such as chemical reagents from proteins and for buffer exchange. Since the molecules to be separated are generally very small, typically < 1 kDa, gels are generally used with an exclusion limit of approx 2–5 kDa. The protein appears in the void volume (V_o) and the reagents and buffer salts are

retained. Because of the distinct mol-wt differences, the columns are shorter than other size-exclusion columns and operated at higher flow rates (e.g., 30 cm/h).
4. In some instances, the length of the column required to obtain a satisfactory separation exceeds that which can be packed into a commerically available column (>1 m). In these cases, columns can be packed in series. The tubing connecting the columns should be as narrow and as short as possible to avoid zone spreading.
5. The majority of protein separations performed using size exclusion are carried out in the presence of aqueous phase buffers. Size exclusion of proteins in organic phases (sometimes called gel permeation) is not normally undertaken, but is sometimes used for membrane protein separations. The agarose- and dextran-based matrices are not suitable for separations with organic solvents. Synthetic polymers, and to a certain extent, silica are suitable for separations in organic phases. For separation in acidified organic solvents, polyacrylamide matrices are particularly suitable. The separation of the proteins may be influenced by the denatured state of the protein in the organic phase. The equipment must be compatible with the solvent system, e.g., glass or Teflon™.
6. Interactions with the matrix are commonly seen when charged proteins are being resolved. Protein–matrix interactions can be minimized by use of ionic strength in excess of $0.1M$. If low ionic strength is necessary, then the risk of interaction can be reduced by manipulating the charge on the protein via the pH of the buffer. This is best achieved by keeping the mobile phase above or below the pI of the protein as appropriate. Protein–matrix interaction is a common cause of protein loss during size exclusion. This may be owing to complete retention on the column or retardation sufficient for the material to elute in an extremely broad dilute fraction, thereby evading detection above the baseline of the buffer system. Another important consideration is when enzymes are detected off line by activity assay. The active enzyme may resolve into inactive subunits. In such cases, a review of the mobile phase is advisable.
7. In addition to determination of the number of theoretical plates (N) described, the performance of the column can be assessed qualitatively by the shape of the eluted peaks (Fig. 5). The theoretically ideal peak is sharp and triangular with an axis of symmetry around the apex. Deviations from this are seen in practice. Some peak shapes are diagnostic of particular problems that lead to broadening and poor resolution. If the down slope of the peak is significantly shallow, it is possible that the concentration of the load was too high or the material has disturbed the equilibrium between the mobile and stationary phases. If the down slope tends to be symmetrical intially,

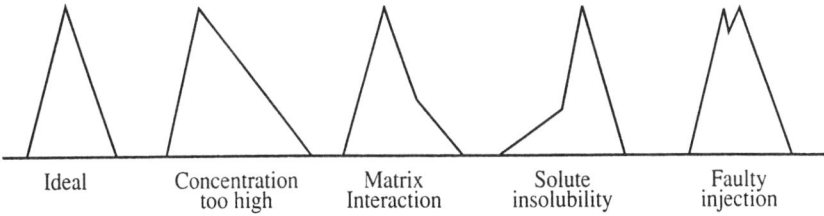

Fig. 5. Diagnosis of column performance by consideration of peak shape.

but then becomes shallow, it is common to assume that there is a poorly resolved component. However it may be suspected that interaction with the matrix is taking place. A shallow up slope may represent insolubility of the loaded material. A valley betwen two closely eluting peaks may suggest poor resolution, but can also be the result of a faulty sample injection.

References

1. Laurent, T. C. (1993) Chromatography classic: history of a theory. *J. Chromatogr.* **633,** 1–8.
2. Hagel, L. (1989) Gel filtration, in *Protein Purification* (Jansen, J.-C. and Ryden, L., eds.), VCH, New York, pp. 63–106.
3. Findlay, J. B. C. (1990) Purification of membrane proteins, in *Protein Purification Applications* (Harris, E. L. V. and Angal, S., eds.), IRL, Oxford, UK, pp. 59–82.
4. Hagel, L. (1993) Size-exclusion chromatography in an analytical perspective. *J. Chromatogr.* **648,** 19–25.
5. Stellwagen, E. (1990) Gel filtration. *Methods Enzymol.* **182,** 317–328.
6. Pohl, T. (1990) Concentration of proteins and removal of solutes. *Methods Enzymol.* **182,** 69–83.
7. Kopaciewicz, W. and Regnier, F. E. (1982) Nonideal size exclusion chromatography of proteins: effects of pH at low ionic strength. *Anal. Biochem.* **126,** 8–16.
8. Dubin, P. L., Edwards, S. L., Mehta, M. S., and Tomalia, D. (1993) Quantitation of non-ideal behavior in protein size-exclusion chromatography. *J. Chromatogr.* **635,** 51–60.
9. Jansen, J.-C. and Hedman, P. (1987) Large scale chromatography of proteins. *Adv. Biochem. Eng.* **25,** 43–97.

CHAPTER 26

Fast Protein Liquid Chromatography (FPLC) Methods

David Sheehan

1. Introduction

High-resolution protein separation techniques critically depend on the availability of column packings of small average particle size. This gives a minimum of peak broadening on the column owing to the direct relationship between the theoretical plate height parameter, H, and particle size (lowest values of H give highest resolution). High-performance liquid chromatography (HPLC) procedures exploit column packings with average diameters of as little as 5–40 µm. However, these are used in high-pressure systems (up to 400 bar) often with organic solvents and are generally limited to rather low sample loadings *(1)*. To provide a more biocompatible high-resolution separation of biopolymers, including (although not exclusive to) proteins, Pharmacia LKB (Uppsala, Sweden) developed fast protein liquid chromatography (FPLC) in 1982 *(2)*.

FPLC provides a full range of chromatography modes, such as ion exchange, chromatofocusing *(3)*, gel filtration, hydrophobic interaction *(4)*, and reverse phase *(5)*, based on particles with average diameter sizes in the same range as those used for HPLC separations. These columns can accommodate much higher protein loadings than conventional HPLC, however, and use a wide range of aqueous, biocompatible buffer systems. Although almost 1000 reports of the use of FPLC have appeared in the literature in the decade since its introduction, two of the most popular FPLC modes are ion exchange and gel filtration. The objective of this chapter is to introduce the reader to these two FPLC chromatography

From: *Methods in Molecular Biology, Vol. 59: Protein Purification Protocols*
Edited by: S. Doonan Humana Press Inc., Totowa, NJ

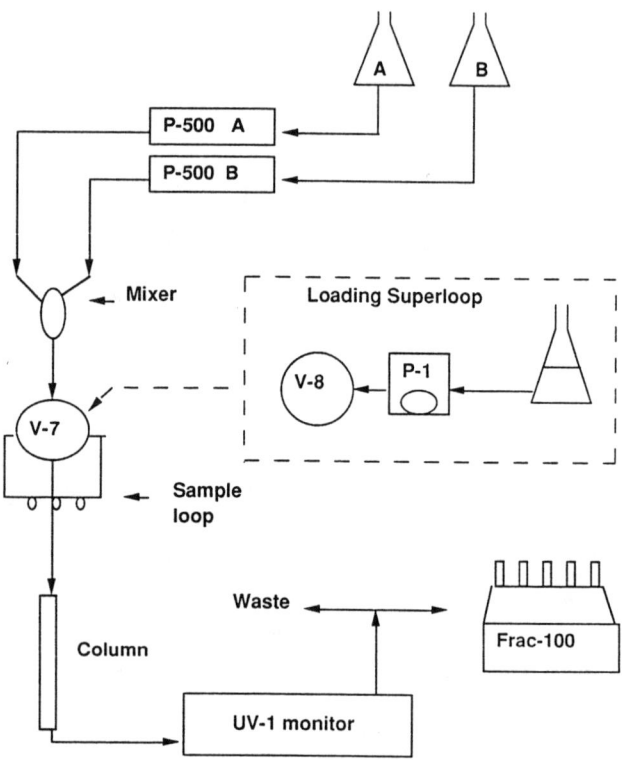

Fig. 1. Format used for FPLC chromatography. Information on other accessories and automation may be obtained from Pharmacia LKB. Superloop may be loaded using the arrangement shown within dashed lines.

modes as used in a typical protein chemistry laboratory. References *3–5* give examples of the use of the other main FPLC chromatography modes, which are not described in detail here.

2. Materials

1. The FPLC system consists of a program controller (LCC 500 Plus), two P-500 pumps (one each for buffers A and B), a mixer, prefilter, seven-port M-7 valve, assorted sample loops (0.025–10 mL), column, UV-1 UV monitor (fitted with an HR-10 flow cell and 280-nm filter), and a Frac-100 fraction collector. Associated equipment includes a superloop attached to an eight-port M-8 valve and a P-1 peristaltic pump for the introduction of samples up to 10-mL volume. All of this equipment is available from Pharmacia LKB. The most commonly used configuration for ion-exchange FPLC is shown in Fig. 1.

FPLC Methods 271

2. Mono Q HR5/5 anion-exchange, Mono S HR5/5 cation exchange, HR10/10 rapid desalting, and Superose 12 HR 10/30 gel-filtration columns are obtained from Pharmacia LKB, and are stored in 20% ethanol. Sephadex G-25 is also obtained from Pharmacia LKB. A glass column (45 × 3 cm in diameter) is required. Filters (0.22 µm) are from Millipore.
3. All reagents for buffer (*see* Notes 3 and 5) preparation are of Analar grade. Buffers should be prepared in HPLC or milli-Q-grade water, and require filtering and degassing immediately before use.
   Buffer A: 10 mM Tris-HCl, pH 7.0.
   Buffer B: 10 mM Tris-HCl, pH 7.0, 1M NaCl.
   Buffer C: 50 mM Tris-HCl, pH 7.0, 100 mM KCl.

3. Methods
3.1. Sample Preparation

Generally, FPLC chromatography is performed on material that has already been subjected to some preliminary chromatography steps (e.g., for ion exchange using cellulose matrices, *see* Chapter 14; for affinity chromatography, *see* Chapter 16). For ion-exchange chromatography, the sample must then be desalted into buffer A (*see* Note 2). This may be accomplished by passing it through a column of Sephadex G-25 (45 × 3 cm in diameter, total bed volume 318 mL).

1. New Sephadex G-25 resin is de-fined by aspiration before packing as a slurry in water into glass column with a no. 1 glass sinter.
2. Column is equilibrated in three to four column volumes of buffer A. Equilibration is complete when the pH and conductivity of eluate are the same as buffer A at the same temperature. The sample (50 mL) is loaded under gravity flow. Buffer A is passed through the column. Protein is monitored by measuring A_{280}, and collected as a single large peak.
3. Conductivity and pH of desalted protein should be the same as those of buffer A at the same temperature.

Sephadex G-25 gel filtration is performed at 4°C in a cold room or cabinet (*see* Note 6). Sample is centrifuged (bench centrifuge, 3–5 min) before application to column.

3.2. Ion-Exchange FPLC

The FPLC system is operated usually at room temperature, taking care to hold samples on ice and to return eluted protein fractions to ice as soon as possible (*see* Note 6).

1. With no column in system (this may be replaced with some tubing), prime P-500 pumps A and B with filtered (0.22-µm filter) and degassed buffers A and B, respectively.

2. Set pressure limits on both P-500 pumps well below the maximum for column in use (*see* Notes 4 and 11).
3. Equilibrate Mono Q column (1-mL vol) with 5 vol buffer A and 10 vol buffer B followed by 5 vol buffer A.
4. Wash loading loop with buffer A.
5. Load 0.5–10 mL sample (approx 1 mg/mL), and wash with buffer A. Collect wash through, and assay for protein of interest.
6. If protein has not bound, immediately replace Mono Q with Mono S, and repeat procedure described in steps 3–5 (generally, at pH 7.0, proteins that do not bind to one resin will bind to the other). If protein still does not bind, follow procedure given in Section 3.3.
5. Mono Q (or Mono S) column is developed with a 0–100% gradient of buffer B. Elution of protein is determined by A_{280} while protein of interest is assayed.
6. Regenerate column by washing with 10 vol buffer B, followed by 5 vol buffer A. Column is now ready to receive another sample.

3.3. Scouting FPLC Methods

The aforementioned procedure describes a basic FPLC ion-exchange chromatography experiment. If the protein of interest has not previously been purified using FPLC, a new method of purification may be needed. Usually, 25–500 µL loadings are used to scout a new procedure. The general approach used is as follows.

1. By varying % B at different time-points, create gradients of varying degrees of shallowness (down to and including isocratic). Knowing % B at which the protein of interest elutes, separation from other proteins can be improved by varying the gradient program around this value *(6)*.
2. Having identified column (either Mono Q or Mono S) to which protein of interest binds, carry out chromatography at different pH values using various buffer systems (Table 1; *see* Notes 3 and 5). By carrying out rechromatography of a single sample at a number of different pH values on a Mono Q or Mono S column, this can often separate two proteins that may elute close to each other at a single pH.
3. Having identified a suitable buffer and gradient program for optimum separation of our protein of interest, the 500-µL sample loading loop may be replaced with the superloop. This allows loading of up to 10 mL (20 mg protein) sample on the column. Fractions may be collected on the Frac-100 for further analysis.
4. Final adjustments to the gradient may be required to allow for the effect of scaling up the loading.

Table 1
Some Buffers Suitable for Ion-Exchange FPLC

pH range	Buffer[a]	Column
3.8–4.3	50 mM Na formate-formic acid	Mono S
4.5–5.2	20 mM Na acetate-acetic acid	Mono S
5.5–6.0	20 mM Histidine-HCl	Mono Q
6.0–7.6	50 mM Na phosphate	Mono S/Mono Q
7.6–7.8	50 mM HEPES	Mono S
7.0–8.0	10 mM Tris-HCl	Mono Q
8.0–9.0	50 mM Tris-HCl	Mono Q
9.0–9.5	20 mM Ethanolamine-HCl	Mono Q

[a]These may be used as buffer A (see Section 3.3.).

3.4. Gel Filtration FPLC

Gel filtration is often used as a final "polishing" step to remove minor contaminants from partially purified protein (see Note 9). It is also especially useful in determining native molecular weights for oligomeric proteins. Run times for samples may be as little as 30 min. Superose 12 has a fractionation range of 1–300 kDa. An identical procedure may be used for Superose 6 (fractionation range: 5–5000 kDa).

1. With no column in system, prime both pumps A and B with buffer C.
2. Set pressure limits on both P-500 pumps to the same value (that for the column in use; see Notes 4 and 11).
3. Connect Superose 12, and operate system at a flow rate of 0.5 mL/min for 90 min to equilibrate column.
4. Wash sample loading loop with buffer C.
5. Load 200 µL sample (maximum) at approx 2 mg/mL.
6. Monitor A_{280}, collect peaks with Frac-100 fraction collector. Note elution volume of peak.

4. Notes

1. Prime pumps with water to remove 20% ethanol in which the system is stored. This avoids any risk of NaCl precipitation on contact of buffer B with 20% ethanol.
2. If the protein of interest is stable, then desalting may also be achieved by dialysis against three changes of 1 L buffer A (minimum of 6 h/change). Centrifuge sample after dialysis to remove any precipitated material. If the protein is particularly unstable then the rapid desalting HR 10/10 column may be included in the system before the ion-exchange column as shown in Fig. 1. This allows for desalting of sample in as little as 4 min.

3. The purpose of using a fairly neutral pH to begin with is to allow assessment of both the anion- and cation-exchange columns at the same pH. It is much quicker to swap columns than to reprime the entire system with new buffers at different pH values. The procedure outlined in Section 3.2. takes < 2 h to complete.
4. Flow rates of 1 mL/min are routinely achievable with newly purchased ion-exchange columns, but even with rigorous cleaning, back-pressure in the system soon rises appreciably after a number of uses. It is often better to accept a comparatively slow flow rate of 0.5 mL/min and modest back-pressures, since this seems to give better chromatography and to extend column life-times.
5. Although it is possible to carry out ion-exchange FPLC across a wide pH range, extremes of pH should be avoided if possible unless it is known that the protein of interest is stable at these pH values.
6. Operation of the FPLC system in a cold room at 4°C may be essential for purification of some particularly labile proteins. However, permanent location in the cold may lead to deterioration of the system. A convenient approach to this problem is to mount the system on a trolley with the components electrically connected to an extension lead on the trolley. This may be wheeled into the cold room, the extension lead connected to a power point, and the system operated once it has cooled to the temperature of the room. After use, the system may be easily wheeled out of the cold room again.
7. It is sometimes desired to operate the system (gel filtration or ion exchange) in the presence of urea. It is not advisable to use high concentrations (e.g., 6–8M) of urea though, because urea may easily precipitate, clogging valves and leading to excessive wear on seals and gaskets. Urea is rarely used at concentrations higher than 6M, and special care has to be taken if the system is being operated at 4°C. It is essential when chromatography in the presence of urea is complete that the entire system is washed extensively with water before storing it in 20% ethanol. Otherwise, urea will precipitate on contact with the ethanol solution. Also, it is important not to allow air (e.g., air bubbles) to come into contact with buffers containing urea, since this will lead to crystal formation. Once formed, crystals are very difficult to remove.
8. It is sometimes advisable to rechromatograph the peak of interest. This is achieved by rapidly desalting the peak into buffer A and reapplying it to the column. This often produces a considerable improvement in purification, since contaminants "move" to one side of the chromatogram away from the peak of interest.
9. Although gel filtration is often used toward the end of a purification procedure, in the particular case of immunoglobulin purification from serum, it may be used with advantage at an early stage of the purification *(7)*.

FPLC Methods 275

10. Calibration of Superose columns should be performed whenever a different chromatography buffer is used. A quick way to achieve this is to load standard proteins with widely separated molecular masses (e.g., cytochrome *c* and ovalbumin) in pairs (i.e., as a single sample).
11. Decrease in flow rate usually improves the resolution of gel-filtration chromatography.
12. When FPLC is complete, wash the system with water. Fill pumps with 20% filtered, degassed ethanol, wash loading loop with 20% ethanol, and place arm of Frac-100 in center of carousel (to minimize risk of damage to optical sensor).

References

1. Boyer, R. F. (1993) Separation and purification of biomolecules by chromatography, in *Modern Experimental Biochemistry*, Benjamin Cummings, Redwood City, CA, pp. 90–102.
2. Richey, J. (1982) FPLC: a comprehensive separation technique for biopolymers. *Am. Lab.* **14,** 104–129.
3. Fagerstam, L. G., Lizana, J., Axio-Fredriksson, U.-B., and Wahlstrom, L. (1983) Fast chromatofocusing of human serum proteins with special reference to α_1-antitrypsin and G_c-globulin. *J. Chromatogr.* **266,** 523–532.
4. Fagerstam, L. G. and Pettersson, G. L. (1979) The cellulytic complex of *Trichoderma reesei* QM9414. An immunological approach. *FEBS Lett.* **98,** 363–367.
5. Jeppson, J.-O., Kallman, I., Lindgren, G., and Fagerstam, L. (1984) A new hemoglobin mutant characterised by reverse phase chromatography. *J. Chromatogr.* **297,** 31–36.
6. Fitzpatrick, P. J. and Sheehan, D. (1993) Separation of multiple forms of glutathione S-transferase from the blue mussel, *Mytilus edulis. Xenobiotica* **23,** 851–862.
7. Havarstein, L. S., Aasjord, P. M., Ness, S., and Endresen, G. (1988) Purification and partial characterisation of an IgM-like serum immunoglobulin from Atlantic salmon *(Salmo salar). Dev. Comp. Immunol.* **12,** 773–785.

CHAPTER 27

Reversed-Phase Chromatography of Proteins

Bill Neville

1. Introduction

Reversed-phase HPLC (RP-HPLC) is now well established as a technique for isolation, analysis, and structural elucidation of peptides and proteins *(1,2)*. Its use in protein isolation and purification may have reached a peak owing to recent developments in high efficiency ion-exchange and hydrophobic interaction supports, which are now capable of equivalent levels of resolution to RP-HPLC without concomitant risk of denaturation and loss of biological activity. Nevertheless, there are many applications in which denaturation may be unimportant and high concentrations of organic modifier can be tolerated (e.g., purity analyses, structural studies, and micropreparative purification prior to microsequencing). It is a widely used tool in Biotechnology for process monitoring, purity studies, and stability determinations.

1.1. Principle of Separation

Application of a solvent gradient is generally superior to isocratic elution because separation is then achievable within a reasonable time frame and peak broadening of later-eluting peaks is reduced and thus sensitivity increased. In gradient elution RP-HPLC, proteins are retained essentially according to their hydrophobic character. The retention mechanism can be considered either as adsorption of the solute at the hydrophobic stationary surface or as a partition between the mobile and stationary phases *(3)*.

From: *Methods in Molecular Biology, Vol. 59: Protein Purification Protocols*
Edited by: S. Doonan Humana Press Inc., Totowa, NJ

In the first case, retention is related to total interfacial surface of the RP packing and is expressed by the adsorption coefficient K_A. Retention is based on a hydrophobic association between the solute and the hydrophobic ligands of the surface *(4)*. By increasing solvent strength of the mobile phase, attractive forces are weakened and the solute is eluted. The process can be regarded as being entropically driven and endothermic, i.e., both ΔS and ΔH are positive *(5)*. Through the relation between the solute capacity factor k' and the mobile and stationary phase properties, solvophobic theory permits prediction of the effect of the organic modifier of the mobile phase, the ionic strength, the ion-pairing reagent, the type of ligand of the RP packing, and other variables on the chromatographic retention of proteins.

The second case assumes a partitioning of the solute between the mobile and stationary phases, the latter being regarded as a hydrophobic bulk phase. This system resembles an *n*-octanol/water two-phase system where the selectivity is expressed by the partition coefficient P of the solute. Both retention principles have their pros and cons *(3)*. In essence, the effective process is largely dependent on the molecular organization of the bonded *n*-alkyl chains of the RP packing in the solvated state and the size and conformation of the solute.

Several methods have been developed for correlating peptide structures with RP-HPLC retention. One method sums empirically derived retention coefficients, representing the hydrophobic contribution of each amino acid residue *(6)*. The assumption behind this approach is that each amino acid residue contacts the adsorbent surface and that the total hydrophobicity of the solute determines RP-HPLC retention. This approach works well for peptides of 20 amino acids or less *(7)*, but generally is not accurate for predicting retention of larger proteins. This is caused by the peculiarities, especially of polar basic moieties, leading to either less retention or irreversible adsorption by residual matrix silanols. Studies with peptides containing amphiphilic helices indicate that secondary structure plays a large role in determining the surface of a polypeptide, which is exposed to the hydrophobic adsorbent *(8)*. Other studies have shown that proteins are less sensitive to changes in the HPLC support or bonded phase than are corresponding small molecules *(9)*. These results imply that protein retention in RP-HPLC is governed mainly by the protein surface and not by the support surface chemistry. Although some proteins show a loss of biological activity during RP-HPLC, others retain full biological activity. Thus, the hydrophobic forces necessary for RP binding compete with those required to maintain the protein's secondary and tertiary structures. Through steric, hydrophobic, and ionic constraints, these structural features define a chromatographic surface or "footprint" of the protein which in turn defines its RP binding. This concept of multipoint attachment of a protein to adsorbent surface is consistent with a relatively large chromatographic footprint for a protein compared with that of a small molecule.

1.2. RP-HPLC Columns

Optimization of chromatographic separation is determined by an appropriate choice of the column matrix, the mobile phase, and the temperature of separation. RP-HPLC has rapidly become the most widely used tool for the separation, purification, and analysis of peptides. For proteins, RP-HPLC has suffered from problems associated with denaturation, including loss of activity, poor recoveries, wide and misshapen peaks, and ghost peaks *(10,11)*. Some of these problems may be column related *(12)* and insurmountable, although others can be overcome by selective optimization of extra-column variables, such as sample pretreatment and mobile phase and hardware considerations *(13)*. As with small-molecule HPLC, there are a large number of RP-HPLC columns for use with peptides and proteins. In trying to select the best column for a new application or for broad applicability, one should consider such column variables as support, bonded phase, pore size, particle size, and column dimensions. These factors are dealt with in the following.

1.2.1. Packing Support

Most commercial RP-HPLC columns for peptides and proteins are silica based, as a consequence of its use for the separation of small molecules over many years. Silica offers good mechanical stability and allows a wide range of selectivities by virtue of the bonding of various phases. Although it is well known that silica-based columns are not stable at basic pH, recent reports suggest that in the acidic mobile-phases typically used, the bonded phase may also be slowly dissolved from the base silica *(14,15)*. Degradation can thus affect not only reproducibility, stability, and lifetime of a column, but also recovery and, in addition, modify selectivity as the silica surface becomes uncovered. An added problem could be contamination of recovered products with silica and bonded-phase material. Because of the limitations associated with silica-based columns, polymer-based columns have gained increased popularity for peptide and protein separations by RP-HPLC. Polymer-based columns, such as divinylbenzene-crosslinked polystyrene are usually stable over the pH range 1.0–14.0, making them more widely applicable with the added advantage that they are easier to clean up after use. Although most polymer-based columns are still inferior to their silica-based counterparts in terms of mechanical strength, selectivity, and efficiency (see Note 1), newer materials, such as PLRP-S (Polymer Laboratories, Church Stretton, Shropshire, UK) show superior performance in complex protein separations (13). Recent studies have shown that the use of polymer-based columns at high pH (>8.0) can offer unique selectivities and thus complement separations with classical acidic mobile phases (16,17). These advantages, coupled with superior stability, longer column life, and better reproducibility make polymer-based columns a good first choice for complex protein separations.

1.2.2. Bonded Phases

A wide variety of bonded phases are available for the separation of peptides and proteins by RP-HPLC, but the most common are n-alkyl bonded phases, namely C_4, C_8, and C_{18}. In general, C_4 and C_8 phases are preferable for more hydrophobic samples and the C_{18} phase for hydrophilic samples. In general, significant differences may be observed in the separations achieved on nominally the same column but from different sources (12). For proteins, a C_8 column is generally a good compromise. If the proteins of interest are too strongly retained a C_4 column could be substituted.

1.2.3. Particle Size

As with small molecules, theory predicts that column performance should increase with decreasing particle size. This is generally the case for most commercially available RP-HPLC columns used for protein separations. Columns with particle sizes of 5–10 μm show little difference in performance. Columns with particle sizes of 2–3 μm are now available and can give much greater resolution (see Note 2).

1.2.4. Pore Size

The standard pore size for separation of small molecules is 80 Å but for the separation of proteins a pore size of 300 Å has become accepted as the norm. For proteins >100 kDa, pore sizes of >1000 Å may be even better *(12)*. For complex samples, if all other column variables are equal, it would be better to use a larger pore size to ensure that restricted diffusion or exclusion from the pores is not encountered by very large proteins.

1.2.5. Column Dimensions

The two factors to consider when selecting a column size are the efficiency and sample loading capacity. In the case of proteins, column length contributes little to efficiency. In addition, longer columns may have an adverse effect on protein recovery. When considering gradient rather than isocratic elution, resolution of a mixture of proteins depends directly on the particle diameter of the packing, but less on the column length and flow rate for gradients of equal time. However peak height, which will determine the detection limit, depends inversely on the particle diameter, flow rate and, to a lesser extent, gradient time. An increase in column length would have little effect on resolution since by the time a protein band reaches the extra column length, it would be overtaken by the gradient and the organic modifier concentration would ensure that the protein spent little time in the adsorbed state. The effect of additional column length is thus largely passive and may even lead to increased band broadening. Long columns might be useful in isocratic protein separations but if extra capacity were needed, a short fat column might be a better choice than a long thin column. Having selected the column length, choice of column diameter should be based on required sample capacity. Columns of 1-mm internal diameter (id) (microbore) to 2.1-mm id (narrowbore) are best suited to submicrogram levels of material thus minimizing sample loss and also increasing sensitivity of detection. For analytical applications in the microgram to low milligram range, columns of 4.6-mm id (analytical) are best.

1.3. Mobile Phase

The mobile-phase composition is the most readily changed variable in an RP-HPLC separation. In normal usage the mobile phase consists of a mixture of water, a miscible organic solvent, and dissolved buffers and salts. A buffer, such as a low concentration of a strong acid or salt, is essential for chromatography. If a protein is adsorbed in the absence of a salt or acid, no increase in the proportion of organic component will elute it. In addition, other components, either solvents or solutes, may be added in order to affect the separation. This may be to accelerate or delay elution, improve peak shape, or adjust the elution position of some components with respect to others thus affecting selectivity. Other important variables are the pH of the mobile phase and the concentration and type of ions present. The ionic strength is a factor in limiting "mixed-mode" retention involving silica hydroxyls, but the choice of ions can also affect the solubility and stability of the protein in the mobile phase and, through ion pairing, the distribution of the protein between stationary and mobile phases. The net effect depends on the combination of column packing and mobile phase.

Numerous studies have examined the effect of mobile phase on the retention and selectivity of proteins in RP-HPLC. Systematic studies by several researchers over the past 10 yr have resulted in the publication of a number of fairly standardized sets of elution conditions. The behavior can be altered by the variation of a number of mobile-phase parameters. Altering the nature of the organic component of the mobile phase can alter column behavior but not usually in a predictable manner. The most successful way to alter column behavior is to manipulate the ionic component of the the solvent system. By altering the nature and strength of the ion-pairing reagent and the pH of the mobile-phase, it is possible to exploit both the acidic and basic character of proteins in a systematic and predictable way. Without prior knowledge of the primary structure of a peptide or protein, one can be certain of some hydrophobic character by virtue of the content of the lipophilic amino acids leucine, iso-leucine, methionine, phenylalanine, tyrosine, tryptophan, and valine. It will also have hydrophilic character largely because of the presence of aspartic acid, glutamic acid, arginine, and lysine. The ratio between hydrophobic and hydrophilic amino acids determines the initial behavior observed for any protein in the most common RP-HPLC solvent system in use today, namely, aqueous acetonitrile containing 0.1% trifluoroacetic acid (TFA). To some extent hydrophilic character of solutes is suppressed at pH 2.0. All carboxylic acid groups are protonated under these conditions and their contribution to hydrophilic character is reduced. This in turn will improve peak shape. Basic amino acids (e.g., arginine, lysine, and histidine) will remain as cations which in turn will lead to a marked reduction in silanophilic solute–matrix interactions.

1.3.1. Organic Component of Mobile Phase

The most popular organic solvents are acetonitrile, 1-propanol, and 2-propanol. Propanols are much more viscous than acetonitrile, giving higher column backpressures (see Note 3) but these are not significant when low flow rates and short columns are used. The acetonitrile concentration needed to elute a protein is considerably higher than the equivalent 2-propanol concentration, perhaps half as high again, and thus proteins are likely to be seriously denatured. There is also evidence that acetonitrile is intrinsically a more powerful denaturant than the alcohols. Recoveries are often higher in propanol-containing mobile phases, especially for the more "difficult" model proteins, such as ovalbumin. Acetonitrile and 2-propanol can be frozen in dry-ice/alcohol mixtures and lyophilized. All solvents should be of the highest quality available (see Note 4).

1.3.2. Minor Mobile-Phase Additives

The most popular minor components of the mobile-phase are the perfluoroalkanoic acids (e.g., TFA and heptafluorobutyric acid). These appear to solubilize proteins in organic solvents. TFA is used in preference to other fully dissociated acids because it is completely volatile and will not corrode stainless steel. TFA also acts as a weak hydrophobic ion-pairing reagent. Most RP columns employed for RP-HPLC are silica-based and apart from the suppression of silanol ionization under acidic conditions (thereby suppressing undesirable ionic interactions with basic residues), silica-based columns are more stable at low pH. As mentioned previously (Section 1.3.), an acid pH is preferred although many polypeptides will lose their tertiary structure at pH 2.0–3.0 (see Note 5).

Other popular additives are pyridinium acetate buffers and buffers based on phosphoric acid, particularly triethylammonium phosphate (TEAP). TEAP is useful in circumstances where tailing is suspected because of nonspecific interactions with surface silanols. Proteins are eluted at a higher percentage of organic component with an equivalent concentration of phosphoric acid. Sulphate, phosphate, perchlorate, and chloride are all transparent in the far UV, while acetate, formate, and fluoroalkanoic acids must be used at concentrations <20 mM if wavelengths below 220 nm are to be monitored. Pyridine-containing eluents cannot be monitored by UV. Ion-pairing reagents, such as heptane and octane sulphonic acids, sodium dodecyl sulphate, or alkyl ammonium salts may be added to the mobile phase to selectively increase the retention of proteins carrying larger charges of opposite sign. Other additives might be needed depending on the nature of the protein being analyzed (see Note 6).

1.3.3. Gradient Elution

The theory of gradient elution is now well established and a good understanding exists concerning the effects of gradient steepness on separation. Geng and Regnier *(18)* developed the stoichiometry factor, Z, which represents the number of solvent molecules displaced during the binding of the protein to the column packing. Thus, the value of Z is a measure of the size of the effective chromatographic footprint of that protein.

For RP-HPLC, Eq. (1) has been derived which shows that Z is proportional to the term S:

$$Z = 2.3 \phi S \quad (1)$$

where ϕ is the volume fraction of the organic in the mobile-phase *(19)*. Under defined gradient conditions, every protein will have an S value described by Eq. (2):

$$\log k' = \log k_w - S\phi \quad (2)$$

where k' is the retention of the solute (capacity factor) and k_w is the value of k' when water is the mobile phase ($\phi = 0$). Values of k_w and S are characteristic of each solute in a sample. Although empirically derived, Eq. (2) is a good approximation of RP-HPLC retention in the gradient mode. For a series of related proteins of nearly identical composition, the hydrophobic contribution to RP retention will be similar and thus observed changes in Z (and S) should reflect changes in the structure of the protein which contacts the adsorbent surface. Examination of Eq. (2) reveals why small changes in the organic composition of RP eluents result in large changes in protein retention as opposed to small molecule separations. Proteins will have a much larger chromatographic footprint (Z and S) than small molecules. Thus, changes in ϕ (% organic) will be amplified by large values of S, resulting in even larger changes in retention (log k'). Many examples of the applicability of Eq. (2) for peptide or protein samples have been reported. Values of S and k_w for each solute can be obtained from two experimental gradient separations of the sample. Retention times in gradient elution then can be predicted as a function of gradient conditions. Two components that elute adjacent to one another in a chromatogram will often show significant changes in band spacing when isocratic solvent strength or gradient steepness is varied. The resolution of such band pairs can usually be accomplished when values of S for the two components differ by 5% or more.

1.4. Temperature

Protein samples will often contain many individual components. A complete separation of such samples poses a real challenge because statistical considerations suggest that one or more peak pairs usually will be resolved poorly. To overcome this dilemma, a systematic variation of separation selectivity would be required. This approach has been widely used for the separation of typical small-molecule samples. In the case of peptide and protein samples, however, the control of selectivity has received little attention. When a change in selectivity is desired, the usual approach is a change of column or mobile phase. The use of elevated temperatures for RP-HPLC of protein samples has been advocated primarily as a means of increasing column efficiency or shortening run time. For samples of this type, a few studies have shown that a change in column temperature can also affect separation selectivity. A combination of low pH and higher temperature can result in a very short life for commonly used alkyl-silica columns which in turn limits the application of temperature optimization. This problem has recently been overcome by the development of "sterically protected" RP packings that are extremely stable at low pH (<2.0) and high temperature (>90°C). The combined use of temperature and gradient steepness would provide an efficient procedure for the control of peak spacing and optimization of separation. At the same time, this approach to selectivity control is more convenient than alternatives, such as change of column or mobile phases because temperature and gradient steepness can be varied via the HPLC system controller.

2. Materials

To illustrate the general points in Section 1., the separation of a mixture of standard proteins is described.

2.1. Equipment

Analysis of proteins by RP-HPLC is suited to most commercially available equipment. Systems require a column packed with a suitable matrix giving adequate retention and resolution, a detector to monitor eluted proteins, a chart recorder or data system to view detector response, and a fraction collector for recovery of eluted proteins, if further analysis is intended. If high sensitivity is a requirement, then narrowbore or microbore chromatography should be considered and specialized equipment may be necessary (*see* Note 7). Proteins are usually monitored by their absorbance in the UV either at 280 nm, which is suitable for proteins with aromatic amino acids, or at 206 nm, which is characteristic of the peptide bond (*see* Note 8). Most modern chromatographic systems now have computers to control operating parameters and to collect and store data. Microfraction collectors are also available with low internal diameter tubing to reduce the volume to a minimum between detector and collection tube in order to more nearly reflect the real time elution situation.

2.2. Column

A wide variety of suitable C_8 or C_{18} columns are commercially available for protein analysis. The price of modern columns is such that little advantage is gained by self-packing a column (*see* Note 9). The column must be able to withstand moderate pressures (250 bar) and be resistant to the mobile phase (pHs typically 1.0–8.0 for silica columns). A guard column (containing the same or a very similar packing material) should be used as a matter of course to extend analytical column lifetime.

2.3. Injector

For most purposes a conventional HPLC manual injection valve (e.g., Valco or Rheodyne) with a fixed volume loop (e.g., 10 µL) is adequate. For microbore analysis, injectors with small internal loops may be better. Many modern chromatographs are now fitted with auto-samplers whereby injection volume can be varied (e.g., 0.1–200 µL) via operating computer.

2.4. Eluents

Use Analar-grade or even Aristar-grade reagents and distilled/deionized water for all solutions (*see* Note 10). All eluents should be filtered through a 0.2-µm filter and degassed under vacuum before use.

1. Eluent A: 0.1% TFA (v/v). To 1 L of water add 1 mL of TFA.
2. Eluent B: acetonitrile:water (70:30) + 0.085% TFA (v/v). To 700 mL of acetonitrile add 300 mL of water and 0.85 mL of TFA (see Note 8).

Eluents should be degassed by a slow stream of helium while on the chromatograph, if at all possible, to counteract the effect of dissolved air. If a ternary system is available, acetonitrile can be programmed into a gradient to fully elute hydrophobic materials which otherwise might be retained by the column packing.

2.5. Detector

By far the most common detector used in chromatography is the variable wavelength UV detector, operated at 210–220 or 280 nm. Most proteins have an absorption maximum at about 280 nm because of the presence of aromatic amino acids, which falls to a minimum at 254 nm, a popular wavelength for fixed-wavelength HPLC detectors. Highest sensitivities are obtained by monitoring the strong absorption bands which peak below 220 nm owing to the "peptide bond" itself (*see* Note 8).

Proteins are usually eluted as fairly broad peaks (*see* Fig. 1, a protein test mixture of insulin, cytochrome *c,* lactalbumin, carbonic anhydrase, and ovalbumin) and thus place fairly modest demands on the dimensions of a normal flowcell (typically 8 µL with pathlength 1 mm). An alternative to UV detection is to monitor intrinsic fluorescence. The usual excitation wavelength is 280 nm with emission monitored at either 320 (for tyrosine) or 340 nm (for tryptophan). The main advantage is a several-fold increase in sensitivity and an improvement in selectivity of detection, particularly in distinguishing protein peaks from "ghost" peaks originating from eluent impurities.

2.6. Protein Sample

A mixture of test proteins, namely insulin (from bovine pancreas), cytochrome *c* (from horse heart), α-lactalbumin (from bovine milk), carbonic anhydrase (from bovine erythrocytes), and ovalbumin (from hen egg whites) can be used as a column test. Two of the proteins, insulin and cytochrome *c,* are difficult to resolve. A partial to complete separation should be achievable when a column with a high enough number of theoretical plates is used. Of the five standard proteins, ovalbumin is not only the most hydrophobic, but also the most difficult to recover. The utility of the column from a recovery perspective will be illustrated by the size of this peak.

3. Method

1. Set up appropriate pump, temperature, injection, detector, and integration methods, as required by chromatograph being used.
2. Install appropriate column (C_8 or C_{18} bonded silica, 300 Å, 5 µm, 10–25-cm length, 2.1 or 4.6 mm id).
3. Equilibrate the column by pumping eluent A at 200 µL/min (narrowbore, 2.1 mm) or 1 mL/min (normalbore, 4.6 mm) for several column volumes. Monitor detector baseline (214 nm) until flat and then zero the detector. Inject 5 µL of HPLC quality water and run a blank gradient to identify any system peaks (*see* Note 10) and to balance baseline (± TFA), if necessary (*see* Note 8).
4. Dissolve protein mixture in HPLC quality water at a concentration of approx 2 µg/µL to produce a stock solution. The stock solution should be

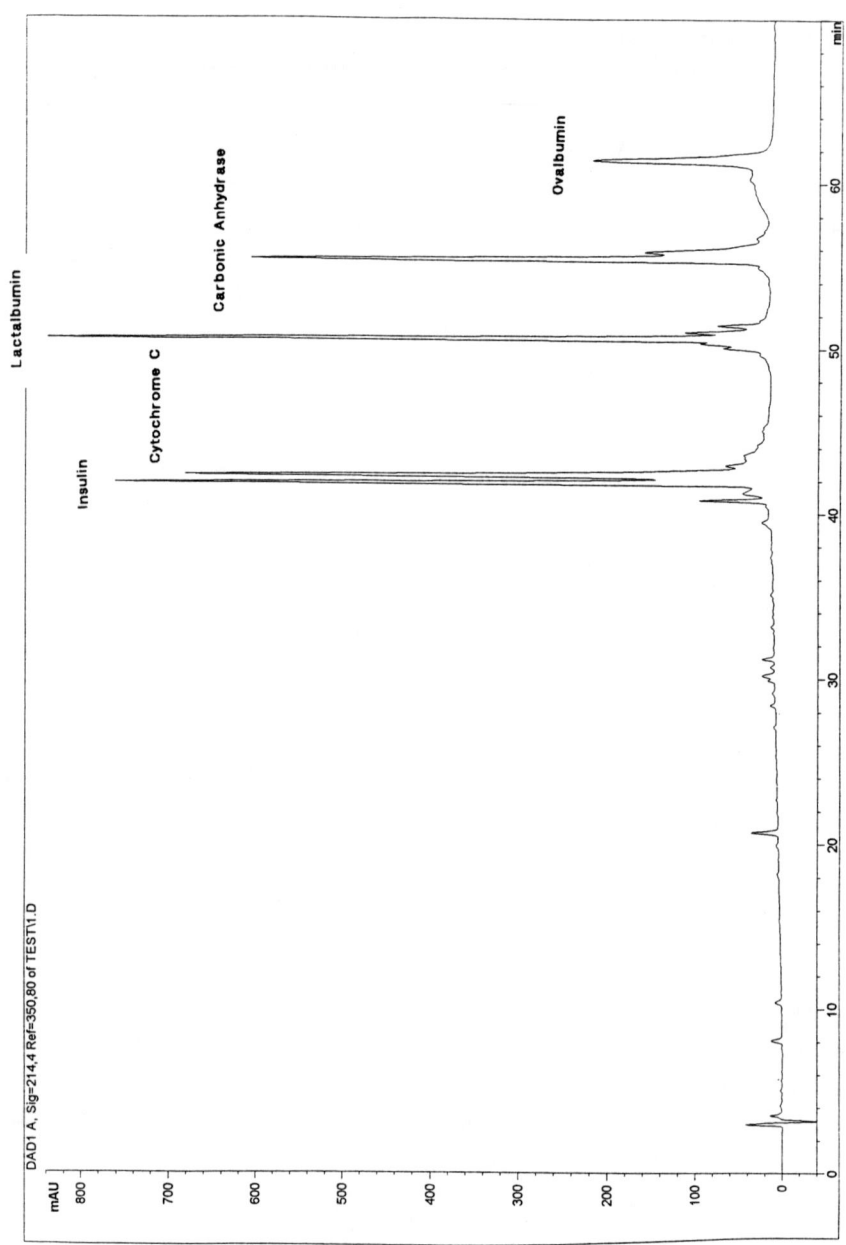

Fig. 1. RP-HPLC of a protein test mixture. The conditions used for chromatography are described in Section 3.

diluted further with eluent A (1:1) to produce a working solution of 1 μg/μL of each protein.
5. Apply the sample (5 μL) at initial gradient conditions (see Note 11).
6. Start gradient to elute proteins.

The elution profile obtained is shown in Fig. 1.

4. Notes

1. Polymer columns unfortunately suffer from a relatively low pressure resistance and, in general, pressure should be kept below 2000 psi. Furthermore, the physicochemical properties are governed by a solvent-dependent swelling and shrinking of the organic matrix which is associated with concomitant changes in pore diameters, leading to changes in mass transfer characteristics of the column. This effect may be encountered during gradient elution starting with an aqueous phase (by which only little or no wetting of the column matrix takes place) and termination with a pure organic solvent (e.g., methanol, acetonitrile, 2-propanol). Pore diameter will continuously change and often may be the cause of poor chromatographic performance. In contrast, silica gel as the starting material for subsequent alkyl silylation offers the advantage that a great number of different alkyl silica substituents can be bound to the silica surface.
2. Any gain in column efficiency can be outweighed quickly by increased column backpressure, susceptibility to plugging, and shorter column lifetime.
3. The pressure of an acetonitrile-containing mobile phase decreases approximately linearly with increasing percentage of organic modifier, whereas aqueous mixtures of methanol, ethanol, 1-propanol, and 2-propanol exhibit a more or less convex pressure dependence on the percentage of organic modifier. Maximum values are reached on average between 40 and 70% and thus a maximum in viscosity. To reduce unacceptably high column pressures, separations may have to be carried out at higher temperatures.
4. Water needs to be free of UV-absorbing organic impurities, which can be adsorbed onto the column during equilibration and the early part of a gradient and which may be eluted later as "ghost" peaks. Water should be freshly distilled and deionized and stored out of contact with plastic thus avoiding dissolution of plasticizers.
5. This may not necessarily be fatal for resultant activity but if the polypeptide consists of subunits stabilized by noncovalent forces, the individual chains will almost certainly be separated during the chromatography.
6. Metal ions, such as calcium, may be included to enhance the stability of a protein and low concentrations of nonionic detergents may improve the

behavior of hydrophobic membrane proteins. Finally, guanidinium hydrochloride or urea, at moderate concentrations, may be added to elute very hydrophobic membrane proteins.

7. In order to obtain good performance from microcolumns, it is very important to have an HPLC system with very low dead volume so that extra-column peak broadening does not destroy the resolution achieved by the column.

    Consideration should be given to the minimization of dead volume at all points in the chromatographic system. If there is a large dead volume between the point at which the gradient is mixed and the injector, then lag will be introduced into the gradient. Also the gradient system must be capable of delivering the gradient accurately and reproducibly at low flow rates. A suitable mixing device (volume <200 µL) may need to be added to the system to enable a reproducible gradient to be delivered at very low flow rates.

8. During a gradient there is usually an associated change in background absorption. This may be caused directly by the increase in the concentration of organic component or by the effect of this concentration change on other mobile-phase components. In addition, refractive index changes may be detected in the flowcell. Usually not much can be done about flowcell design but careful choice of wavelength and pH can minimize the change in absorption spectrum of mobile-phase components. For TFA and acetonitrile, the optimum wavelength is 214 nm. Alternatively, the concentration of other components in the mobile phase can be adjusted to compensate or a small amount of a UV absorbing solute can be added to one of the eluent reservoirs. For example, in a TFA/water/acetonitrile system monitored at 214 nm, the rise in background can be balanced by using 0.1% (v/v) aqueous TFA as eluent A and 70% aqueous acetonitrile, containing 0.085% (v/v) TFA as eluent B.

9. It is generally accepted that Vydac columns are the "industry standard," against which other manufacturers' columns are compared. Packing columns is based on know-how. Packing being commercially important, know-how is not readily imparted. Packing LC columns is still not fully understood. Commercial columns are usually supplied with a test and a guarantee to replace the column if performance does not match that specified. Although packed columns cannot be expected to last indefinitely, a usual lifetime of several months and several hundred injections is a reasonable objective.

10. HPLC quality water is commercially available from a number of suppliers if access to high-quality distilled or distilled/deionized water is not possible. A definition of HPLC quality is when 40 mL water is pumped onto the column and eluted with a linear water/acetonitrile gradient at 5%/min, at a flow rate of 2 mL/min, and absorbance of largest eluted peak, monitored at 254 nm, is <0.002 AU.

11. Proper gradient conditions are critical in optimizing separations of proteins by RP-HPLC. The gradient time, range, and shape are all important and must be optimized for a given sample. A typical starting gradient would be a 2%/min linear increase in acetonitrile over perhaps 70 min. A washing step of 100% eluent B then should be included before returning to initial conditions, allowing sufficient equilibration before the next injection.

Column re-equilibration time is an extremely important variable. The time must be sufficient to allow the initial mobile phase to equilibrate in the pores of the column packing (at least 3 column volumes) and should be exactly repeatable if precise retention times are required. A linear gradient of the above type (0.5–5%/min increase in acetonitrile) should allow resolution of the proteins in the test mixture. For increased resolution of certain of the proteins, a number of steps may have to be included in the gradient profile.

References

1. Hancock, W. S. and Harding, D. R. K. (1984) Review of separation conditions, in *CRC Handbook of HPLC for the Separation of Amino Acids, Peptides and Proteins,* vol. 2 (Hancock, W. S., ed.), CRC, Boca Raton, FL, pp. 303–312.
2. Regnier, F. E. (1987) Peptide mapping. *LC/GC* **5(5),** 392–395.
3. Dill, K. A. (1980) The mechanism of solute retention in reversed-phase liquid chromatography. *J. Phys. Chem.* **91,** 1987–1992.
4. Melander, W. and Horvath, Cs. (1980) Reversed-phase chromatography, in *HPLC—Advances and Perspectives,* vol. 2 (Horvath, Cs., ed.), Academic, New York, p. 114.
5. Hearn, M. T. W. (1980) HPLC of peptides, in *HPLC—Advances and Perspectives,* vol. 3 (Horvath, Cs., ed.), Academic, New York, p. 99.
6. Guo, D., Mant, C. T., Parker, J. M. R., and Hodges, R. S. (1986) Prediction of peptide retention times in reversed-phase HPLC: I. Determination of retention coefficients of amino acid residues of model synthetic peptides. *J. Chromatogr.* **359,** 499–517.
7. Meek, J. L. and Rossetti, Z. L. (1981) Factors affecting retention and resolution of peptides in HPLC. *J. Chromatogr.* **211,** 15–28.
8. Heinitz, M. L., Flanigan, E., Orlowski, R. C., and Regnier, F. E. (1988) Correlation of calcitonin structure with chromatographic retention in HPLC. *J. Chromatogr.* **443,** 229–245.
9. O'Hare, M. J., Capp, M. W., Nice, E. C., Cooke, N. H., and Archer, B. G. (1982) Factors influencing chromatography of proteins in short alkylsilane-bonded large pore-size silicas. *Anal. Biochem.* **126,** 17–128.
10. Pearson, J. D., Lin, N. T., and Regnier, F. E. (1982) Separation of proteins, in *HPLC of Peptides and Proteins* (Hearn, M. T. W., Wehr, C. T., and Regnier, F. E., eds.), Academic, New York, p. 81.
11. Hearn, M. T. W. (1986) Reversed-phase chromatography, in *HPLC—Advances and Perspectives,* vol. 3 (Horvath, Cs., ed.), Academic, New York, pp. 87–91.

12. Burton, W. G., Nugent, K. D., Slattery, T. K., and Summers, B. R. (1988) Separation of proteins by reversed-phase HPLC: I. Optimising the column. *J. Chromatogr.* **443**, 363–379.
13. Nugent, K. D., Burton, W. G., Slattery, T. K., and Johnson, B. F. (1988) Separation of proteins by reversed-phase HPLC: II. Optimising sample pre-treatment and mobile phase conditions. *J. Chromatogr.* **443**, 381–397.
14. Glajch, J. L., Kirkland, J. J., and Kohler, J. (1987) Effect of column degradation on the reversed-phase HPLC of peptides and proteins. *J. Chromatogr.* **384**, 81–90.
15. Sagliano, J., Floyd, T. R., Hartwick, R. A., Dibussolo, J. M., and Miller, N. T. (1988) Studies on the stabilisation of reversed-phases for liquid chromatography. *J. Chromatogr.* **443**, 155–172.
16. De Vos, F. L., Robertson, D. M., and Hearn, M. T. W. (1987) Effect of mass loadability, protein concentration and *n*-alkyl chain length on the reversed-phase HPLC behaviour of bovine serum albumin and bovine follicular fluid inhibin. *J. Chromatogr.* **392**, 17–32.
17. Guo, D., Mant, C. T., and Hodges, R. S. (1987) Effects of ion-pairing reagents on the prediction of peptide retention in reversed phase HPLC. *J. Chromatogr.* **386**, 205.
18. Geng, X. and Regnier, F. E. (1984) Retention model for proteins in reversed-phase liquid chromatography. *J. Chromatogr.* **296**, 15–30.
19. Kunitani, M., Johnson, D., and Snyder, L. R. (1986) Model of protein conformation in the reversed-phase separation of interleukin-2 muteins. *J. Chromatogr.* **371**, 313–333.

CHAPTER 28

Extraction of Membrane Proteins

Kay Ohlendieck

1. Introduction

The fluid mosaic model of biomembrane structure, proposed in 1972 by Singer and Nicolson *(1)*, defines two classes of membrane proteins that are associated, to varying degrees, with the phospholipid bilayer. In addition to peripheral and integral membrane proteins, a third class of membrane proteins is represented by lipid-anchored proteins. As well as these protein–membrane interactions, certain components of the membrane cytoskeleton and the extracellular matrix are also directly or indirectly associated via binding proteins or receptor molecules with membranes. Treatment of biological membranes with salt solutions or change in pH usually dissociates peripheral proteins, since these extrinsic membrane proteins interact with the membrane surface mostly via electrostatic and hydrogen bonds. Integral membrane proteins that possess hydrophobic surfaces are more strongly associated with the bilayer, and these intrinsic proteins extend across or are partially inserted into the lipid bilayer. Extraction of integral membrane proteins is commonly accomplished by solubilizing the protein containing membrane fraction using a variety of detergents *(2)* (Fig. 1).

1.1. Peripheral Membrane Proteins

Since peripheral membrane proteins are extracted by relatively mild treatments, care must be taken not to partially release these membrane proteins accidentally during subcellular fractionation procedures prior to purification. Isolation of a soluble form of an extrinsic membrane protein is usually achieved by a single technique or combinations of mild

From: *Methods in Molecular Biology, Vol. 59: Protein Purification Protocols*
Edited by: S. Doonan Humana Press Inc., Totowa, NJ

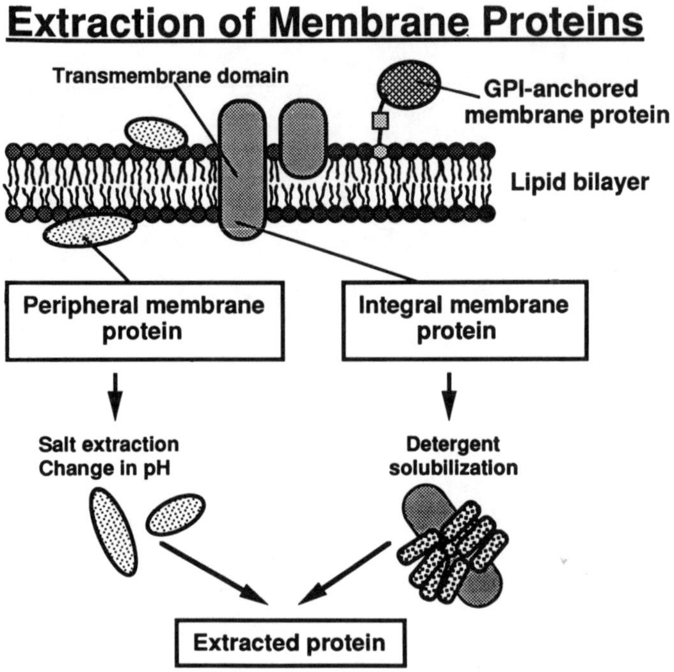

Extraction of Membrane Proteins

Fig. 1. Diagram of the different kinds of membrane proteins and the most commonly employed methods to extract them from biological membranes.

extraction procedures, depending on the particular properties of the protein under investigation and the starting material, i.e., a particular subcellular fraction enriched in the peripheral protein *(3–5)*. Most extraction procedures aim at the disordering of the structure of water, the disruption of weak electrostatic interactions and hydrogen bonds, and occasionally weak hydrophobic interactions in order to break the interactions between the extrinsic proteins and the membrane. Following extraction for 10–30 min at 40°C, the remaining membrane bilayer and its associated integral proteins are separated by centrifugation (30–60 min at 100,000g), and the released peripheral membrane proteins are recovered in the supernatant. The following list summarizes commonly used procedures for the extraction of peripheral membrane proteins:

1. Treatment with alkaline buffers (pH 8.0–12.0).
2. Treatment with acidic buffers (pH 3.0–5.0).
3. Use of metal chelators (10 mM EGTA or EDTA).
4. Treatment with high-ionic-strength (1M NaCl or KCl).

5. Treatment with denaturing agents (i.e., urea).
6. Treatment with organic solvents (i.e., butanol).
7. Sonication of membrane fractions.

Usually extraction procedures employing high-ionic-strength NaCl or KCl, alkaline, or acidic buffers and metal chelators result in a relatively distinct separation between solubilized peripheral proteins and membrane-associated integral membrane proteins. However, the nonspecific association of soluble proteins with membrane fractions and/or the entrapment of cytosolic proteins in membrane vesicles during subcellular fractionation, as well as the partial release of integral membrane proteins even under mild conditions of extraction and the existence of certain proteins in both integral and soluble isoforms, makes the interpretation of these kinds of differential extraction procedures sometimes difficult *(3–5)*.

1.2. Integral Membrane Proteins

Extraction of integral membrane proteins is most conveniently achieved by the use of detergents. Detergents are amphipathic molecules that contain both hydrophobic and hydrophilic moities, and the preferred form of detergent aggregation in water is the formation of micelles. Detergent micelles are characterized by a unique critical micelle concentration (CMC), and below the specific CMC value, individual detergent molecules predominate in solution. CMC values differ quite significantly between individual classes of detergents, i.e., the CMC for octylglucoside is 25 mM, whereas Triton X-100 exhibits a CMC of 0.3 mM *(2,6,7)*. With respect to micelle size, certain detergents form relatively large micelles, i.e., Triton X-100 forms micelles with approx 150 detergent molecules/micelle and a molecular mass of 90–95 kDa *(2)*. A detailed listing of detergents used in the biological sciences and a description of their individual properties can be found in a review on detergent structure by Neugebauer *(6)*. Ideally, the detergent of choice should not only be available in pure form and be relatively inexpensive for its usage in large-scale preparation, but it should also sufficiently solubilize the membrane protein under investigation without irreversibly denaturing it. Furthermore, easy removal of excess detergent from the solubilized protein fraction is an additional criterion for the choice of detergent, which will be discussed in Chapter 29. For comprehensive information on the availability, purity, unit prices, and sources of commonly used detergents for the solubilization of biological membranes, *see* the review article by Jones et al. *(7)*.

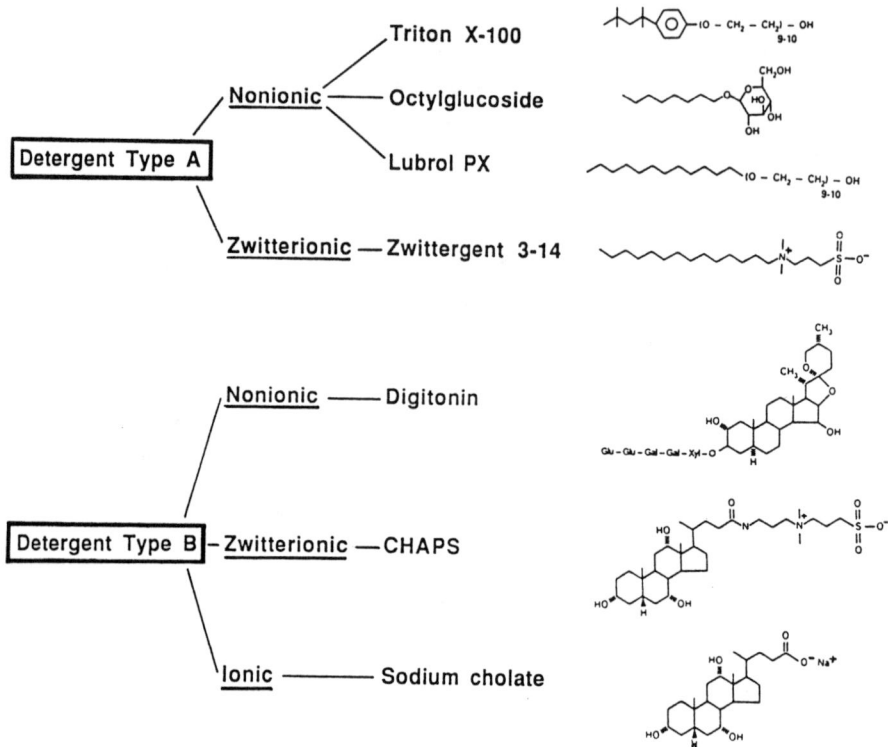

Fig. 2. Commonly used detergents employed in the solubilization of biological membranes.

A good initial selection of suitable detergents for pilot experiments on the extraction of novel integral membrane proteins with little knowledge of their physical and chemical properties is summarized in Fig. 2. As extensively reviewed by Helenius and Simons (2), detergents may be classified by their overall chemical structure as Type A or Type B detergents, and further subdivided according to their electric charge as nonionic, ionic, or zwitterionic detergents. Although Type A detergents exhibit flexible hydrophobic tails and hydrophilic head groups, Type B detergents are more rigid and are cholesterol-based structures with amphiphilic properties (Fig. 2). Commonly used Type A detergents are the nonionic components Triton X-100, octylglucoside, and Lubrol PX, as well as the zwitterionic detergent sulfobetaine 14 (Zwittergent 3–14). The nonionic detergent digitonin, the zwitterionic component CHAPS, and the ionic bile salt sodium cholate are typical Type B detergents extensively used in biological research *(2,6,7)*. Many other detergents exist that are closely related to the described Type A and Type B detergents, and are listed in recent review articles on membrane solubilization *(6,7)*. The ionic detergent sodium dodecyl sulfate, although widely used in analytical and preparative polyacrylamide gel electrophoresis techniques *(8)*, is not usually employed in the initial extraction of membrane proteins and therefore is not further discussed in this chapter.

1.3. Experimental Design

Although the design of an extraction experiment may partially be based on the physical and chemical properties of the specific detergent used and the biological characteristics of the particular integral membrane protein and subcellular fraction under investigation, often a series of small-scale trial and error experiments is a good start to determining conditions for optimum solubilization. In pilot experiments, aliquots of the particular biological membrane are incubated with different concentrations of a variety of commonly used detergents. Initially, standard conditions are used with respect to buffer composition, salt solutions, temperature, and incubation time. In order to optimize the extraction process, these conditions may be varied. Most importantly, the denaturing properties of individual detergents, compatibility of detergents with divalent cations, the pH dependency of detergent solubility, spectral properties of detergents, and electrostatic and temperature effects on detergent behavior should be taken into account when choosing the appropriate buffer conditions and the ratio of detergent to protein *(6)*. Effective solubilization is usually achieved in a well-buffered, physiological pH environment and the addition of $0.15M$ NaCl to the extraction mixture for optimum ionic strength. Membrane preparations are used at a protein concentration of 1–5 mg/mL and are solubilized by detergent concentrations of 0.1–5% (v/v). By convention, retention of a membrane protein in the supernatant following centrifugation for 60 min at $100,000g$ after solubilization defines this protein operationally as soluble *(6)*. However, in large-scale preparations, shorter spins and lower g-forces might be sufficient to extract a high yield of a particular membrane protein. To determine if the integral protein of interest was sufficiently solubilized by a particular concentration of a detergent, a relatively simple and fast assay procedure should be choosen. Enzyme assays or ligand binding assays, as well as cell biological test systems are certainly very convenient to evaluate the efficiency of the extraction protocol, and might also be useful in determining the potential loss of biological activity during the solubilization procedure. If a specific labeled probe or antibody is available to the integral membrane protein to be solubilized, overlay assays with immobilized protein fractions and immunoblot analysis using minigel systems can be performed reasonably quickly and with large numbers of samples.

To illustrate the use of commonly employed methods and reagents for the extraction of peripheral and integral membrane proteins, this chapter covers the initial isolation of a typical extrinsic protein, calsequestrin, and that of a typical intrinsic protein, Ca^{2+}-ATPase. Both proteins are major constituents of the sarcoplasmic reticulum from skeletal muscle. The Ca^{2+}-ATPase is localized to the longitudinal tubules and is responsible for the resequestration of calcium into the lumen of the sarcoplasmic reticulum following muscle contraction (9,10). Calsequestrin represents a high-capacity, intermediate affinity calcium-binding protein localized to the luminal site of the junctional sarcoplasmic reticulum and increases the calcium holding capacity of the terminal cisternae in muscle (11,12). These properties make both proteins key components of calcium homeostasis and excitation–contraction coupling in muscle (13).

2. Materials

2.1. Preparation of Sarcoplasmic Reticulum from Rabbit Skeletal Muscle

1. Dissection kit: good quality scissors for fine mincing of muscle tissue.
2. Ice bucket or tray, sturdy glass plate (size of the ice-containing bucket or tray).
3. Skeletal muscle from a New Zealand white rabbit (100 g).
4. Waring blender.
5. Buffer A (homogenization buffer): 10% (w/v) sucrose, 20 mM histidine, pH 7.0, 0.5 mM EDTA, 0.23 mM PMSF, 0.83 mM benzamidine (see Note 1).
6. Refrigerated high-speed centrifuge.
7. Cheese cloth.
8. Ultracentrifuge with fixed-angle rotor.
9. Small glass homogenizer (for resuspension of pellets).
10. Buffer B: 0.6M KCl, 30 mM histidine, pH 7.0, 0.23 mM PMSF, 0.83 mM benzamidine.
11. Buffer C (storage buffer): 10% (w/v) sucrose, 30 mM histidine, pH 7.4, 0.23 mM PMSF, 0.83 mM benzamidine.

2.2. Extraction of the Extrinsic Calcium-Binding Protein Calsequestrin from Sarcoplasmic Reticulum

1. Buffer D (extraction buffer): 100 mM sodium carbonate, pH 11.4.
2. Ultracentrifuge and fixed-angle rotor.
3. Magnetic stirrer.
4. Concentrated HCl and a pH meter.
5. NaCl.
6. Morpholinopropanesulfonic acid (MOPS).
7. Dithiothreitol (DTT).

2.3. Assay of Calcium Binding to Calsequestrin

1. Standard dialysis tubing (with a mol-wt cutoff of approx 10,000).
2. Buffer E (to pretreat dialysis membrane): 50 mM EDTA, 100 mM sodium carbonate.
3. Heat plate and large glass beaker (to boil dialysis tubing).
4. Buffer F (calcium-binding buffer): 5 mM Tris-HCl, pH 7.5, 0.1mM ^{45}CaCl$_2$ (40,000 cpm/mL).
5. Scintillation fluid.
6. Scintillation spectrometer.

2.4. Extraction of the Intrinsic Enzyme Ca^{2+}-ATPase from Sarcoplasmic Reticulum

1. Buffer G: 10% (w/v) sucrose, 50 mM sodium phosphate, pH 7.4, 1M KCl, 0.23 mM PMSF, 0.83 mM benzamidine.
2. Ultracentrifuge and fixed-angle rotor.
3. Magnetic stirrer.
4. 10% (w/v) Sodium deoxycholate.

2.5. Assay of Ca^{2+}-ATPase Activity

1. Stock solutions for assay buffer: 100 mM histidine, pH 7.4, 1M KCl, 50 mM MgCl$_2$, 1 mM CaCl$_2$, 10 mM EGTA, 25 mM ATP.
2. Stock solutions for phosphate determination:
   a. Ammonium molybdate: 28.6 g of ammonium molybdate are dissolved in 500 mL of 6M HCl.
   b. Polyvinyl alcohol: 11.6 g of polyvinyl alcohol are dissolved in 500 mL of boiling water and are allowed to cool.
   c. Malachite green: 0.81 g of malachite green are dissolved in 1 L of distilled water.
   d. Mixed reagent: 2 vol of malacite green are mixed with 1 vol each of polyvinyl alcohol and ammonium molybdate, and 2 vol of distilled water. Allow the reagent to stand at room temperature until it turns a golden-yellow color, which takes approx 30 min.

3. Potassium hydrogen phosphate standard (10–100 nmol P_i/mL).
4. Spectrophotometer.

3. Methods

Isolation of membrane vesicles derived from sarcoplasmic reticulum of skeletal muscle were based on the method of Eletr and Inesi *(14)*. Ca^{2+}-ATPase activity and calcium binding was assayed according to Meissner et al. *(15)* and MacLennan *(16)*, respectively. Inorganic phosphate was determined by the method of Chan et al. *(17)*, and measurement of protein concentration was performed according to Hartree *(18)* using bovine serum albumin as a standard. Extraction procedures for the extrinsic membrane protein calsequestrin and the intrinsic membrane protein Ca^{2+}-ATPase were based on the protocols of Cala and Jones *(19)* and Warren et al. *(20)*, respectively. All preparative steps are performed at 4°C unless otherwise stated.

3.1. Preparation of Sarcoplasmic Reticulum from Rabbit Skeletal Muscle

1. Mince 100 g of trimmed white rabbit skeletal muscle (*see* Note 2) with fine scissors on a sturdy glass plate positioned on top of a tray of crushed ice. Disperse the tissue pieces in 300 mL of ice-cold buffer A.
2. Homogenize the muscle tissue in a Waring blender at full speed four times for 15 s.
3. Centrifuge the homogenate for 20 min at 15,000g, filter the supernatant through three layers of washed cheesecloth, and recentrifuge at 40,000g for 90 min.
4. Resuspend the pellet in 50 mL of buffer B, incubate the suspension for 40 min on ice, and then centrifuge at 15,000g for 20 min. Recentrifuge the supernatant at 40,000g for 90 min, and resuspend the pellet in buffer C. At this stage, the crude sarcoplasmic reticulum preparation may be quick-frozen in liquid nitrogen, and stored at –70°C until future usage (*see* Note 3).

3.2. Extraction of the Extrinsic Calcium-Binding Protein Calsequestrin from Sarcoplasmic Reticulum

1. Pellet freshly thawed sarcoplasmic reticulum, prepared as described in Section 3.1., at 100,000g for 30 min, and resuspend in ice-cold buffer D to a protein concentration of 1.5 mg/mL.
2. Incubate the suspension on ice for 30 min, and then centrifuge at 100,000g for 30 min yielding a pellet and the carbonate-extracted supernatant fraction containing calsequestrin.
3. Stir the supernatant magnetically, and make the suspension to 50 mM MOPS, 0.5M NaCl, and 1 mM DTT by the addition of solid reagents, and adjust the pH to 7.0 by the dropwise addition of concentrated HCl (*see* Note 4).

Table 1
Calcium Binding to Calsequestrin Extracted from Sarcoplasmic Reticulum

Step	Total protein, mg	Specific Ca^{2+} binding, nmol Ca^{2+}/mg protein
Sarcoplasmic reticulum	80	175
Carbonate solubilized supernatant	27	280
Purified calsequestrin	–	713

3.3. Assay of Calcium Binding to Calsequestrin

1. Boil a sufficiently long piece of thin dialysis tubing for 10 min in buffer E, followed by boiling for 10 min in distilled water, and then rinse the tubing extensively in distilled water (*see* Note 5).
2. Place a 0.4-mg protein sample in 1-mL vol and dialyze against 100 mL of buffer F for 24 h in a cold room at 4°C. Use all the necessary precautions when handling the radioactive isotope (*see* Note 6).
3. Dissolve 0.1-mL samples from both inside and outside of the dialysis bag each in 10 mL of scintillation fluid, and count the specific radioactivity in a scintillation spectrometer.
4. Assuming that the dialysis reached an equilibrium, calculate Ca^{2+} binding/mg of protein from the increased radioactivity within the dialysis bag (Table 1).

3.4. Extraction of the Intrinsic Enzyme Ca^{2+}-ATPase from Sarcoplasmic Reticulum

1. Centrifuge freshly thawed sarcoplasmic reticulum, prepared as described in Section 3.1., at 100,000g for 30 min, and resuspend the pellet in ice-cold buffer G.
2. Add sodium deoxycholate from a 10% (w/v) detergent stock solution slowly under magnetic stirring and on ice to the membrane suspension at a ratio of 0.4 mg detergent/mg protein.
3. Centrifuge the resulting mixture at 100,000g for 60 min, and collect the clear supernatant containing the solubilized Ca^{2+}-ATPase (*see* Note 7).

3.5. Assay of Ca^{2+}-ATPase Activity

1. Mix in a microcuvet 10 µL of protein sample with 390 µL of distilled water and with 100 µL each of the assay stock solutions, including histidine, KCl, $MgCl_2$, as well as calcium (to measure total ATPase activity) or EGTA (to measure basal ATPase activity).
2. Initiate the enzyme reaction by the addition of 200 µL of ATP stock solution.

Table 2
Enzyme Activity of Ca^{2+}-ATPase Extracted from Sarcoplasmic Reticulum

Sample	Protein, mg	Ca^{2+}-ATPase activity, μmol P$_i$/mg protein/min
Sarcoplasmic reticulum	52	2.1
Detergent solubilized supernatant	29	3.9
Purified Ca^{2+}-ATPase	–	8.1

3. After an incubation time of 5 min, terminate the reaction by the addition of 2 mL of mixed reagent (malachite green-molybdate-polyvinyl alcohol reagent) (*see* Note 8).
4. Immediately after the addition of the mixed reagent, measure color development at 630 nm against a mixed reagent blank.
5. Compare your measurements with a potassium dihydrogen phosphate standard graph.
6. Calculate the Ca^{2+}-ATPase activity by subtracting the basal ATPase activity (in the presence of EGTA) from the total ATPase activity (in the presence of calcium) (*see* Note 9) (Table 2).

4. Notes

1. No general composition of a protease inhibitor cocktail suitable for all homogenization or extraction procedures can be given. To prevent proteolytic degradation of calsequestrin and the Ca^{2+}-ATPase during isolation from rabbit skeletal muscle, the addition of EDTA, PMSF, and benzamidine to the isolation buffer appears to be sufficient. Other widely used protease inhibitors include antipain, aprotinin, bestatin, chymostatin, E-64, leupeptin, pefabloc, pepstatin, and phosporamidon. Some companies, i.e., Boehringer-Mannheim (Mannheim, Germany), sell protein inhibitor sets for small-scale pilot experiments in order to determine the minimum concentration and composition of a protease inhibitor cocktail sufficient for preventing proteolytic degradation in a novel purification protocol.
2. Muscle tissue should be quickly removed from the freshly killed animal and trimmed of fat tissue. Avoid contaminating the isolation buffer and blender with rabbit hair. Fine mincing of muscle tissue should preferentially be performed in a cold room at 4°C with the muscle positioned on top of an ice-cold glass plate.
3. To purify the sarcoplasmic reticulum further, the crude preparation of Section 3.1. may be centrifuged through a 26–40% sucrose gradient for 150 min at 100,000g. The upper half of the sucrose density gradient contains the purified sarcoplasmic reticulum vesicles with considerably lower

amounts of contaminating mitochondrial membranes and vesicles derived from the transverse tubular membrane system.
4. Solubilized calsequestrin, extracted by alkaline treatment as described in Section 3.2., can be very simply purified to homogeneity using Phenyl-Sepharose chromatography *(19)*. Bound calsequestrin is eluted from the column by 10 mM $CaCl_2$, and this results in a homogeneous protein preparation.
5. Alternatively, gentler washing of dialysis tubing may be performed at 60°C by incubation for 2 h in buffer E, followed by extensive washing in distilled water. To avoid time-consuming washing of dialysis tubing in order to remove impurities and heavy metals, more expensive brands of tubing can be purchased that have been prewashed and sterilized (Spectrum, Houston, TX). However, irrespective of the brand of tubing, handle dialysis tubing at all times with gloves to avoid contamination of protein samples with proteases and other impurities from your hands.
6. When using $^{45}CaCl_2$ solutions, handle the radioactive isotope with extreme caution. Work exclusively in an area of the cold room designated for radioactive work, and confine radioactive experiments to a minimum space. Wear at all times double plastic gloves and protective clothing when handling radioactive solutions, and clearly mark all radioactive glassware and equipment. When removing labeled sample from the inside of the tubing following equilibrium dialysis, ask someone to help you in order to avoid spillage of radioactive solution when transferring it to the scintillation cocktail.
7. Solubilized Ca^{2+}-ATPase may be purified to homogeneity by centrifuging the solubilized preparation (Section 3.4.) through a linear 20–60% sucrose density gradient at 100,000g for 24 h *(20)*. The peak fractions with the highest specific Ca^{2+}-ATPase activity are usually found in the lower part of the sucrose gradient and are greatly depleted of excess detergent.
8. An alternative way to determine Ca^{2+}-ATPase activity is a coupled enzyme assay. Instead of measuring the production of inorganic phosphate, the increase of ADP is measured by coupling the ATPase reaction to the widely employed pyruvate kinase/lactate dehydrogenase system. The production of NAD can conveniently be measured at 340 nm.
9. Contaminating mitochondrial ATPase activity, possibly present in crude membrane preparations, can be inhibited by the addition of 5 mM NaN_3. The presence of inorganic phosphate in the protein sample prior to initiation of the enzyme reaction or nonenzymatic hydrolysis of ATP can be accounted for by control assays using sample but no ATP, or just ATP with no sample.

References

1. Singer, S. J. and Nicholson, G. L. (1972) The fluid mosaic model of the structure of cell membranes. *Science* **175**, 720–731.

2. Helenius, A. and Simons, K. (1975) Solubilization of membranes by detergents. *Biochim. Biophys. Acta* **415**, 29–79.
3. Penefsky, H. S. and Tzagoloff, A. (1971) Extraction of water-soluble enzymes and proteins from membranes. *Methods Enzymol.* **22**, 204–219.
4. Thomas, T. C. and McNamee, M. G. (1990) Purification of membrane proteins. *Methods Enzymol.* **182**, 499–520.
5. Scopes, R. K. (1994) *Protein purification: Principles and Practice, 3rd ed.* Springer-Verlag, New York.
6. Neugebauer, J. M. (1990) Detergents: an overview. *Methods Enymol.* **182**, 239–252.
7. Jones, O. T., Earnest, J. P., and McNamee, M. G. (1987) Solubilization and reconstitution of membrane proteins, in *Biological Membranes: A Practical Approach* (Findlay, J. B. C. and Evans, W. H., eds.), IRL, Oxford, UK, pp. 139–172.
8. Laemmli, U. K. (1970) Cleavage of structural proteins during the assembly of the head of bacteriophage T4. *Nature (London)* **227**, 680–685.
9. MacLennan, D. H. (1970) Purification and characterization of an adenosine triphosphate from sarcoplamic reticulum *J. Biol. Chem.* **245**, 4508–4518.
10. MacLennan, D. H., Brandl, C. J., Korczak, G., and Green, N. M. (1985) Amino-acid sequence of a $Ca^{2+} + Mg^{2+}$-dependent ATPase from rabbit sarcoplasmic reticulum, deduced from its complementary DNA sequence. *Nature (London)* **316**, 696–701.
11. MacLennan, D. H. and Wong, P. T. S. (1971) Isolation of a calcium-sequestrating protein from sarcoplasmic reticulum. *Proc. Natl. Acad. Sci. USA* **68**, 1231–1235.
12. Meissner, G. (1975) Isolation and characterization of two types of sarcoplasmic reticulum vesicles. *Biochim. Biophys. Acta* **389**, 51–68.
13. Mortonosi, A. N. (1994) Regulation of calcium by the sarcoplasmic reticulum, in *Myology,* 2nd ed. (Engel, A. G. and Franzini-Armstrong, C., eds.), McGraw-Hill, New York, pp. 553–584.
14. Eletr, S. and Inesi, G. (1972) Phospholipid orientation in sarcoplasmic membranes: spin-label ESR and proton NMR studies. *Biochim. Biophys. Acta* **282**, 174–179.
15. Meissner, G., Conner, G. E., and Fleischer, S. (1973) Isolation of sarcoplasmic reticulum by zonal centrifugation and purification of Ca^{2+}-pump and Ca^{2+}-binding proteins. *Biochim. Biophys. Acta* **298**, 246–269.
16. McLennan, D. H. (1975) Isolation of proteins of the sarcoplasmic reticulum. *Methods Enzymol.* **32**, 291–302.
17. Chan, K. M., Delfert, D., and Junger, K. D. (1986) A direct colorimetric assay for Ca^{2+}-stimulated ATPase activity. *Anal. Biochem.* **157**, 375–380.
18. Hartree, E. F. (1972) Determination of protein: a modification of the Lowry method that gives a linear photometric response. *Anal. Biochem.* **48**, 422–427.
19. Cala, S. E. and Jones, L. R. (1983) Rapid purification of calsequestrin from cardiac and skeletal muscle sarcoplasmic reticulum vesicles by Ca^{2+}-dependent elution from Phenyl-Sepharose. *J. Biol. Chem.* **258**, 11,932–11,936.
20. Warren, G. B., Toon, P. A., Birdsall, N. J. M., Lee, A. G., and Metcalfe, J. C. (1974) Reconstitution of a calcium pump using defined membrane components. *Proc. Natl. Acad. Sci. USA* **71**, 622–626.

CHAPTER 29

Removal of Detergent from Protein Fractions

Kay Ohlendieck

1. Introduction

The solubilization of biological membranes by detergents plays an important role in the identification, characterization, and extraction of integral membrane proteins. The most commonly used detergents in the biological sciences are described in Chapter 28, and several review articles exist on solubilization procedures (1–3). Since the initial extraction of intrinisic membrane proteins usually involves high detergent concentrations, excess detergent has to be removed or exchanged for another type of detergent at later stages of preparative or analytical procedures involving the solubilized protein fraction. The efficiency of certain chromatographic procedures is often significantly improved by lowering the overall detergent concentration or by exchanging one type of detergent for another. Lectin chromatography, a powerful technique to affinity-purify subsets of glycoproteins (see Chapter 18), appears to be especially sensitive to high concentrations of a variety of detergents (4). Furthermore, since high concentrations and/or certain types of detergent interfere with many physical and chemical analyses and detergents also exhibit undesirable side effects on highly sensitive cell biological assays, detergent removal is in many cases of central importance for retaining the biological activity of isolated membrane proteins. Another important area of detergent removal is reconstitution studies with hydrophobic membrane proteins that exhibit vectorial transport (3).

From: *Methods in Molecular Biology, Vol. 59: Protein Purification Protocols*
Edited by: S. Doonan Humana Press Inc., Totowa, NJ

Table 1
Techniques Used for the Removal
or Exchange of Detergents from Protein Fractions[a]

Equilibrium dialysis
Batch or column chromatography
Hydrophobic adsorbtion
Ion-exchange chromatography
Gel filtration
Affinity chromatography
Lectin chromatography
Sucrose density gradient centrifugation
Protein precipitation procedures
Phase partioning
Electroelution techniques

[a]Suitability of individual techniques for detergent removal depends on the CMC, aggregation number, and HLB of the individual detergent (see Section 1.).

This chapter summarizes some of the techniques employed in removing or exchanging detergents used in the solubilization of biological membranes. For a more in-depth discussion, see recent reviews on procedures of detergent removal (5–8). Technical bulletins from companies selling commercially available resins for detergent removal, i.e., Biobeads SM-2 (Bio-Rad, Hercules, CA) or Extracti-Gel D (Pierce, Rockford, IL), usually contain a list of references relevant to conditions recommended for detergent removal. It is well worth studying these general guidelines and recommendations prior to attempting to remove or exchange detergent from a precious, novel protein sample. Detergents are amphipathic molecules, and their preferred form of aggregation in water is the formation of micelles whose size and molecular weight may vary considerably between different types of detergent (1–3). Thus, the most important properties of a detergent with respect to removal is its unique critical micelle concentration (CMC), its hydrophile-lipophile balance (HLB), and its micellar molecular weight determined by the aggregation number of detergent molecules (7).

The most commonly employed techniques for detergent removal from protein fractions are summarized in Table 1, and are based on physical and chemical differences between protein–detergent complexes and detergent micelles. Suitability of individual techniques depends on the unique properties of the detergent used, and the choice of procedure is furthermore strongly dependent on the concentration range of the protein

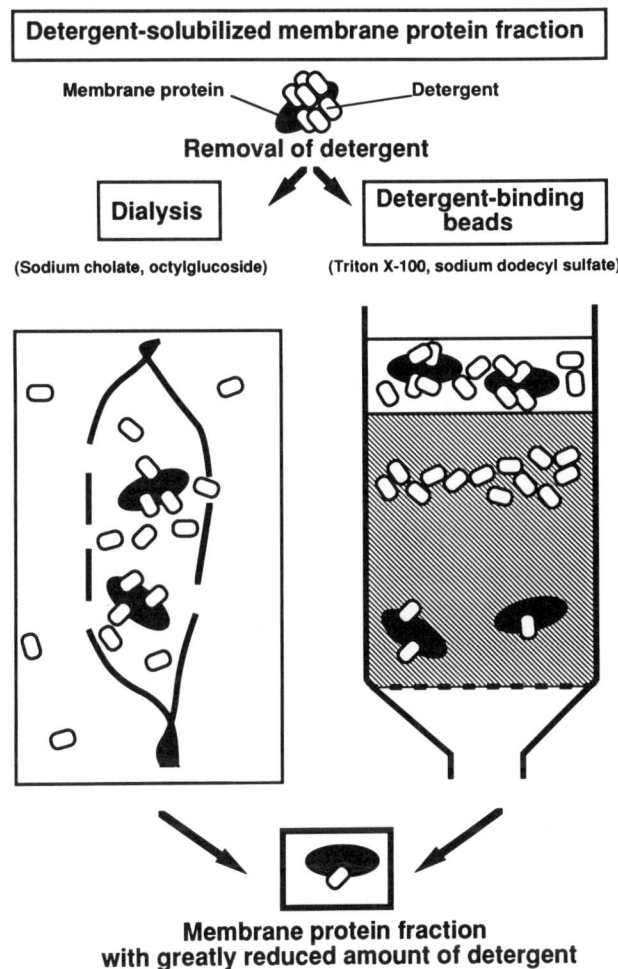

Fig. 1. Commonly used methods to remove detergent from protein fractions.

fraction to be depleted of detergent. Probably the simplest and least work-intensive method to remove detergent from protein fractions is dialysis (Fig. 1). Using large external volumes and frequent changes of dialysis medium, ionic detergents with a relatively high CMC can quite successfully be removed. However, it can take considerable time to reach an equilibrium, and this can lead to undesirable side effects, such as protein degradation. An excellent example of detergent removal, routinely performed by dialysis, is reconstitution experiments *(3)*. Alternatively, incorporation of hydrophobic membrane proteins into lipid vesicles can be achieved by gel-filtration or dilution procedures.

A variety of chromatographic techniques are available to remove or exchange detergents. These are more work-intensive than dialysis, but take less overall time and thereby tend to keep protein degradation during detergent removal to a minimum. Hydrophobic adsorption chromatography is certainly a very convenient way to exchange different classes of detergent (Fig. 1). The exchange of alkyl detergents, i.e., octyl glucoside and dodecyl sulfate for Triton X-100 type detergents using Phenyl-Sepharose (Pharmacia), was reported by Robinson et al. *(9)*. Reasonably inexpensive matrices for the specific binding of detergents are commercially available from Bio-Rad (Bio-Beads SM-2) and Pierce (Extracti-Gel D). Both adsorbents, however, require a high enough protein concentration to avoid losses in recovering protein samples during detergent removal. The preincubation of columns with bulk carrier proteins in order to saturate nonspecific protein binding sites might at least partially solve this problem. Alternatively, protein–detergent complexes could be bound to an affinity matrix and eluted following washing and/or exchange with a different detergent. Affinity matrices with immobilized ligands for receptor binding or a variety of lectin columns highly specific for binding to subsets of glycosylated membrane proteins are suitable for these kinds of procedures. However, high detergent concentrations often adversely affect ligand–protein interactions, and dilution prior to application to the affinity column is beneficial. Another way to bind charged membrane proteins and remove excess detergent by extensive washing is ion-exchange chromatography. Elution of the bound protein fraction can be achieved by increasing the ionic strength or addition of an ionic detergent *(5,7)*. Furthermore, if the difference in size between detergent micelles and protein–detergent complexes is large enough, gel-filtration chromatography can be employed to exchange detergents in a protein fraction *(5,7)*. The use of sucrose gradient centrifugation in the removal of excess detergent was demonstrated by Warren et al. *(10),* who could successfully separate excess deoxycholate from an integral membrane protein by this method. Precipitation of solubilzed membrane proteins from aqueous solutions can be achieved by treatment with polyethylene glycol, and phase partitioning can also be exploited to precipitate hydrophobic integral proteins *(5,7)*. Finally, electroelution is a widely used method to recover protein samples separated by sodium dodecyl sulfate-polyacrylamide gel electrophoresis (see Chapter 33), and many reasonably priced electroelution units are now commercially available.

To illustrate the practical aspects involved in the removal of detergent from protein fractions, equilibrium dialysis and detergent adsorbtion chromatography are described in more detail in Section 3. For a typical protocol of detergent-exchange chromatography, see the article by Robinson et al. (9). Since the unique properties of individual classes of integral membrane proteins and their interaction with a variety of ionic, nonionic, and zwitterionic detergents cannot adequately be predicted, the procedures described are only general outlines and do not describe the removal of detergent from a specific solubilized membrane protein. Refer to quoted research papers and review articles for details on specific requirements with respect to individual membrane proteins.

2. Materials
2.1. Dialysis

1. Dialysis tubing with a mol-wt cutoff of approx 10,000.
2. Wash buffer: 100 mM NaHCO$_3$, 50 mM EDTA.
3. Hot plate and large glass beaker.
4. Reliable, leak-proof plastic clamps for closing dialysis tubing.
5. Dialysis buffer: 20 mM Tris-HCl, pH 7.4, 0.15M NaCl.
6. Large beaker (4–6 L).
7. Small plastic funnel.
8. Magnetic stirrer and suitable large stir bar.

2.2. Detergent-Adsorbtion Chromatography

1. Small columns (1–5 mL bed volume).
2. Detergent adsorbtion matrix, i.e., macroporous Bio-Beads SM-2 (Bio-Rad), Extracti-Gel D (Pierce), or any other commercially available detergent-adsorption matrix.
3. Blocking buffer: 0.1% (w/v) bovine serum albumin in 50 mM Tris-HCl, pH 7.4, 0.15M NaCl.
4. Washing buffer: 50 mM Tris-HCl, pH 7.4, 0.15M NaCl.
5. Small peristaltic pump and suitable tubing.

3. Methods

All procedures are performed in a cold room at 4°C, unless stated otherwise.

3.1. Dialysis

1. Take a sufficiently long piece of standard dialysis tubing, and boil it for 10 min in washing buffer, followed by boiling for 10 min in distilled water and

extensive washing in distilled water. Alternatively, use prewashed dialysis tubing (*see* Note 5, Chapter 28).
2. Transfer the solubilized membrane protein fraction with the aid of a small funnel into the dialysis tubing, which is securely closed at the lower end (*see* Note 1).
3. Close the dialysis tubing after removal of any air bubbles possibly introduced during transfer of the detergent-containing suspension, and allow for a small increase in volume during equilibrium dialysis. Generally, dialysis tubing is very sturdy, and leakage is not a problem as long as the tubing is tightly closed.
4. Dialysis should be performed with large external volumes (4–6 L) and adequate stirring, as well as frequent exchanges of the external solution (*see* Note 2).
5. At the end of the dialysis, wash the outside of the tubing, and carefully remove the dialyzed protein fraction (*see* Note 3).

3.2. Detergent-Adsorbtion Chromatography

1. The protein fraction to be treated with respect to detergent removal or detergent exchange should have a relatively high protein concentration (*see* Note 4) and be of large enough molecular weight to avoid entrapment in the pores of the affinity matrix (*see* Note 5).
2. Wash the detergent-removing column matrix first with distilled water, then equilibrate it thorougly with blocking buffer (*see* Note 6), and wash with two column volumes of washing buffer.
3. Apply the protein sample, preferentially dissolved in the washing buffer or another suitable buffer for optimum detergent binding, to the equilibrated column.
4. Collect 0.5–1 mL fractions, and combine the protein peak fractions. The protein concentration can conveniently be measured by microprotein assays to avoid substantial losses owing to assaying. Peak fractions may then be concentrated by ultrafiltration prior to subsequent analytical or preparative procedures.

4. Notes

1. Dialysis tubing should be carefully closed by tight knotting and preferentially secured by special leak-proof plastic clamps (Spectrum, Los Angeles, CA).
2. Since reaching equilibrium in dialysis procedures can be quite time consuming, the placement of a suitable matrix to bind detergent outside the dialysis tubing might accelerate the process. Although nonionic detergent might be trapped by a hydrophobic adsorbtion matrix, ionic detergents might be removed by the use of an ion-exchange matrix.

3. Following equilibrium dialysis, avoid losing dialyzed samples when removing them from the tubing. Dialysis bags can be expanded following extensive dialysis, and should be carefully opened surrounded by a larger, clean glass beaker to avoid any accidental spillage.
4. Generally, a very important requirement for avoiding substantial losses of protein during procedures of detergent removal is a high enough protein concentration of the starting material. Especially, chromatographic techniques and precipitation procedures might result in a severe loss of proteins from very dilute solutions. Ideally, protein concentrations >1 mg/mL should be used to avoid these problems. Thus, concentrating a protein fraction prior to detergent removal by a suitable type of ultrafiltration is recommended, although these techniques are usually also not without danger of losing protein samples owing to nonspecific binding to membrane filters. These kinds of problems have to be worked out with every new class of solubilized membrane protein and no general strategy with respect to optimizing overall protein recovery can be given.
5. Using hydrophobic adsorbtion chromatography to remove or exchange detergent, the molecular weight of the protein to be recovered should be large enough to avoid entrapment of smaller peptides within the pores of the support matrix. The chromatographic matrix of commercially available columns exhibits exclusion limits between 2 and 10 kDa.
6. If dilute starting material cannot be concentrated, the detergent-exchange columns should be pretreated with solutions of bulk protein to avoid substantial protein losses. Bovine serum albumin is usually very useful in blocking nonspecific binding sites for protein on affinity matrices. This precaution should significantly lower the loss of dilute protein samples during detergent removal or exchange.

References

1. Helenius, A. and Simons, K. (1975) Solubilization of membranes by detergents. *Biochim. Biophys. Acta* **415**, 29–79.
2. Neugebauer, J. M. (1990) Detergents: an overview. *Methods Enzymol.* **182**, 239–253.
3. Jones, O. T., Earnest, J. P., and McNamee, M. (1987) Solubilization and reconstitution of membrane proteins, in *Biological Membranes: A Practical Approach* (Findlay, J. B. C. and Evans, W. H., eds.), IRL, Oxford, UK, pp. 139–177.
4. Lotan, R., Beattie, G., Hubbel, W., and Nicolson, G. L. (1977) Activities of lectins and their immobilized derivitaives in detergent solutions. Implications on the use of lectin affinity chromatography for the purification of membrane glycoproteins. *Biochemistry* **16**, 1787–1794.
5. Hjelmeland, L. M. (1990) Removal of detergents from membrane proteins. *Methods Enzymol.* **182**, 277–282.
6. Furth, A. J., Bolton, H., Potter, J., and Priddle, J. D. (1985) Separating detergents from proteins. *Methods Enzymol.* **104**, 318–328.

7. Furth, A. J. (1980) Removing unbound detergent from hydrophobic proteins. *Anal. Biochem.* **109**, 207–215.
8. Moriyama, H., Nakashima, H., Makino, S., and Koga, S. (1984) A study on the separation of reconstituted proteoliposomes and unincorporated membrane proteins by use of hydrophobic affinity gels, with special reference to band 3 from bovine erythrocyte membranes. *Anal. Biochem.* **139**, 292–297.
9. Robinson, N. C., Wiginton, D., and Talbert, L. (1984) Phenyl-Sepharose-mediated detergent-exchange chromatography: its application to exchange of detergent bound to membrane proteins. *Biochemistry* **23**, 6121–6126.
10. Warren, G. B., Toon, P. A., Birdsall, N. J. M., Lee, A. G., and Metcalfe, J. C. (1974) Reconstitution of a calcium pump using defined membrane components. *Proc. Natl. Acad. Sci. USA* **71**, 622–626.

CHAPTER 30

Purification of Membrane Proteins

Kay Ohlendieck

1. Introduction

In contrast to soluble proteins, the isolation of a peripheral or integral membrane protein requires extraction of the biological membrane containing the protein of interest prior to purification. The most commonly used extraction procedures to isolate membrane proteins are described in Chapter 28. Following initial solubilization in high concentrations of detergent *(1)*, intrinsic proteins can then be purified to homogeneity using a variety of biochemical techniques (*see* Chapters 14–24) in the presence of relatively low concentrations of detergent. Solubilized membrane proteins from diverse sources and with different properties have successfully been purified by a combination of standard techniques, such as density gradient centrifugation, ion-exchange chromatography, gel filtration, lectin chromatography, and different forms of affinity chromatography methods. Usually separation techniques based on biological differences, such as affinity chromatography using specific ligands, probes, or antibodies highly specific for a membrane protein, result in a higher yield and purity of the isolated protein than chromatographical methods based on physical differences, such as size and charge of the membrane protein *(2,3)*. An especially powerful separation technique for the initial purification of membrane-associated glycoproteins is lectin affinity chromatography. As described in more detail in Chapter 18, this method explores the highly specific interaction between N- and O-linked oligosaccharide chains on glycosylated proteins with immobilized lectins, which usually results in a remarkable enrichment of integral glycoproteins *(4)*.

Although the ultimate experiments to be carried out with an isolated membrane protein may influence the overall strategy of purification, it is desirable that the molecules of interest are obtained in their native, biologically active form in high yield and purity. However, biochemical, biophysical, or physiological experiments with homogeneous membrane proteins might require a gentler purification scheme than, for example, the preparation of only partially purified antigen samples for the production of monoclonal antibodies. Once a suitable source, such as a subcellular membrane fraction enriched in the membrane protein of interest, is found (*see* Chapters 6 and 7), small-scale pilot experiments should be performed to test the effects of various detergents on the biological activity of the membrane protein to be purified (*see* Chapter 28). With respect to the choice of detergent, a reagent with a high critical micelle concentration (i.e., cholates) should be tried first, since these detergents can be more easily removed following purification than detergents with a lower critical micelle concentration (*see* Chapter 29).

To illustrate the basic strategy for purifying an integral membrane protein and to monitor its purification and analyze the purity of the final product, this chapter addresses the purification of the sea urchin egg receptor for sperm. This highly glycosylated, integral surface protein represents a novel class of cell recognition molecules and exists in its native configuration as a disulfide-bonded homo-tetrameric complex *(5,6)*. Each sea urchin egg contains approx 1.25×10^6 receptor molecules whose subunits exhibit an apparent mol wt of 350 kDa *(7)*. Based on the previous biochemical characterization of proteolytic receptor fragments *(8)* and information on the primary structure deduced from its cDNA sequence *(9)*, a strategy to purify the sperm receptor to homogeneity was worked out. The purification protocol, typical for the isolation of integral membrane proteins, includes cell homogenization, subcellular fractionation, solubilization of membrane fractions, lectin affinity chromatography, ion-exchange chromatography, dialysis, ultrafiltration, and lyophilization *(7)*. Since this membrane protein does not exhibit enzyme activity, purification was monitored using immunoblot analysis with an antibody raised to a recombinant protein representing the extracellular sperm binding domain of the receptor molecules. The purity of the final product was analyzed by gradient SDS-PAGE in combination with sensitive silver staining *(5,7)*.

Thus, the techniques described in Section 3. and elaborated on in Section 4. give a good overview of how to approach the purification of a

Purification of Membrane Proteins

novel protein based on limited information on its physical and biological properties. The flowchart in Fig. 1 summarizes the different steps in the purification of the sea urchin sperm receptor typical for the isolation of an integral membrane protein. Other excellent examples of detailed descriptions of the methods employed in the purification of membrane proteins, i.e., rat liver 5'nucleotidase and rabbit skeletal muscle Ca^{2+}-ATPase appeared previously in this series *(10,11)*.

2. Materials
2.1. Isolation of Crude Egg Surface Membrane Complex

1. Sea urchin eggs (50 mL of settled, dejellied, and washed eggs from *Strongylocentrotus purpuratus*) (*see* Note 1).
2. Small plastic beakers (for collection of gametes).
3. 6°C Water bath for storage of sturdy 2-L glass beakers.
4. 120-µm Nitex nylon membrane (Tekton, Inc., Elmsford, NY).
5. 0.5M KCl.
6. Buffer A (artificial sea water): 0.48 M NaCl, 10 mM KCl, 27 mM $MgCl_2$, 29 mM $MgSO_4$, 11 mM $CaCl_2$, 2 mM $NaHCO_3$, pH 7.5 (*see* Note 2).
7. Buffer B (homogenization buffer): 0.5M NaCl, 10 mM KCl, 25 mM $NaHCO_3$, 63 mM NaOH, 25 mM ethylene glycol-bis(β-aminoethyl ether) N,N,N',N'-tetracetic acid (EGTA), 63 mM NaOH, pH 8.0; supplemented with 1 mM of each of aprotinin, soybean trypsin inhibitor, antipain, leupeptin, benzamidine, and phenylmethanesulfonyl fluoride (Sigma, St. Louis, MO) (*see* Note 3).
8. Buffer C: 20 mM Tris-HCl, 0.15M NaCl, pH 7.4, supplemented with the same protease inhibitor cocktail as described for buffer B.
9. Hand-operated Potter-Elvehjem homogenizer (25-mL vol).
10. Low-speed bench centrifuge with 4 × 200 mL rotor.
11. Standard bright-field microscope.
12. Water aspirator.

2.2 Solubilization of Sperm Receptor Complex

1. 4% (w/v) Octylglucoside (Boehringer-Mannheim, Indianapolis, IN) in buffer C.
2. Ultracentrifuge with 30-mL fixed-angle rotor and appropriate tubes.

2.3. Lectin Affinity Chromatography

1. Wheat germ agglutinin (WGA)-agarose (EY Labs, San Mateo, CA) matrix with a bed volume of approx 10 mL (*see* Note 4).
2. Wash buffer: 0.1% (w/v) octylglucoside in buffer C.

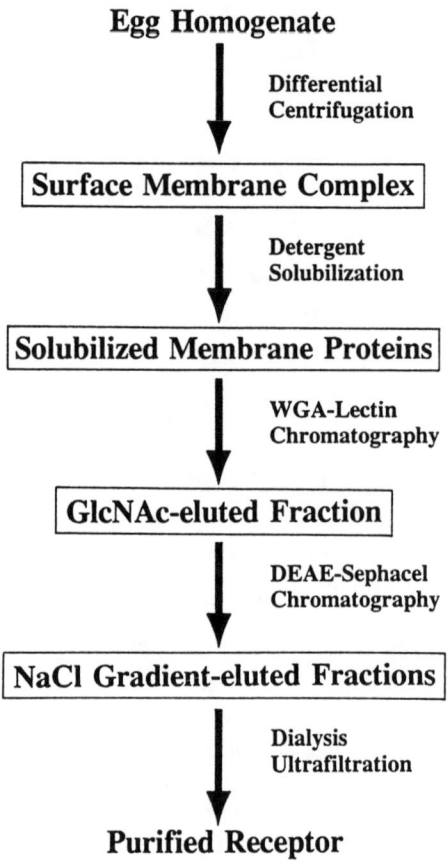

Fig. 1. Flowchart of the receptor purification protocol. The sea urchin egg receptor for sperm is purified to homogeneity from *S. purpuratus* egg homogenates. Following isolation of a crude surface membrane complex by subcellular fractionation, membranes are solubilized, and a subset of glycoproteins purified by lectin affinity chromatography. The eluted glycoprotein fraction is then bound at low ionic strength to an ion-exchange matrix, and eluted with a linear salt gradient. The peak fractions containing homogeneous receptor are dialyzed, concentrated and lyophilized, resulting in a preparation of purified sperm receptor.

3. Elution buffer: $0.5M$ N-acetylglucosamine, 0.1% (w/v) octylglucoside in buffer C.
4. End-over-end mixer.

2.4. Ion-Exchange Chromatography

1. DEAE-Sephacel column (Pharmacia, Uppsala, Sweden) with a bed volume of approx 10 mL.
2. Wash buffer: 0.1% (w/v) octylglucoside in buffer C.
3. Elution buffer A: 0.5M NaCl, 0.1% (w/v) octylglucoside, 20 mM Tris-HCl, pH 7.4, supplemented with the protease inhibitor cocktail as described for buffer B.
4. Elution buffer B: 4M NaCl, 0.1% (w/v) octylglucoside, 20 mM Tris-HCl, pH 7.4, supplemented with the protease inhibitor cocktail as described for buffer B.
5. 30-mL linear gradient maker, small magnetic pellet, and magnetic stirrer.
6. Small peristaltic pump and suitable elastic tubing.
7. Fraction collector (optional).

2.5. Dialysis, Ultrafiltration, and Lyophilization

1. Standard dialysis tubing (with a mol wt cutoff of approx 10,000).
2. Ultrafiltration filter, i.e., pre-equilibrated Amicon PM-30 membrane filter.
3. Ultrafiltration apparatus, i.e., 50-mL vol Amicon concentrator.
4. Lyophilizer.

3. Methods

Perform all manipulations of eggs at 6°C and subsequent homogenization and chromatographical purification steps in a cold room at 0–4°C.

3.1. Isolation of Crude Egg Surface Membrane Complex

1. To release mature gametes, treat 10–20 adult sea urchins with intracoelomic 0.5M KCl injection, and collect eggs in small plastic beakers of artificial sea water. After microscopical examination of individual batches of eggs, combine batches of morphological integrity, and wash them three times by settling in 2 L of artificial sea water.
2. Dejelly washed eggs by 10 slow passages through a 120-µm Nitex membrane, and then wash twice in filtered artificial sea water and once in 2 L of buffer B (see Note 5).
3. Resuspend the dejellied, washed, and settled eggs in 100 mL of ice-cold buffer B, supplemented with the described protease inhibitor cocktail, and homogenize by approx 10 gentle strokes using a hand-operated glass homogenizer and Teflon™ pestle. When breakage of eggs is adequate, the majority of eggs can be seen to release cytoplasmic organelles, such as yolk platelets, when viewed under a light microscope (see Note 6).
4. After 10-fold dilution, centrifuge the ghost membranes for 2 min at 1000g, carefully remove the yellowish supernatant with the help of a water aspirator, and combine the white membrane pellets.

5. Wash the pellet twice in buffer B, and finally resuspend it in 10 mL of buffer C (*see* Note 7).
6. If the egg surface complex cannot be processed immediately, it may be quick-frozen in liquid nitrogen, and stored at −80°C until usage.

3.2 Solubilization of Sperm Receptor Complex

1. Solubilize the egg surface complex membranes at a final protein concentration of 1 mg/mL by adding an equal volume of 4% (w/v) octylglucoside in buffer C, and incubate this mixture for 60 min on ice.
2. Remove insoluble material by centrifugation at 105,000g for 60 min, and decant the supernatant containing the solubilized membrane protein fraction.
3. Dilute the supernatant 1:4 (v/v) with buffer C.

3.3. Lectin Affinity Chromatography (see *Note 8*)

1. Equilibrate the WGA matrix by settling five times in 20 vol of buffer C.
2. Add the diluted, solubilized egg surface membrane complex suspension to the settled lectin matrix, and incubate gently for 3 h or overnight in a cold room on an end-over-end mixer.
3. Allow the lectin matrix to settle on ice and carefully remove the supernatant. Wash the matrix three times with 10 vol of wash buffer.
4. Add 1.5 vol of elution buffer to the lectin matrix and incubate the mixture for 20 min in a cold room on a mixer.
5. Allow the matrix to settle on ice, carefully recover the supernatant containing the eluted glycoprotein fraction, and proceed immediately to Section 3.4. For regenerating the WGA matrix, *see* Note 9.

3.4. Ion-Exchange Chromatography

1. Apply the glycoprotein fraction to 10 mL of DEAE-Sephacel ion-exchange resin, which has been pre-equilibrated with wash buffer. Gently shake this mixture for 2 h in a cold room.
2. Allow the ion-exchanger matrix to settle, remove the supernatant, and wash the resin three times with 10 vol of wash buffer.
3. Pour the resin into a clean glass column, and elute bound protein with a linear NaCl gradient of increasing ionic strength. The gradient is formed in a small gradient maker under adequate stirring using elution buffers A and B.
4. Collect 3-mL fractions, and store them on ice prior to analysis.
5. Analyze the fractions by immunoblot and/or lectin blot analysis. The sea urchin egg receptor for sperm exhibits a characteristic brownish color in silver-stained SDS polacrylamide gels, indicative of highly glycosylated proteins. In addition, the receptor protein band of approx 350 kDa is strongly labeled by peroxidase-conjugated WGA lectin (*see* Fig. 2).

Purification of Membrane Proteins

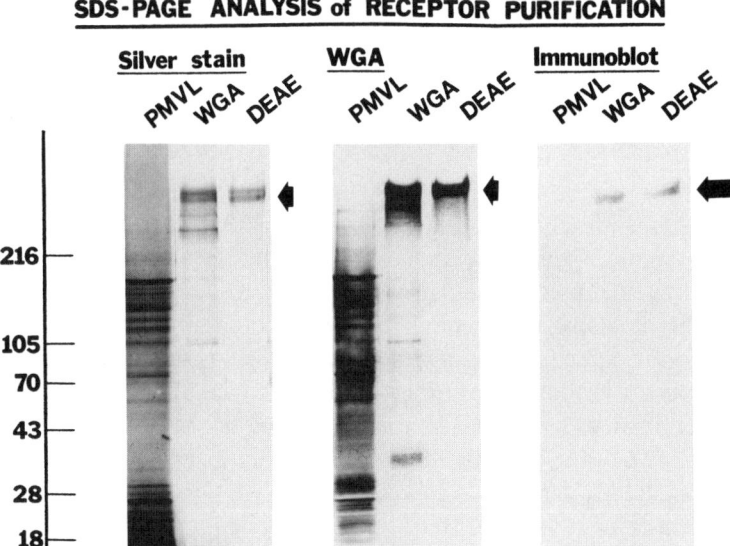

Fig. 2. Analysis of sperm receptor purification. Shown is the SDS-PAGE analysis of the different steps in the purification of the receptor using subcellular fractionation of the crude egg surface (PMVL; plasma membrane/vitelline layer), followed by lectin affinity chromatography (WGA) and finally ion-exchange chromatography (DEAE-Sephacel). The silver-stained and the WGA lectin-blotted 350-kDa protein band corresponds to the receptor band visualized by immunoblotting. The double protein band in silver staining might reflect a partial proteolytic degradation of the high-mol-wt receptor molecule or the existence of two isoforms with slightly different molecular weights. This phenomena is currently under investigation. Mol-wt standards ($M_r \times 10^{-3}$) are indicated on the left.

3.5. Dialysis, Ultrafiltration, and Lyophilization

1. Combine the peak fractions containing homogeneous sperm receptor, and dialyze them overnight against distilled water. For the preparation of dialysis tubing, see Chapter 11 or 29.
2. Concentrate the dialyzed receptor preparation using ultrafiltration with a PM-30 membrane. Following removal of the concentrated protein solution from the ultrafiltration chamber, rinse the filter twice with 1 mL of buffer to elute some of the protein possibly sticking to the membrane.
3. Quick-freeze the concentrated receptor preparation in liquid nitrogen and lyophilize the sample. Resuspend the freeze-dried receptor preparation in a small volume of buffer suitable for the subsequent analytical procedures,

i.e., filtered artificial sea water for fertilization inhibition assays or sample buffer for SDS-PAGE.
4. Figure 2 shows a silver-stained SDS-PAGE and corresponding immunoblot and lectin blot of the different preparative steps involved in the isolation of the integral receptor glycoprotein (*see* Note 10).

4. Notes

1. Sea urchins are common marine organisms that can be held in captivity under simple conditions, and mature gametes can be obtained in large quantities. In mature animals, intracoelomic KCl injection may release 10^7 eggs during a single spawning. The most commonly used sea urchin is *S. purpuratus*, which can be obtained from Marinus, Inc. (Long Beach, CA). This species produces mature gametes from late autumn to early summer.
2. Artificial sea water should be made from high-quality water, and filtered through 0.22-μm filters (Millipore, Bedford MA) prior to use on isolated gametes. Ready-made mixtures of artificial sea water (i.e., "Instant Ocean" from Aquarium Systems, Mentor, OH) may be purchased from a pet shop.
3. The composition of the protease inhibitor cocktail used during the purification of the sea urchin egg receptor for sperm may not be suitable for other purification protocols. Since certain protease inhibitors are relatively expensive, small-scale pilot experiments should be performed to determine the lowest concentration of protease inhibitors necessary to prevent protein degradation during preparative steps (*see* Note 1, Chapter 28).
4. Immobilized WGA is an expensive affinity matrix and can alternatively be prepared quite simply from crude wheat germ purchased in a health store. Following extraction, WGA can be purified to homogeneity by affinity chromatography using *N*-acetylglucosamine agarose (Sigma, St. Louis, MO) *(12)*. Purification of the lectin can be monitored by SDS-PAGE analysis combined with immunoblot analysis using a polyclonal antibody to WGA commercially available from Sigma. Purified WGA can then be immobilized by standard procedures using cyanogen bromide-activated agarose (Pharmacia).
5. If suitable nylon membranes are not available, dejellying of sea urchin eggs can also be achieved by acid treatment. A 10% solution of eggs is treated for 2 min at pH 5.5 with the addition of $0.1M$ HCl under gentle stirring with a plastic spatular. The solution is then neutralized by the addition of $2M$ Tris, pH 8.0, and washed three times in filtered artificial sea water.
6. It is important to monitor visually the generation of ghost membranes following homogenization. Figure 2 in ref. *5* illustrates the typical appearance of ghost membranes. Depending on the type of hand-operated homogenizer used, the breakage of the majority of eggs (>95%) can vary from batch to batch. On the one hand, the breakage of sea urchin eggs

should be adequate to release the cytoplasmic content, but on the other hand, the surface membrane should not be greatly fragmented, since this leads to loss in yield of suitable ghost membranes.
7. If a small yellow pellet is seen at the bottom of the final membrane complex pellet, this is owing to eggs that were inadequately broken. Avoid resuspending this part of the otherwise whitish egg surface membrane complex.
8. Alternative to the described batch method, binding and elution of the glycoprotein fraction may be performed by column chromatography. However, elution of the bound glycoprotein fraction using a linear gradient of N-acetylglucosamine does not result in a major separation of WGA binding proteins. Since the batch method is technically simpler, this procedure is used for routine purification of the sperm receptor. However, the disadvantage of batch methods is a possible gradual loss of matrix when removing the eluted supernatant fraction. This can be a problem with expensive affinity resins, and in that case, column chromatography would be advantageous.
9. Regeneration of a lectin matrix is usually achieved by salt washes. Treat the WGA matrix, following elution with N-acetylglucosamine, with three 1.5M NaCl washes. Then re-equilibrate the WGA matrix with incubation buffer or phosphate-buffered saline complemented with 5 mM NaN$_3$ for storage at 4°C.
10. For a description of the standard techniques used in the analysis of the sperm receptor, see ref. 7. Gel electrophoretic procedures combined with analytical blotting techniques are ideal for monitoring the purification of a membrane protein, since they are not only comparatively inexpensive and simple methods, but also relatively fast and reliable. Silver staining of proteins separated by gel electrophoresis is a very useful method to estimate the purity of the final preparation, since it is highly sensitive in the detection of contaminating proteins.

Acknowledgments

The author thanks W. J. Lennarz, State University of New York, in whose laboratory experiments purifying the sea urchin sperm receptor were worked out. This research was funded by the National Institutes of Health (HD 18590-W. J. Lennarz).

References

1. Helenius, A. and Simons, K. (1975) Solubilizationn of membranes by detergents. *Biochim. Biophys. Acta.* **415**, 29–79.
2. Scopes, R. K. (1994) *Protein Purification: Principles and Practice,* 3rd ed. Springer-Verlag, New York.
3. Thomas, T. C. and McNamee, M. G. (1990) Purification of membrane proteins. *Methods Enzymol.* **182**, 499–519.

4. Lis, H. and Sharon, N. (1986) Lectins as molecules and as tools. *Annu. Rev. Biochem.* **55**, 35–67.
5. Ohlendieck, K., Partrin, J. S., and Lennarz, W. J. (1994) The biologically active form of the sea urchin egg receptor for sperm is a disulfide-bonded homo-multimer. *J. Cell Biol.* **125**, 817–824.
6. Ohlendieck, K. and Lennarz, W. J. (1995) Role of the sea urchin egg receptor for sperm in gamete interactions. *Trends Biochem. Sci.* **20**, 29–33.
7. Ohlendieck, K., Dhume, S. T., Partin, J. S., and Lennarz, W. J. (1993) The sea urchin egg receptor for sperm: isolation and characterization of the intact, biologically active receptor. *J. Cell Biol.* **122**, 887–895.
8. Foltz, K. R. and Lennarz, W. J. (1992) Identification of the sea urchin egg receptor for sperm using an antiserum raised against a fragment of its extracellular domain. *J. Cell Biol.* **116**, 647–658.
9. Foltz, K. R., Partin, J. S., and Lennarz, W. J. (1993) Sea urchin egg receptor for sperm: sequence similarity of binding domain and hsp 70. *Science (Wash. DC)* **259**, 1421–1425.
10. Luzio, J. P. and Bailyes, E. M. (1993) Isolation of a membrane-bound enzyme, 5'nucleotidase, in *Methods in Molecular Biology, Vol. 19: Biomembrane Protocols: I. Isolation and Analysis* (Graham, J. M. and Higgins, J. A., eds.), Humana, Totowa, NJ, pp. 229–242.
11. East, J. M. (1994) Purification of a membrane protein (Ca^{2+}/Mg^{2+}-ATPase) and its reconstitution into lipid vesicles, in *Methods in Molecular Biology, Vol. 27: Biomembrane Protocols: II. Architecture and Function* (Graham, J. M. and Higgins, J. A., eds.), Humana, Totowa, NJ, pp. 87–92.
12. Vretblad, P. (1976) Purification of lectins by biospecific affinity chromatography. *Biochim. Biophys. Acta.* **434**, 169–176.

CHAPTER 31

Lyophilization of Proteins

Ciarán Ó Fágáin

1. Introduction

Lyophilization, or freeze-drying, is a method for the preservation of labile materials in a dehydrated form. It can be particularly suitable for high-value biomolecules, such as proteins. The process involves the removal of bulk water from a frozen protein solution by sublimation under vacuum with gentle heating (primary drying). This is followed by controlled heating to more elevated temperatures for removal of the remaining "bound" water from the protein preparation (secondary drying). Residual moisture levels are often <1% *(1)*. If the freeze-drying operation is carried out correctly (*see* Section 3.), the protein will preserve all or most of its initial biological activity in the dry state. This dry state offers many advantages for long-term storage of the protein in question.

An isolated protein in an aqueous system can suffer various adverse reactions over time owing to physical, chemical and biological factors. Typical physical phenomena are aggregation *(2)* and precipitation. There are many deleterious chemical reactions involving the side chains of amino acid residues, notably asparagine, aspartic acid, and cysteine/cystine *(2,3)*, or the glycation of lysine residues with reducing sugars (the Maillard reaction) *(4,5)*. Biological deterioration can result from loss of an essential cofactor or from the action of proteolytic enzymes, either endogenous or arising from microbial contamination. Water participates directly in many of the chemical reactions and in proteolysis. In any case, it provides a medium for molecular movement and interactions. For these reasons, removal of water effectively prevents deterioration of the protein. The freeze-dried preparation will be much less bulky than the original solution and can conveniently be stored in a laboratory freezer or refrigerator (or perhaps even at room temperature). When one wishes to use the protein preparation, one can rehydrate it simply by addition of an appropriate volume of pure water or suitable buffer solution.

At very low temperatures, a liquid may behave in one of two ways. Eutectic solutions undergo a sharp liquid–solid freezing transition over a very narrow temperature range, whereas amorphous liquids are characterized by a glass transition in which viscosity increases dramatically with cooling and the solution takes on the macroscopic properties of a solid, even though it has not crystallized. Below this glass transition temperature (T_g'), virtually no adverse chemical or biological reaction can take place. Above T_g', however, the viscous, rubbery material is very prone to deterioration *(6) (see* Fig. 1). Understanding these low-temperature features of liquids is important for effective freeze-drying. Eutectic and glass transitions will greatly influence the freeze-drying protocol (the way the freeze-dryer is run), and the choice of substances used as preservatives or excipients in the product formulation subjected to lyophilization. Understanding of freeze-drying as a process has grown in recent years *(6,8),* and knowledge of the effects of additives has also grown *(9,10).* Together these advances have allowed lyophilization to be undertaken on a more rational and less strictly empirical basis. These matters are discussed in more detail in Section 3. Reference *11* contains a useful treatment that touches on the underlying theory.

Effective freeze-drying is time consuming and cannot be accomplished in a hurry. Process times of 72 h or greater are usual, depending on the nature of the product formulation and the properties of its constituents. The complex lyophilizer apparatus will have high capital and running costs (the latter owing to its power consumption and the time required). Freeze-drying is usually reserved, therefore, for high-value proteins, or is used in cases where alternative product presentations (such as ammonium sulfate precipitates) are unsuitable or give insufficient shelf lives.

A typical freeze-dryer comprises a set of shelves that can be cooled to about –40°C or heated above ambient temperature, a condenser usually cooled to –60°C, and a high-performance vacuum pump. The shelves allow rapid freezing of the protein solution and, later, gentle programmed heating for primary and secondary drying. The vacuum pump reduces the pressure within the system to allow sublimation of the bulk water

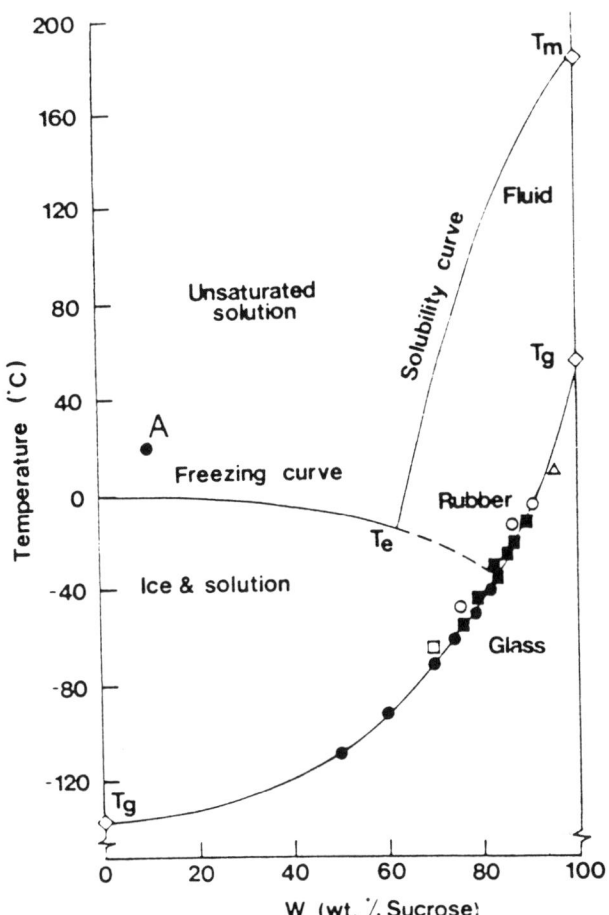

Fig. 1. Solid–liquid state diagram for the sucrose–water system, showing the glass transition T_g/concentration profile, the equilibrium solubility curve, running between the melting point and the eutectic temperature, T_e, and the region of metastability (supersaturation) beyond T_e (broken line). Symbols represent experimental T_g data taken from different sources. Point A represents a typical dilute solution that is to be dried to a stable solid product. From ref. 7.

(i.e., primary drying), whereas the condenser acts as a lower temperature trap for the sublimed water that collects in the form of ice (i.e., it provides a temperature gradient within the apparatus). This explains why the condenser temperature must be considerably less than that of the shelves.

FREEZING

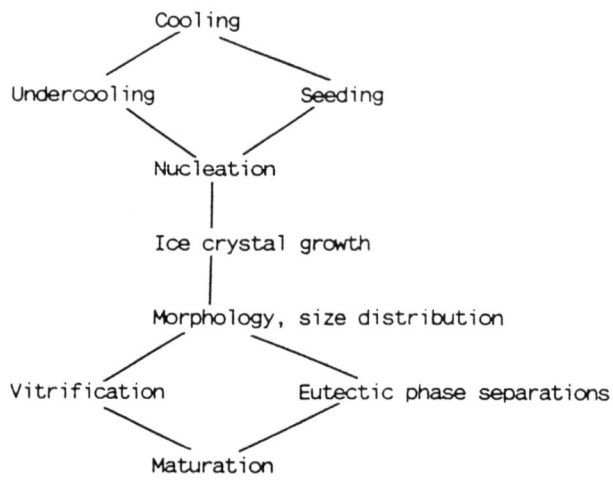

DRYING

Ice sublimation (Product Temperature, Chamber pressure)

Desorption (Temperature profile, Residual moisture)

STORAGE   Residual moisture/temperature control

Scheme 1. Physical principles of freeze-drying. From ref. 7.

A complex interplay of chemical and physical phenomena takes place during freeze-drying (*see* Scheme 1). The product yield (that is, the percentage recovery of the initial active protein) depends on the formulation in which the protein is placed prior to lyophilization *(6,10),* whereas its ease of rehydration and its stability on long-term storage (or "shelf life") are influenced by the processing regime *(6).* Section 3.1. describes, in broad terms, the operation of a typical freeze-dryer apparatus. Some critical factors concerning the operations of freezing, and primary and secondary drying are outlined in Sections 3.2.–3.4.

2. Materials

Freeze-dryers differ greatly in their specifications depending on the model or manufacturer. Regardless of the type, one must use it at all times in accordance with the manufacturer's instructions. The method discussed in Section 3.1. assumes that the reader studies this chapter together with the user's handbook.

The most basic lyophilizer equipment will consist of a condenser/vacuum pump unit to which one may attach a centrifugal test tube holder or a manifold for the drying of multiple round-bottomed flasks. Freezing is accomplished in a separate cooling bath, usually filled with an alcohol. Such equipment can be used successfully for small-volume samples, but fine and reproducible control of the overall process may not be possible. Neither can one attain continuous recording of sample temperature. Higher grade equipment, with temperature-programmable shelves and a number of temperature probes, is to be preferred. Shelf-equipped freeze-dryers are especially suitable for use with rubber-capped pharmaceutical vials or with microtiter plates. Often an externally operated screw press will be present, allowing one to seal vials (partially closed with rubber stoppers so as not to restrict gaseous movement) under vacuum before releasing air into the chamber.

Obviously, one should thoroughly familiarize oneself with the machine to be used. The user should know the locations of the main components and indicators together with appropriate settings for the meters, controls, and valves.

The characteristics of the vials (usually glass) in which product is lyophilized will influence the process. The vial diameter, glass type, and bottom shape and thickness will all affect the rate of heat transfer from the shelves to product. The vials should withstand freezing and pressure changes, and should be uniform with respect to internal diameter and bottom thickness. The vial bottoms should be completely flat to make good contact with the shelves. The same model of vial should be used consistently, since a change of vial type will introduce a variable into an optimized process *(8)*. Ensure that any other vessels used (e.g., round-bottomed flasks) are of sufficient quality to withstand the temperatures and pressures associated with freeze-drying.

3. Methods

3.1. Operation of Freeze-Dryer

These directions for freeze-drying are of necessity very general. Exact details will depend on the material and apparatus in question, and on the user's requirements. In pilot or manufacturing operations, Good Manufacturing Practice, ISO 9000, or regulatory authorities will likely impose many procedural disciplines that are beyond the scope of this chapter.

3.1.1. Start-Up

Ensure that the valve connecting the vacuum to the drying chamber is closed. Start the vacuum pump and allow it to evacuate. Observe the decrease in pressure on the vacuum indicator. It is important to start the pump first so as to reach a steady-state high vacuum long before evacuation of the main chamber. It is also important that the pump warms up thoroughly before the condenser or shelves are cooled: water in any form reaching the pump can cause damage. A warm-up time of 30 min will usually suffice. Close the condenser drain outlet, which should always be left open when the freeze-dryer is not in use. (If not, open it, allow any water to drain completely, and then close tightly once again.) Switch on the condenser, and allow to cool to –60°C. At this point, it may be convenient to cool the shelves slightly: this will aid the overall chilling process later. Do not cool the shelves more than a few degrees below ambient temperature, however, to prevent atmospheric condensation. Condensation will add to the water load, which must be removed during the lyophilization process.

3.1.2. Filling and Loading

Now loading of the shelves with the preparation to be lyophilized can begin (*see* Note 1). Fill only minimal amounts of material into each container to ensure a high ratio of surface area to volume: This will aid effective freeze-drying. The product matrix will tend to inhibit sublimation of water vapor from the surface of the ice crystal. This resistance depends on the depth of liquid and on the solids content of the product *(8)*. Also, a good head space in the vial or ampule will allow easier and better gaseous movement. One can conveniently load filled vials into metal trays at the bench and then place the trays into the drying chamber. The trays should have level bottoms to make good contact with the freeze-dryer shelves. Many vials have narrow necks, which one closes with special rubber stoppers. The design of these stoppers allows one to cap the vials partially while maintaining contact between their contents and the atmosphere. (Sealing of the vials takes place later under vacuum.) If one wishes to use such closures, one should partially stopper the vials before loading them into the freeze-dryer. If using microtiter plates, place them within special metal frames, which then come into contact with the shelves.

The freeze-dryer will likely have a number of flexible temperature probes; carefully clean each of these. Place some of the probes so as to monitor the temperatures of the shelves. Dip others directly into the material to be lyophilized. A temperature record of the actual solution (as distinct from the shelf underneath it) is well worth the loss of a small number of product vials. Arrange the temperature probes in different locations throughout the chamber: in the centers of shelves and also at the edges. Conditions will not be homogeneous across all vials, and monitoring should be as complete as equipment will allow. Insert the temperature probes right to the bottom of vials. Ensure that the probes remain in contact with the intended sample or shelf, and do not become displaced during product loading or manipulation or during lyophilization.

3.1.3. The Freeze-Drying Operation

Once loading is complete, close and secure the door of the drying chamber. Now the freezing process may begin. Bring the material well below its freezing temperature (or glass transition temperature) as quickly as possible. This is particularly important for a number of reasons (*see* Section 3.2.). Allow the shelf/product temperature to fall further to a steady −40°C before drawing the vacuum. Check that the drying chamber door is properly closed and sealed; then evacuate the chamber by opening the vacuum valve. The vacuum gage reading will indicate atmospheric pressure as the valve is opened and may take a few minutes to show a vacuum once again as the air within the drying chamber evacuates. Soon, however, the pressure within the drying chamber will decrease, and a steady-state high vacuum will result.

There now exists a high vacuum within the drying chamber together with a temperature gradient from the shelves/product (−40°C) to the condenser (−60°C). These conditions permit sublimation of the bulk water in the product over a period of hours as the shelves are gently heated. The sublimed water will collect as ice on the condenser. Sublimation, or primary drying, removes only the bulk water in the system; it is insufficient to remove the "bound" water closely associated with the protein molecules. Removal of this bound water, or secondary drying, requires heating to elevated temperatures. This is usually applied through the shelves on which the product rests (and has previously been frozen). One can program the shelves to heat to a particular temperature at a defined rate appropriate for the product undergoing lyophilization. One must select the heating regime with particular care (*see* Sections 3.3. and 3.4.).

3.1.4. Termination of Run and Removal of Product

One may terminate the freeze-drying process when the cake has a good appearance and the product has reached a sufficiently low steady-state percentage of moisture (determined in a separate series of experiments). Seal vials, partially capped with rubber stoppers, before insertion into the chamber, by operating the special screw press. This operation will insert the stoppers fully before air is admitted into the drying chamber. Sealing of vials while still under vacuum has an important advantage over the use of an inert, moisture-free gas, such as nitrogen. Air will rush in when a vial, sealed under vacuum, is uncapped. The sound of the in-rushing air will be absent from a defective vial that has failed to seal or where the seal has broken down. In this way, the user will immediately know of the defect and be aware that the vial contents may have deteriorated *(1)*.

Close the valve to the vacuum pump firmly. Slowly release the vacuum within the drying chamber by a minimal opening of the air inlet valve: air admission should be as gentle as possible to prevent undue sudden stresses on sealed vials, and also to prevent disturbance or upset of chamber contents by a strong jet of incoming air. The vacuum/pressure gage will show a rise in drying chamber pressure; this will eventually equalize to atmospheric, and one can then open the chamber door. Vials of freeze-dried product can be removed for storage (preferably at refrigerated temperatures). Store microtiter plates immediately in a desiccator or (better) vacuum-sealed within foil packets containing desiccant sachets.

3.1.5. Shutting Down

If the freeze-dryer is to be reused immediately, one must be certain that the condenser's ice capacity will withstand the accumulation of ice from two runs (*see* Note 1). If there is no more material for lyophilization, shut down the apparatus carefully. Remove any material spilled in the drying chamber and clean the shelves carefully according to the manufacturer's instructions. Leave the chamber door slightly ajar to allow circulation of air, and to prevent sticking and compression of the door seals. Switch off the condenser, and open its drain outlet. (The drain should remain open until the freeze-dryer is next used.) Over a period of hours, the ice on the condenser surfaces will melt and drain away. Once the condenser temperature has returned to ambient and all the melted ice has drained away, allow the vacuum pump to run for a further 3 h before shutting it down. This is to prevent any damage to the pump owing to occurrence or accumulation of condensation. To maintain good pump performance, change the oil often according to the handbook's directions.

3.1.6. Using Simpler Freeze-Dryers

Use of simpler apparatus with manifold or centrifuge accessories is carried out in much the same stepwise fashion as above. In these cases, aliquots of the product are frozen in individual opennecked flasks or tubes. Switch on the vacuum pump and condenser, and allow to run as noted. Freeze flask contents by immersion in an alcohol cooling bath or in liquid nitrogen (follow the normal precautions). Swirl or rotate the flask during the freezing step to effect even distribution of the product over the widest possible surface area. This will minimize the depth of material through which water loss must occur (*see* Section 3.1.2. and ref. *8*). Connect the frozen material directly to the manifold assembly (or load into the centrifugal tube dryer), and draw a vacuum in the chamber as quickly as possible so as to minimize back-melting of ice under atmospheric pressure. The rapid reduction of pressure by the vacuum pump will aid sublimation and help minimize melting to water. Although control of temperature and heating rates are more problematic with such accessories, visual inspection of the material and of the dried cake is much easier than in an enclosed chamber.

3.2. Freezing

In freezing, liquid water crystallizes to yield solid ice. Water, in common with other substances, including many solutes, is an eutectic material with a sharp transition between the liquid and solid states. Noncrystalline materials undergo a glass transition and become rigid at a certain low temperature (T_g'). This is the end result of a notable increase in viscosity with decreasing temperature. The material becomes increasingly rubbery until rigidification occurs at T_g' and liquid movement effectively ceases (*see* Note 2). Even for eutectics, ice formation usually does not occur at the thermodynamic freezing point: the actual temperature of freezing is generally 10–15°C lower than this, so that supercooling (or undercooling) is required *(6,8)*. It is generally satisfactory to cool the product to just below 0°C initially, but not so low as to induce crystallization. When all the product has cooled to this set temperature, reduce the shelf temperature sharply to crystallize bulk water to ice. Typical temperatures lie in the range –20 to –30°C. Allow time for ice to form in all containers before proceeding to the next stage, which involves further shelf cooling below the lowest eutectic temperature (for crystalline materials) or below the glass transition temperature (T_g', for amorphous substances) (see Note 3). A usual final temperature is –40°C. The product is reliably solidified only below these temperatures, and only now may primary drying begin. It is vital to maintain the freezing temperature below T_g' before and during primary drying. If the temperature rises above this value, the material ceases to be a solid and becomes a viscous rubber very prone to deleterious reactions, with resulting losses of activity. The rubber will also undergo a mechanical collapse, take on a different appearance, and be difficult to rehydrate (6). Collapsed product is always unacceptable even if reasonable biological activity remains.

Other notable changes occur in the product during freezing, many of which can lead to significant protein denaturation. As the bulk water freezes to ice, the amount of liquid water remaining naturally decreases. This leads to great increases in the concentrations of solutes, such as salts. Such freeze-concentration can have far-reaching effects. Concentration will increase the rates of unwanted chemical reactions, such as oxidations. Buffer components may crystallize differentially, leading to pronounced pH shifts; also, pK_a values are temperature-dependent. For phosphate buffers, one should use potassium or mixed salts in preference to sodium salts (6,12). For all of these reasons, one should accomplish the freezing steps (crystallization of bulk water and cooling below the eutectic or T_g' temperatures) in any lyophilization process as quickly as possible.

Suitable excipients or protective substances can lessen or overcome some of these damaging freezing effects: they can particularly influence melting or collapse temperatures (13). Excipients may be classed as bulking agents, tonicity modifiers, buffers, and cryoprotectants/lyoprotectants (13). One should use bulking agents where the product's solids content is low. They help to ensure the development of a plug of dried material, which will be of good appearance. Bulking agents can also prevent blowout: where the material has a very low solids content of the order of 1%, disruption of cake structure by the issuing water vapor can result in loss of the dried product from the container along with the vapor (10). Typical bulking agents include polyols and certain sugars, mostly nonreducing.

Buffer substances must be chosen with great care. One must consider possible variations with temperature in pK_a and solubility as well as compatibility with the protein(s) in question and other constituents. Useful information can be found in ref. *14*, and in monographs and articles dealing with the handling of proteins.

Many bulking agents will double as tonicity modifiers in preparations for administration in vivo to ensure that the preparation is isotonic with body fluids. Any agent used for this purpose must, of course, be pharmaceutically acceptable.

Cryoprotectants and lyoprotectants are substances that stabilize a protein against the deleterious effects of freezing and of lyophillization/storage, respectively. These protecting additives appear to be preferentially excluded from the protein surface. This results in a strengthening of the water "shell" surrounding the protein *(15)*. Polyols and certain salts and sugars can be beneficial. In general, nonreducing sugars are preferable, since, unlike reducing sugars, they cannot participate in Maillard reactions. Xylitol's T_g' is −46.5°C in a freeze concentrate of 42.9 wt% water, whereas T_g' values for sorbitol, sucrose, and trehalose are −43.5, −32, and −29.5°C at wt% water values of 18.7, 35.9, and 16.7, respectively *(6)*. Glycerol has a notably low T_g' value of −65°C at 46 wt% water *(6)*. This means that glycerol-containing formulations must reach very low temperatures to become glassy. Glycerol, therefore, is not an ideal lyophilization excipient, since such temperatures may be difficult to achieve. Useful discussions and examples of cryoprotectants occur in refs. *9,15,16; see also* Note 2 and Section 3.3. of Chapter 32 in this volume.

3.3. Primary Drying

Primary drying is the sublimation under vacuum of bulk ice from the product to the much colder condenser typically held at −60°C. Sublimation will be faster at higher shelf temperatures, so heat the shelves to a few degrees below T_g' to quicken the process (*see* Note 2). **Never** allow the shelf temperature to equal or exceed T_g', however. If it does, collapse and product deterioration may occur. A usual safety margin is 2–5°C below the eutectic or collapse temperature *(8)*. (The term "collapse temperature" is usually equivalent to T_g' for an amorphous substance.) The rate of sublimation depends on the vacuum and on the condenser temperature, and therefore is largely determined by the characteristics of the lyophilizer apparatus *(6)*. Drying often becomes easier as the temperature approaches T_g' or the eutectic temperature: The sublimation rate can increase by about 13% for each 1°C rise in temperature *(8)*. Resistance to drying also decreases with decreasing product thickness (i.e., filling height) and with increasing vial diameter (which influences the area of the drying surface) *(8)*. It is important that the temperature of the actual product remains constant (and therefore is monitored) throughout primary drying. The sublimation rate will decrease as primary drying proceeds, and therefore, the degree of product cooling owing to sublimation will decrease also. The purpose of shelf warming is to counter this sublimation cooling. Accordingly, be sure to adjust the heat input to the shelves as the product dries to prevent a net rise in product temperature. An unchecked rise in product temperature could lead to collapse. One can use chamber pressure to control the length of the primary drying cycle *(8)*. Even under uniform conditions, primary drying times may vary by up to 10% for a given process. Make sure to include a delay period (ascertained empirically and occupying a period of hours) at the end of the primary drying cycle to ensure that all ice has sublimed satisfactorily and to avoid collapse *(8)*. *See* Note 4 for a reference to a quantitative model that has been used to predict drying times.

3.4. Secondary Drying

Secondary drying removes the remaining unfrozen, bound water from the lyophilizing material yielding a final product, low in residual moisture, which will be stable for an extended period without deterioration. The unfrozen water is removed by heating the shelves on which the primarily dried product rests. The rule of thumb for initiation of secondary drying is the equivalence of product and shelf temperatures. For safety, however, one should include a delay period at the end of primary drying in any freeze-drying protocol to prevent collapse of any vials that have not quite finished sublimation *(8)* (*see also* Section 3.3.). The partial pressure of water within the drying chamber drops at the end of primary drying as the last of the ice sublimes. If it can be monitored, this drop in water partial pressure can be a good indicator of the completion of primary drying. Even during secondary drying, with much of the original water gone, the preparation's temperature should never rise above T_g', something that is seldom appreciated *(6)*. T_g', however, rises as the residual water content drops *(8)*. One can, therefore, increase the temperature (within limits) during secondary drying. The vacuum need not be exhaustive during secondary drying; indeed, one should use a pressure in the region of 0.2 torr for secondary drying *(8)*. There is little evidence that one can harm a protein by overdrying it *(8)*.

3.5. Quality Indices

Most freeze-dryers will have a built-in chart recorder to provide a profile of shelf temperatures (and perhaps other parameters, such as vacuum) during the lyophilization run. Study this carefully, and retain with other batch information. Note the cake shape and texture and any variations between vials, especially those located at different positions within the chamber. Choose a representative number of vials from different chamber locations for scrupulous moisture determination (*see* Note 5). Measure the yield or recovery (%) of the initial biological activity by appropriate assay following rehydration of a representative number of vials. While waiting to assay, note the time required for complete product rehydration. Also note whether any turbidity remains on rehydration or after what interval turbidity appears in a clear sample *(6)*. Persistence of the rehydrated biological activity can be measured at suitable or convenient time intervals. Accelerated degradation methods can predict shelf lives of the lyophilized preparation at temperatures of interest for long-term storage. These methods involve extrapolation of an Arrhenius plot. Accelerated degradation protocols can be of value, provided the activity decay is first-order at each of the temperatures tested and all data are scrupulously accurate and precise. Reference *17* gives guidelines for the proper use of accelerated methods; *see also* Section 3.6. of Chapter 32 in this volume. Note that the formulation used greatly influences yield, whereas the process parameters affect ease of rehydration and shelf life *(6)*.

4. Notes

1. The volumes of bulk liquid subjected to freeze-drying must never exceed the manufacturer's recommendations. If the condenser's ice capacity is reached or exceeded, the degree of product drying will be insufficient and many problems can result.
2. The glass transition temperature, T_g', and the product water content are critical parameters in the freeze-drying process. It is essential to prevent or minimize damage to the protein of interest during freezing. Inclusion of excipients (additives) with high T_g' values in the protein formulation to be freeze-dried can be very useful. The mixture will form a glass at relatively low temperatures, minimizing freezing damage. A high T_g' will also allow

the use of higher temperatures during primary drying with less danger of product collapse. Any constituent that will lower the unbound water content of the freeze concentrate will help shorten the secondary drying operation *(6)*, but uncrystallized salts will decrease T_g' since any salt will bring about a depression of freezing point. Thus, the salts content of the product formulation should be as low as is practicable *(6)*. The optimum temperature for freezing and primary drying depends on the ratio protein:protectant additive:salt in the freeze concentrate rather than in the initial solution *(6,10)*. The ratio protein:other solids in the freeze concentrate influences T_g' *(6)*.
3. The glass transition temperature can be determined by differential scanning calorimetry, although a sensitive instrument is required *(6)*. Electrical resistance measurements can sometimes be useful, but microscopic observation of freeze-drying over a range of temperatures is arguably the most direct, sensitive, and unambiguous method for determination of the collapse temperature *(7)*.
4. Reference *18* contains a brief quantitative treatment of freeze-drying, discussing it in terms of heat and mass fluxes to derive an equation for the prediction of drying times. The model successfully predicted times for removal of 65–90% of the total initial water. However, the actual times required for removal of the remaining 10–35% of the water were greater than those predicted.
5. The residual moisture content of, and its distribution throughout, the lyophilized preparation will dictate its long-term stability. Uneven moisture distribution between vials often leads to biphasic activity loss profiles on extended storage *(6)*. Each 1% of moisture can depress T_g' by more than 10°C *(11)*.

Acknowledgments

The author thanks F. Franks for permission to reproduce figures from his work and P. K. Walsh for a critical reading of the manuscript.

References

1. Thuma, R. S., Giegel, J. L., and Posner, A. H. (1987) Manufacture of quality control materials, in *Laboratory Quality Assurance* (Howanitz, P. J. and Howanitz, J. H., eds.), McGraw-Hill, New York, pp. 101–123.
2. Liu, W. R., Langer, R., and Klibanov, A. M. (1991) Moisture-induced aggregation of lyophilized proteins in the solid state. *Biotech. Bioeng.* **37,** 177–184.
3. Volkin, D. B. and Middaugh, C. R. (1992) The effect of temperature on protein structure, in *Stability of Protein Pharmaceuticals, Part A: Chemical and Physical Pathways of Protein Degradation* (Ahern, T. J. and Manning, M. C., eds.), Plenum, New York, pp. 215–247.
4. Hageman, M. J. (1992) Water sorption and solid-state stability of proteins, in *Stability of Protein Pharmaceuticals, Part A: Chemical and Physical Pathways of*

Protein Degradation (Ahern, T. J. and Manning, M. C., eds.), Plenum, New York, pp. 273–309.
5. Quax, W. J. (1993) Thermostable glucose isomerases. *Trends Food Sci. Technol.* **4,** 31–34.
6. Franks, F. (1990) Freeze-drying: from empiricism to predictability. *Cryo-Lett.* **11,** 93–110.
7. Franks, F. (ed.) (1993) *Protein Biotechnology: Isolation, Characterization, and Stabilization,* Humana, Totowa, NJ.
8. Pikal, M. J. (1990) Freeze-drying of proteins. Part 1: process design. *BioPharm.* **3,** 18–27.
9. Carpenter, J. F. and Crowe, J. H. (1988) The mechanism of cryoprotection of proteins by solutes. *Cryobiology* **25,** 244–255.
10. Pikal, M. J. (1990) Freeze-drying of proteins. Part 2: formulation selection. *BioPharm.* **3,** 26–30.
11. Franks, F., Hatley, R. H. M., and Mathias, S. F. (1991) Materials science and the production of shelf-stable biologicals. *Pharm. Technol. Int.* **3,** 24–34.
12. Franks, F. and Murase, N. (1992) Nucleation and crystallization in aqueous systems during drying: theory and practice. *Pure Appl. Chem.* **64,** 1667–1672.
13. Hanson, M. A. and Rouan, S. K. R. (1992) Formulation of protein pharmaceuticals, in *Stability of Protein Pharmaceuticals, Part B, In Vivo Pathways of Degradation and Strategies for Protein Stabilisation* (Ahern, T. J. and Manning, M. C., eds.), Plenum, New York, pp. 209–233.
14. Blanchard, J. S. (1984) Buffers for enzymes. *Methods Enzymol.* **104,** 404–415.
15. Timasheff, S. N. (1992) Stabilisation of protein structure by solvent additives, in *Stability of Protein Pharmaceuticals, Part B, In Vivo Pathways of Degradation and Strategies for Protein Stabilization* (Ahern, T. J. and Manning, M. C., eds.), Plenum, New York, pp. 265–285.
16. Schein, C. H. (1990) Solubility as a function of protein structure and solvent components. *Biotechnology* **8,** 308–317.
17. Kirkwood, T. B. L. (1984) Design and analysis of accelerated degradation tests for the stability of biological standards III. Principles of design. *J. Biol. Stand.* **12,** 215–224.
18. Geankoplis, C. J. (1983) *Transport Processes and Unit Operations,* 2nd ed., Allyn and Bacon, Boston, MA, pp. 554–557.

CHAPTER 32

Storage of Pure Proteins

Ciarán Ó Fágáin

1. Introduction

There is often a need to store an isolated or purified protein for varying periods of time. If the protein in question is to be studied, it will take some time to characterize the properties of interest. If the protein is an end product or is for use as a tool in some procedure, it will likely be used in small quantities over a period of time. It is vital, therefore, that the protein retains as much as possible of its original, postpurification, biological (or functional) activity over an extended period of storage. This storage period or "shelf life" may vary from a few days to more than 1 yr. Shelf life can depend on the nature of the protein and on the storage conditions. This chapter explores the means by which activity losses occur on storage, and discusses a range of measures to prevent or lessen the inactivating events. The chapter also describes the use of accelerated storage (accelerated degradation) testing for the prediction of shelf lives at particular temperatures.

Apart from extremes of temperature and pH (which will, naturally, be avoided as conditions for routine or long-term storage), a variety of factors can lead to loss or deterioration of a protein's biological activity. These include proteolysis, aggregation, and certain chemical reactions (1). Proteolysis may arise from microbial contamination or be an intrinsic hazard in the case of proteolytic enzymes. A protein may lose an essential cofactor, or the subunits of oligomeric proteins may dissociate from each other, resulting in each case in loss of activity (1). Adsorption to surfaces may also lead to inactivation (1,2).

From: *Methods in Molecular Biology, Vol. 59: Protein Purification Protocols*
Edited by: S. Doonan Humana Press Inc., Totowa, NJ

The purely chemical reactions are few and well defined. Deamidation of glutamine and asparagine can occur at neutral to alkaline pH values, whereas peptide bonds involving aspartic acid undergo cleavage under acidic conditions. Cysteine is prone to oxidation, as are tryptophan and methionine. Alkaline conditions lead to reduction of disulfide bonds, and this is often followed by β-elimination or thiol-disulfide exchange reactions *(1)*. Where reducing sugars are present with free protein amino groups (N-termini or lysine residues), there may be destructive glycation of the amino functions by the reactive aldehyde or keto groups of the sugar (the Maillard reaction) *(3)*. Elevated temperatures favor all of these reactions, but it is important to note that aggregation and deleterious chemical reactions can occur at moderate temperatures also. Virtually complete aggregation of lyophilized bovine serum albumin occurred over 24 h at 37°C following addition of just 3 µL of physiological saline. The degree of aggregation was less, but still significant, at lower temperatures. The underlying cause was formation of an intermolecular disulfide bond by a thiol-disulfide exchange. Ovalbumin, glucose oxidase, and β-lactoglobulin underwent similar aggregation under the same conditions, but by a somewhat different mechanism *(4)* (*see* Note 1). These findings underline how important it is to ascertain correct storage conditions for the protein of interest. Simple reliance on the laboratory refrigerator to minimize activity losses may not suffice over an extended period.

Other factors may work against the refrigerator. Certain proteins are more stable at room temperature than in the refrigerator and are said to be cold labile. This cold denaturation has been well characterized for myoglobin and a few other proteins *(5,6)*. It is a property of the protein itself and is distinct from freezing inactivation (*see* Section 3.2. of Chapter 31). This phenomenon arises from the fact that it is thermodynamically possible for a protein to unfold at low as well as at high temperatures (*see* ref. *7* for a summary of the notable features of cold denaturation). Although cold denaturation is completely reversible *(7)*, it is inconvenient and can waste valuable time as the protein refolds (as well as causing laboratory scientists some unpleasant shocks).

Storage concerns a protein's long-term or kinetic stability. Kinetic stability is distinct from (and need not correspond with) thermodynamic stability. Thermodynamic stability refers to a protein's conformational stability in terms of the change in heat capacity or the Gibbs (free) energy on reversible unfolding. (Unfolding is often reversible in vitro, but other

Storage of Pure Protein

events occurring subsequently can lead to irreversible inactivation.) Kinetic stability measures the persistence of activity with time (or, to put it another way, the progressive loss of function). It can be represented by the scheme:

$$N \xrightarrow{k_{in}} I \qquad (1)$$

where N is the native, functional protein, I is an irreversibly inactivated form, and k_{in} is the rate constant for the inactivation process. The equation:

$$V_{in} = -d[N]/dt = k_{in}[N] \qquad (2)$$

describes the process mathematically, where V_{in} is the experimentally observed rate of disappearance of the native form *(1)*. Often the activity loss will be first-order, although more complex inactivation patterns are well documented *(8)*. Note that an apparently unimolecular first-order time-course of inactivation may mask a more complex set of inactivating molecular events *(8)*.

This chapter does not consider the special requirements of pharmaceutical regulatory authorities, good manufacturing practice (GMP), or ISO 9000 specifications. Readers needing to meet these directives and standards should consult materials produced by the appropriate authorities in order to ensure compliance. Finally, mention of suppliers names does not imply endorsement of any particular product(s).

2. Materials

1. All containers used for storage of pure proteins should be of good quality, and should tolerate temperatures as low as –20°C or even –70°C if freezer storage is desired or necessary. A number of manufacturers (such as Sarstedt, Nümbrecht, Germany, or Nunc, Roskilde, Denmark; there are many others) supply presterilized screw-cap plasticware with good mechanical and low-temperature properties. Clean glassware exhaustively, and sterilize it by dry heat. Screw caps or rubber stoppers that will not withstand dry heat can be autoclaved.
2. Membrane or cartridge filters, of pore size 0.22 µm for sterile filtration, are available from such companies as Gelman (Ann Arbor, MI) or Sartorius (Göttingen, Germany), with or without a Luer lock attachment for extra secure attachment to a hand-held syringe.
3. Use an ordinary domestic refrigerator for storage at temperatures of a few degrees Celsius. Modern machines are often combined with a freezer unit that can easily maintain temperatures as low as –20°C. Storage at –70°C or below will require a specialized low-temperature freezer.

4. A number of standard constant-temperature laboratory incubators will be required if accelerated storage testing is to be performed.
5. Highly purified forms of the following chemicals will be useful:
   a. Antimicrobials: sodium azide, thiomersal.
   b. Protease inhibitors: phenylmethylsulfonyl fluoride (PMSF).
   c. Chelators: ethylenediamine tetra-acetic acid (EDTA).
   d. Salts: ammonium sulfate; other ammonium, citrate, sulfate, acetate, and phosphate salts.
   e. Osmolytes: sucrose, glucose, trehalose, xylitol, glycerol, polyethylene glycol (PEG).
   f. Reducing agents: dithiothreitol, 2-mercaptoethanol.

3. Methods

3.1. Prevention of Bacterial Contamination

3.1.1. Antimicrobials

Microbial contamination can lead to significant losses of a pure protein by proteolysis. Contamination can be detected by standard microbiological plating techniques, but the aim must always be to avoid it in the first place. Even if one can achieve successful elimination or removal of contaminating microorganisms, the protein of interest may already have lost at least some activity or may have deteriorated in ways difficult to detect. Addition of antimicrobial compounds, such as sodium azide or thiomersal (sodium merthiolate, a mercury-containing compound), can prevent microbial growth. (Both of these compounds are poisonous: handle them with care.) Add sodium azide to a final concentration of 0.1% (w/v) or thiomersal to a final concentration of 0.01% (w/v). Note that azide will inactivate oxygen-binding proteins, such as hemoglobin or peroxidase (*see also* Note 2).

3.1.2. Filtration

Where one wishes to ensure sterility or to avoid use of the antimicrobials discussed in Section 3.1.1., filtration offers a useful alternative. (One can use both strategies together, of course.) A filter of pore size 0.22 μm will exclude all bacteria; indeed, this method is used in industry to sterilize labile materials that cannot be autoclaved or irradiated. Disposable filter cartridges are widely available in a variety of configurations; most have very low protein-binding capacities. Typically, one draws the solution to be sterilized into a syringe, and then removes the needle or tube. Next, connect the filter to the syringe nozzle, ensuring it is firmly mounted. Uncap a suitable sterile storage container directly beneath the filter outlet (using standard aseptic manipulations to avoid contamination of container or cap), and depress the syringe plunger to force the protein solution through the sterilizing filter into the container. Recap immediately. It will not be possible to flame plastic containers in a Bunsen burner as part of aseptic technique: it is much better practice to perform filtration operations of this sort in a Class 2 laminar flow microbiological safety cabinet, the design of which prevents contamination of the sample. Following the manufacturer's instructions closely, turn on the cabinet's fans, and allow to run for at least 10 min to allow adequate filtration of cabinet air. Open and remove the front door. Swab the cabinet's internal surfaces and the outer surfaces of storage containers brought inside the cabinet with 70% alcohol, and allow to evaporate. Carry out the filtration maneuvers, remove the storage containers, and dispose of waste materials appropriately. Swab the internal surfaces of the cabinet with alcohol once again, replace the front cover and allow to run for 10 min (or according to the user's handbook) before shutting down. It is not always possible to use a filter as fine as 0.22 µm directly (*see* Note 3).

3.2. Avoidance of Proteolysis

It can be difficult to remove proteases completely during purification of a target protein. Unless the object protein is completely pure (homogeneous), even tiny amounts of contaminating proteolytic enzymes can cause serious losses of activity during extended storage periods. The molecular diversity of proteases complicates the situation: there are exopeptidases, which remove amino acid residues from the N- or C-termini, and endopeptidases, which cleave internal peptide bonds within protein chains. In addition, there are four types of proteases classified by their molecular reaction mechanisms: the serine, cysteine (or thiol), acid, and metalloproteases *(9)*. Use EDTA in the concentration range 2–5 mM to complex the divalent metal ions essential for metalloprotease action. Pepstatin A is a potent, but reversible inhibitor of acid proteases. It is used at concentrations of around 0.1 µM, as are similar protease inhibitors. The compound PMSF reacts irreversibly with the essential serine in the active site of serine proteases, inactivating them. It can also act on some thiol proteases. Use it at a final concentration of 0.5–1 mM, having dissolved it first in a solvent, such as acetone (it is poorly soluble in water) *(10)*. One must, of course, ensure before addition that none of these compounds will adversely affect the protein of interest (*see also* Note 4).

If the protein of interest is itself a proteolytic enzyme, use of protease inhibitors is not feasible. One may need to store such a protein in dried form (Section 3.5.) or as a freeze-dried preparation (Chapter 31). Alternatively, one can place it in a solution with a pH value far removed from the protease's optimum. Trypsin, for example, is most active at mildly alkaline pH values. Stock solutions of trypsin are often prepared in 1 mM HCl, where the very acid pH value renders the enzyme effectively incapable of catalysis. This helps prevent autolysis during the course of the experiment. The enzyme molecule does not inactivate under these conditions and is fully active on dilution into a suitable assay solution *(11)*.

3.3. Use of Stabilizing Additives

3.3.1. Background

It has long been known that inclusion of low-mol-wt substances, such as glycerol or sucrose, in protein solutions can greatly stabilize the critical protein's biological activity. However, it took some time for the exact mechanism of this stabilization to be ascertained. Timasheff and Arakawa *(12)* showed that these types of substances are preferentially excluded from the vicinity of the protein molecules, since their binding to the protein is thermodynamically unfavorable. The protein molecule is preferentially hydrated by the solvent water. Loss of the protein's compact, properly folded structure (denaturation) will increase the protein–solvent interface. This, in turn, will tend to increase the degree of thermodynamically unfavorable interaction between the additive and the protein molecule. The result is that the protein molecule is stabilized by the additive *(12,13)*. This preferential exclusion means that there is less of the solute (additive) immediately surrounding the protein than there is in the bulk solution; it does not necessarily mean that no solute molecules can penetrate the protein molecule's hydration shell *(14)*.

It is important to note that the additives discussed in the following sections are generally applicable as stabilizing agents for proteins, but a given substance may not be effective for a particular protein. Both sucrose and PEG, for instance, are good stabilizers of invertase, but have denaturing effects on lysozyme; the same additive has contrary effects on the two enzymes *(15)*.

3.3.2. Addition of Salts

Certain salts can significantly stabilize proteins in solution. The effect varies with the constituent ions' positions in the Hofmeister lyotropic series,

Storage of Pure Protein

which relates to ionic effects on protein solubility *(16,17)*. This series ranks both cations and anions in order of their stabilizing effects. In the series below, the most stabilizing ions are on the left, whereas those on the right are actually destabilizing:

$$(CH_3)_4N^+ > NH_4^+ > K^+, Na^+ > Mg^{2+} > Ca^{2+} > Ba^{2+}$$
$$SO_4^{2-} > Cl^- > Br^- > NO_3^- > ClO_4^- > SCN^-$$

The stabilizing ions force protein molecules to adopt a tightly packed, compact structure by salting out hydrophobic residues. This helps prevent the unfolding, which is the initial event in any protein deterioration process (*see* Section 1.). Most stabilizing ions seem to act via a surface tension effect *(12,13)*. Ions can also stabilize proteins by shielding surface charges and can act as osmolytes by affecting the bulk properties of water *(18)*. Note that ammonium sulfate, which is widely used as a stabilizing additive and as a noninactivating precipitant, comprises two of the most stabilizing ions from this list, the NH_4^+ cation and the SO_4^- anion. To stabilize proteins in solution while avoiding precipitation, add ammonium sulfate to a final concentration in the range 20–400 mM *(18)*. One can do this by adding a minimal volume of a stock solution of ammonium sulfate of known molarity or by the careful addition of solid ammonium sulfate. Sprinkle the solid salt, a few grains at a time, into the protein solution. Ensure that each portion of ammonium sulfate added dissolves fully before addition of the next lot. This procedure will prevent accumulation of high local salt concentrations that are undesirable. In addition to ammonium sulfate, salts containing citrate, sulfate, acetate, phosphate, and quaternary ammonium ions are generally useful *(18)*. Note, however, that the nature of the counterion will influence the overall effect of such compounds on protein stability *(12,13)*.

This discussion assumes that ions used or added do not act as substrates, activators, or inhibitors of the enzyme or protein in question, and that added ions do not interfere with or precipitate possibly essential ions already in solution. Ammonium ion, for instance, is actually a substrate for the enzyme glutamate dehydrogenase.

3.3.3. Use of Osmolytes

Osmolytes are a diverse group of substances comprising such compounds as polyols, mono and polysaccharides, neutral polymers (e.g., PEG), and amino acids and their derivatives. They are not strongly charged and have little effect on enzyme activity below 1M concentration. In general, they affect water's bulk solution properties and do not interact directly with the protein *(18)*.

Use polyols and sugars at high final concentrations: Typical figures range from 10–40% (w/v) *(13,18)*. Sugars are reckoned to be the best stabilizers, but reducing sugars can react with protein amino groups leading to inactivation *(3,13)*. This problem can be avoided by using nonreducing sugars or the corresponding sugar alcohols. Glycerol is a very widely used low-mol-wt polyol. Its advantages include its ease of removal by dialysis and its noninterference with ion-exchange chromatography *(18)*. However, glycerol suffers from two significant disadvantages: It is a good bacterial substrate *(18)* and it greatly lowers the glass transition temperature (T_g') of materials to be preserved by lyophilization (*see* Section 3.2. of Chapter 31) or drying (Section 3.5., this chapter). The five-carbon sugar alcohol, xylitol, can often replace glycerol; it can be recycled from buffers and is not a convenient food source for bacteria *(18)*.

Polymers, such as PEG, are generally added to a final concentration of 1–15% (w/v). They increase the viscosity of the single-phase solvent system and so help prevent aggregation. Note, however, that higher polymer concentrations will promote the development of a two-phase system. The protein of interest will concentrate in one of these phases, and this may actually lead to aggregation *(18)*.

Amino acids with no net charge, notably glycine and alanine, can act as stabilizers if used in the range 20–500 mM *(18)*. Amino acids and derivatives occur as osmolytes in nature *(13)*. Some related compounds, such as γ-amino butyric acid (GABA) and trimethylamine N-oxide (TMAO), can be good stabilizers in the 20–500 mM range used for amino acids *(13,18)*.

3.3.4. Substrates and Specific Ligands

Addition of specific substrates, cofactors, or competitive (reversible) inhibitors to purified proteins can often exert great stabilizing effects. (Indeed, their inclusion may be necessary where an essential metal ion or coenzyme is only loosely bound to the apoprotein.) Occupation of the target protein's binding or active site(s) by these substances leads to minor, but significant conformational changes in the polypeptide backbone. The protein adopts a more tightly folded conformation, reducing any tendency to unfold *(19)* and (sometimes) rendering it less prone to proteolytic degradation. Occlusion of the protein's active site(s) by a bound substrate molecule or reversible inhibitor will protect those amino acid side chains

Storage of Pure Protein

that are critical for function. A starch-degrading amyloglucosidase enzyme (from an *Aspergillus* species) stored in the presence of 14% (w/v) partial starch hydrolysate was 80% more stable over a 24-wk period at ambient temperature than the corresponding enzyme preparation stored in the hydrolysate's absence *(20)*.

Note that dialysis (or some other procedure for the removal of low-molecular-mass substances) may be necessary to avoid carryover effects of the substrate or inhibitor when the protein is removed from storage for use in a particular situation where maximal activity is desired.

3.3.5. Use of Reducing Agents and Prevention of Oxidation Reactions

The thiol group of cysteine is prone to destructive oxidative reactions. One can prevent or minimize these by using reducing agents, such as 2- (formerly β-) mercaptoethanol (a liquid with an unpleasant smell) or dithiothreitol (DTT or Cleland's reagent, a solid with little odor). Add 2-mercaptoethanol to reach a final concentration of 5–20 mM, and then keep the solution under anaerobic conditions. To achieve these anaerobic conditions, gently bubble an inert gas, such as nitrogen, through the solution and fill it to the brim of a screw-cap container to minimize headspace and the chances of gaseous exchange. DTT is effective at lower concentrations: usually 0.5–1 mM will suffice *(10)*. Indeed, Schein advised that the DTT concentration should not exceed 1 mM: It can act as a denaturant at higher temperatures and is not very soluble in high salt *(18)*. Note that these reducing agents are themselves prone to oxidation. (This is why solutions containing them must be stored under anaerobic conditions.) DTT oxidizes to form an internal disulfide that is no longer effective, but will not interfere with protein molecules *(10)*. 2-Mercaptoethanol, on the other hand, participates in intermolecular reactions and can form disulfides with protein thiol groups *(10)*. Such thiol–disulfide exchanges are highly undesirable and may actually lead to inactivation or aggregation. It is probably best to add reducing agents only in situations where they are known (or can be demonstrated) to be beneficial *(16)*.

Much of the oxidation of thiol groups is mediated by divalent metal ions, which can activate molecular oxygen. Complexation of free metal ions (where they are not themselves essential for activity) can prevent destructive oxidation of thiol groups. *See* Section 3.2. regarding the use of EDTA to complex metal ions.

3.3.6. Extremely Dilute Solutions

Very dilute protein solutions are highly prone to inactivation. This is especially true of oligomeric proteins where dissociation of subunits can occur at low concentration. The individual polypeptide chains comprising the oligomer may lack activity alone and/or may denature with consequent loss of activity. Protein solutions of concentration <1–2 mg/mL should be concentrated as rapidly as possible *(10)* by ultrafiltration or sucrose concentration *(see* Note 5).

Where rapid concentration is not possible, inactivation may be prevented by addition of an exogenous protein, such as bovine serum albumin (BSA), typically to a final concentration of 1 mg/mL. Alkaline phosphatase from *E. coli* is unstable at room temperature at concentrations <10 μg/mL, but can be stabilized by addition of BSA *(21)*. Scopes discussed possible reasons for the undoubted benefits of BSA addition *(10)*. It may seem foolish to add an exogenous, contaminating protein, such as BSA, deliberately to a pure protein preparation, but occasionally this may be the price to be paid in order to avoid inactivation.

3.4. Low-Temperature Storage

Refrigeration at 4–6°C is often sufficient for the preservation of a protein's biological activity provided the hints in Sections 3.1.–3.3. of this chapter are followed judiciously. Many proteins are supplied commercially in 50% glycerol or as slurries in approx 3*M* ammonium sulfate. Freezing of such preparations is not necessary and should be avoided. Occasionally one may observe the phenomenon of cold denaturation; this has been discussed briefly in Section 1.

Some proteins can deteriorate at refrigerator temperatures and require storage at temperatures <0°C. Usually, temperatures between –18 and –20°C, attainable by a domestic freezer, will allow for stable storage *(see* Note 6). Sometimes, however, it may be necessary to use temperatures below –20°C. In these cases, it is normal to use a low-temperature laboratory freezer designed to maintain temperatures in the range of –70 to –80°C *(see* Note 7).

Most protein solutions will undergo freezing to a solid at temperatures below 0°C. (Mixtures containing high concentrations of glycerol will remain liquid at –20°C; *see* Section 3.3.3. in this chapter and Section 3.2. of Chapter 31.) The events occurring on freezing of a protein-containing mixture or biological system are much more complex than the simple macroscopic phase change would suggest. Differential freezing of particular components of the mixture can lead to enormous concentration effects and to dramatic changes of pH at low temperatures. These chemical processes can lead, in turn, to a notable degree of protein inactivation. The subject of freezing damage and its avoidance is discussed in Section 3.2. of Chapter 31. Note that the problem can often be minimized by judicious choice of stabilizing additives as discussed in Section 3.3.; *see also* ref. *14*.

Prevention of freezing will, of course, avoid freezing damage. It is possible to undercool liquids without freezing by preventing the nucleation of ice crystals. This means that proteins can be stored well below 0°C in the liquid phase. The preparation of protein-containing aqueous-organic emulsions, which can maintain complete biological activity in the liquid state over extended periods at –20°C, has been described *(22); see* Note 8. The method is very useful for small volumes of valuable proteins, avoids the need to use additives, and is more economical than freeze-drying. The same process is used for many different proteins, and one can remove portions of a sample without effect on the activity of the remainder. The actual storage temperature matters little, provided the upper temperature is <4°C and the lower temperature remains > –40°C, the nucleation temperature for ice crystal formation *(23)*.

3.5. Drying for Stable Storage

The advantages of water removal as a protein storage/stabilization strategy have been set out in Section 1. of Chapter 31. Lyophilization can remove more than 95% of water from a protein preparation, but there is the risk of freezing damage (Section 3.2. of Chapter 31). It is possible to design protein-compatible formulations with T_g' values typically as high as 37°C *(24)*. With these high T_g' values, controlled evaporative drying can be used in place of lyophilization to stabilize proteins in the solid state. Worthwhile evaporation rates will occur below these high T_g' values at reduced pressure. Evaporation is faster, less costly, and more easily controlled than freeze-drying *(23,24)*. The high T_g' values also mean that one can sometimes store the resulting dried product at ambient temperature: If room temperature is less than Tg' the protein formulation will not undergo a glass/rubber transition during storage at room temperature. The glass-forming compounds are typically carbohydrates; maltose and maltohexose are particularly valuable, whereas sucrose can be useful if the moisture content can be reduced to 2% or less *(24)*. Reconstitution of the solid protein preparation is accomplished simply by rehydration with added water or buffer. The method has been patented and is described in ref. 25. An alternative formulation for vacuum-drying of proteins, involving the use of a cationic, soluble polymer (e.g., diethylaminoethyl dextran) and the sugar alcohol, lactitol, has also been patented (26) and published (27).

3.6. Stability Analysis and Accelerated Degradation Testing

3.6.1. Background

The distinction between conformational and kinetic stability has already been drawn in Section 1. Kinetic stability is usually measured at elevated temperatures (1), but the inactivating event(s) at high temperatures may not mirror that/those at the much lower temperatures used for storage. It is not feasible, however, to monitor stability in real time at the actual storage temperature; the experiment would take too long. Inaccuracy may result over shorter intervals, since only minimal losses, virtually indistinguishable from the starting activity, would be apparent.

Fortunately, there is a methodology that can in many cases overcome these difficulties, namely accelerated degradation (or accelerated storage) testing. This involves the periodic assay of samples incubated at different temperatures and use of the Arrhenius equation to predict shelf lives at temperatures of interest. The Arrhenius equation is conveniently expressed in logarithmic form as:

$$\ln k = -E_a/R \cdot T + \ln A \tag{3}$$

where k is the first-order rate constant of activity decay, E_a is the activation energy, R is the gas constant, and T is the temperature in Kelvin. This log form of the Arrhenius equation yields a straight-line plot of $\ln k$ against $1/T$ with slope $-E_a/R$ (28). Extrapolation of this plot can give the rate constant (and hence the useful life) at a particular temperature; see Note 9. Accelerated storage testing has been used as a practical means of quality assurance for biological standards (29) and has been employed in some scientific investigations (30,31).

3.6.2. Setting Up a Test

An accelerated storage test must use the minimum amount of the (often precious) test protein that is compatible with achievement of precise and accurate results. Since each experiment will occupy a period of weeks, it is important that every test yields a meaningful outcome. Accordingly, take great care in setting up an accelerated degradation test run.

It is essential to prevent microbial contamination of test samples (*see* Section 3.1.), especially since one assumes that no (or very little) sample degradation takes place at the reference temperature during the period of the experiment. Place test samples at a series of elevated temperatures (e.g., 48, 45, 42, 37, 33, 30°C) and at a suitable lower reference temperature, e.g., 4°C (*see* Note 6). Ensure that the reference temperature will not cause freezing of liquid samples. (The liquid–solid phase change will introduce a further variable and may also lead to freezing damage, as discussed in Section 3.2. of Chapter 31; *see also* Note 10.) Remove samples at intervals from the various incubation conditions, bring them to the same temperature, and assay for activity under standardized, optimal conditions. Ensure that stock solutions used in assays are carefully prepared and standardized: it is likely that different batches of assay solutions will be needed over the period of the accelerated storage test, so variations must be minimized. Be equally scrupulous with regard to procedural details of the assay and the performance of the instruments used.

Kirkwood made some practical recommendations for successful accelerated storage testing *(32)*, that are summarized here. Use at least three elevated temperatures plus a low reference temperature. Set up 10 or more samples at each of these temperatures. The following schedule avoids waste of material owing to testing at inappropriate intervals. It also allows checking of the order of reaction. An acceptable result is very likely if enough material is placed on test at the beginning.

1. At intervals, test samples stored at the highest temperatures against the low-temperature reference samples. Ignore intermediate temperatures until a loss of 25% or greater (against the low-temperature reference sample) has occurred at the highest temperature. Now test a second sample to confirm the first result. If the results disagree, continue this stage 1 testing. If they agree, move on to stage 2.
2. Test the next two highest temperatures against a reference sample. Fit all available data to the Arrhenius equation. This procedure will not give a final precise result, but it will facilitate progress to stage 3.
3. Use the approximate result from stage 2 to ascertain further periods of storage at all temperatures such that measurable activity losses will have occurred at the intermediate temperatures. Assay samples from all available temperatures, and check all available data for their ability to fit the Arrhenius equation. This should result in a reasonable estimate of the low-temperature degradation rate, especially if multiple assays have been performed. If too little degradation has occurred to yield a precise result, one can repeat stage 3 after a longer time *(32)*.

3.6.3. Analysis of Results

It is clear that raw experimental data must undergo a number of transformations for use in the Arrhenius plot, notably conversion to natural log or reciprocal forms. Error relationships are often significantly affected by these sorts of transformations. Use a computer for all statistical fitting to minimize such errors. Use of good-quality replicate results is very important.

The activity decay must be first-order at all of the temperatures used in the Arrhenius plot. Verify this by fitting the time course of activity loss at each test temperature to a first-order exponential decay. Assess goodness of fit by inspection of the graphic fit and of parameters, such as standard errors or chi-square values. First-order exponentials yield straight-line plots when transformed into semilogarithmic form, unlike higher order functions (i.e., a plot of ln[Activity] against time is linear for a first-order decay). Deviations from a first-order function are more likely to occur at higher temperatures *(29)*.

Using the *k*-values determined at different temperatures, plot the Arrhenius graph (ln*k* is the ordinate, 1/T the abscissa). A linear plot of negative slope results. Extrapolate the line to a temperature of interest (e.g., 0, 4, or 25°C; these are respectively 273, 277, and 298 K) and estimate the value of *k* at this temperature. It is easy to estimate a true half life at this temperature by using the equation:

$$t_{1/2} = 0.693/k \tag{4}$$

Specially designed programs for the analysis of accelerated storage data are available (*see* Note 11).

4. Notes

1. Loss or decrease of the protein's biological or functional activity will be the main and most important index of deterioration. Often, however, the degree or time course of activity loss will not give any indication of the underlying molecular cause (although aggregation may be readily visible). Reference *1* provides a useful table of methods to identify the molecular changes leading to inactivation of the protein.
2. Thiomersal and azide are totally unacceptable in any product for administration. Do not discard azide compounds or azide-containing solutions down laboratory sinks. Not only is azide toxic, but it can accumulate in old lead piping, leading to the formation of potentially explosive compounds.
3. Some biological matrices, particularly sera, will not filter effectively through a 0.22-µm filter alone. One may need to prefilter the material ini-

tially through a coarser 0.45-μm filter to which the desired 0.22-μm filter is connected in series. Alternatively, one can accomplish the finer filtration as a separate operation. One can best filter larger volumes (hundreds of milliliters or liters) using a stack of filters clamped in a special filtration unit. A filter as coarse as 1 μm may be used directly in contact with the solution of interest, the stack comprising progressively finer filters until the sterilizing 0.22-μm filter is encountered at the bottom of the stack. Technical representatives of filtration manufacturers can give specialist advice for individual cases.

4. Many suppliers offer specific inhibitors of proteases or classes of protease. These inhibitors are often peptides or proteins, e.g., aprotinin, soybean trypsin inhibitor. A cocktail of protease inhibitors is available in tablet form from Boehringer Mannheim (Mannheim, Germany) under the trade name Complete. The product is stated to give effective inhibition of serine, cysteine, and metalloproteases during protein extractions from a variety of tissues and sources.

5. A variety of vessels and membranes for laboratory-scale ultrafiltration, with a range of defined mol-wt cutoffs, are commercially available. These may comprise permanent stirred pressure cells with replaceable membranes (for volumes in the range of 10–500 mL) or disposable centrifugal concentrators (for volumes up to 10 mL) *(18)*. The Amicon and Millipore corporations are prominent suppliers of such apparatus. Schein gave some useful observations on ultrafiltration and suggested some other means of achieving protein concentration *(18)*.

Sucrose concentration is an effective and rapid means of concentrating a dilute protein solution. Place the solution of interest into a suitably treated, softened dialysis tube, and secure the ends. Tear off a piece of aluminum foil such that the dialysis tube will rest on the foil with roughly 5–6 cm to spare all around. Shake some solid sucrose onto the foil, and then rest the dialysis tube on top of the sucrose. Shake more sucrose on top of the dialysis tube, wrap the foil around the sucrose and dialysis tube to form a parcel, and place in the refrigerator. Water from the dilute protein solution will move by osmosis through the pores of the dialysis tube to the surrounding solid sucrose, leading to concentration of the protein. Examine the dialysis tubing every 15–20 min. The surrounding sucrose will gradually form a viscous liquid, which can be removed periodically and replaced with fresh solid material. Volume reduction can take place quite quickly. The method has the drawback that sucrose will enter the dialysis tube in amounts not readily calculable. (The sucrose will help to stabilize the protein, of course.) If the presence of sucrose is undesirable, gently pull the dialysis tube between finger and thumb to force its contents into one end. Knot or clamp the dialysis tube as close as possible to the concentrated solution, and then dialyze the shortened dialysis tube against a suitable buffer to remove the sucrose. Note that the dialysis tube will swell in dilute buffer as water moves by osmosis into the protein solution, which will have a high sucrose concentration. The dialysis tube must be clamped as short as possible to prevent undue redilution of the sucrose-concentrated protein solution.

6. It can be a good idea to place a maximum/minimum thermometer inside the refrigerator, freezer, or incubator(s) close to the containers of interest in order to record any significant variations of temperature that may occur over an extended period. (Ensure first that the thermometer will withstand the low or high temperatures!)
7. Low temperature freezers typically function at –70 to –80°C. These temperatures are extremely cold and can inflict a cold burn on exposed skin. Always wear insulating or autoclave gloves when handling low-temperature items: latex laboratory gloves are not sufficient.
8. Special fluids for the preparation of undercooling emulsions are available commercially under the trade name U-COOL.
9. The Arrhenius equation is empirical but a similar equation results from the applications of Eyring's transition-state theory: *see* ref. *28* for a fuller discussion of this point.
10. Amorphous solid preparations will follow Arrhenius kinetics provided they remain in the glassy state. However, if any of the elevated temperatures used exceeds the glass transition temperature (T_g'), the product will become rubbery and will no longer obey the Arrhenius equation. (In fact, it will deteriorate much faster than predicted by the Arrhenius equation.) *See* ref. *24* for a description of this phenomenon and of an alternative kinetic scheme. Other situations in which deviations from Arrhenius kinetics may occur are outlined in ref. *33; see also* Note 5 of Chapter 31.
11. There follows a brief description, for information only, of two programs for analysis of accelerated degradation data. Their underlying statistical methodology is beyond the scope of this chapter, and the author makes no endorsement or judgment of their merits or claims. Each user must determine the suitability of any package for his or her purposes.

    The DEGTEST program carries out a three-stage analysis of data. First, the % activities at each temperature (relative to the low-temperature reference sample) are used to give estimates of relative degradation rates, assuming first-order decay. Second, these rates are fitted to the Arrhenius equation, and the goodness of fit is ascertained. Third, the decay rate at the proposed storage temperature is predicted from the Arrhenius fit, and the statistical precision of this value is calculated. The program can fit to the Arrhenius equation or to the more rigorous Eyring equation *(34)*. In practice, however, little difference is observed between the two equations *(35)*.

The POTENCYLOSS program is based on a linear model for stability prediction that uses a double logarithmic primary plot together with an altered secondary plot corresponding to the Arrhenius plot. The program can distinguish between zero- and first-order data and gives loss rate and shelf life parameters at a desired designated temperature. In addition, one can obtain estimates of activity values at any temperature or future time period *(36)*.

References

1. Mozhaev, V. V. (1993) Mechanism-based strategies for protein thermostabilization. *Trends Biotechnol.* **11,** 88–95.
2. Sluzky, V., Klibanov, A. M., and Langer, R. (1992) Mechanism of insulin aggregation and stabilization in agitated aqueous solutions. *Biotechnol. Bioeng.* **40,** 895–903.
3. Quax, W. J. (1993) Thermostable glucose isomerases. *Trends Food Sci. Technol.* **4,** 31–34.
4. Liu, W. R., Langer, R., and Klibanov, A. M. (1991) Moisture-induced aggregation of lyophilized proteins in the solid state. *Biotechnol. Bioeng.* **37,** 177–184.
5. Privalov, P. L. (1990) Cold denaturation of proteins. *Crit. Rev. Biochem. Mol. Biol.* **25,** 281–305.
6. Franks, F. and Hatley, R. H. M. (1991) Stability of proteins at subzero temperatures: thermodynamics and some ecological consequences. *Pure Appl. Chem.* **63,** 1367–1380.
7. Franks, F. (1993) Conformational stability of proteins, in *Protein Biotechnology: Isolation, Characterization, and Stabilization* (Franks, F., ed.), Humana Press, Totowa, NJ, pp. 395–436.
8. Sadana, A. (1988) Enzyme deactivation. *Biotechnol. Adv.* **6,** 349–446.
9. Beynon, R. J. and Bond, J. S. (1987) *Proteolytic Enzymes: A Practical Approach.* IRL, Oxford, UK.
10. Scopes, R. K. (1994) Maintenance of active enzymes, in *Protein Purification: Principles and Practice,* 2nd ed., Springer-Verlag, Berlin, pp. 317–324.
11. Erlanger, B. F., Kokowsky, N., and Cohen, W. (1961) The preparation and properties of two new chromogenic substrates of trypsin. *Arch. Biochem. Biophys.* **95,** 271–278.
12. Timasheff, S. N. and Arakawa, T. (1989) Stabilization of protein structure by solvents, in *Protein Structure: A Practical Approach* (Creighton, T. E., ed.), IRL Press, Oxford, UK, pp. 331–345.
13. Timasheff, S. N. (1992) Stabilization of protein structure by solvent additives, in *Stability of Protein Pharmaceuticals, Part B* (Ahern, T. J. and Manning, M. C., eds.) Plenum, New York, pp. 265–285.
14. Carpenter, J. F. and Crowe, J. H. (1988) The mechanism of cryoprotection of proteins by solutes. *Cryobiology* **25,** 244–255.
15. Combes, D., Yoovidhya, T., Girbal, E., Willemot, R.-M., and Monsan, P. (1987) Mechanism of enzyme stabilization. *Ann. NY Acad. Sci.* **501,** 59–62.
16. Tombs, M. P. (1985) Stability of enzymes. *J. Appl. Biochem.* **7,** 3–24.
17. Klibanov, A. M. (1983) Stabilisation of enzymes against thermal inactivation. *Adv. Appl. Microb.* **29,** 1–25.

18. Schein, C. H. (1990) Solubility as a function of protein structure and solvent components. *Biotechnology* **8,** 308–317.
19. Volkin, D. B. and Klibanov, A. M. (1989) Minimizing protein inactivation, in *Protein Function: A Practical Approach* (Creighton, T. E., ed.), IRL Press, Oxford, UK, pp. 1–24.
20. Shah, N. K., Shah, D. N., Upadhyay, C. M., Nehete, P. N., Kothari, R. M., and Hegde, M. V. (1989) An economical, upgraded, stabilized and efficient preparation of amyloglucosidase. *J. Biotechnol.* **10,** 267–276.
21. Reid, T. W. and Wilson, I. W. (1971) *E. coli* alkaline phosphatase, in *The Enzymes* (Boyer, P. D., ed.), Academic, London, pp. 373–415.
22. Hatley, R. H. M., Franks, F., and Mathias, S. F. (1987) The stabilization of labile biomolecules by undercooling. *Proc. Biochem.* **22,** 169–172.
23. Franks, F. (ed.) (1993) Storage stabilization of proteins, in *Protein Biotechnology: Isolation, Characterization, and Stabilization,* Humana, Totowa, NJ, pp. 489–531.
24. Franks, F., Hatley, R. H. M., and Mathias, S. F. (1991) Materials science and the production of shelf-stable biologicals. *Pharm. Technol. Int.* **3,** 24–34.
25. Franks, F. and Hatley, R. H. M. (1992) US Patent no. 5,098,893.
26. Gibson, T. D. and Woodward, J. R. (1991) *Enzyme Stabilization.* PCT/GB91/00443, Publ. No. Wo91/14773. World Intellectual Property Organization, Geneva.
27. Gibson, T. D., Higgins, I. J., and Woodward, J. R. (1992) Stabilisation of analytical enzymes using, a novel polymer-carbohydrate system and the production of a stabilised, single reagent for alcohol analysis. *Analyst* **117,** 1293–1297.
28. Eisenberg, D. S., Crothers, I., and Donald, M. (1979) Chemical and biochemical kinetics, in *Physical Chemistry with Applications to the Life Sciences.* Benjamin/Cummings, Menlo Park, CA, pp. 239–245.
29. Jerne, N. K. and Perry, W. L. M. (1956) The stability of biological standards. *Bull. Wld. Hlth. Org.* **14,** 167–182.
30. Malcolm, B. A., Wilson, K. P., Matthews, B. W., Kirsch, J. F., and Wilson, A. C. (1990) Ancestral lysozymes reconstructed, neutrality tested, and thermostability linked to hydrocarbon packing. *Nature* **345,** 86–89.
31. Ó Fágáin, C., O'Kennedy, R., and Kilty, C. (1991) Stability of alanine aminotransferase is enhanced by chemical modification. *Enzyme Microb. Technol.* **13,** 234–239.
32. Kirkwood, T. B. L. (1984) Design and analysis of accelerated degradation tests for the stability of biological standards III. Principles of design. *J. Biol. Stand.* **12,** 215–224.
33. Franks, F. (1994) Accelerated stability testing of bioproducts: attractions and pitfalls. *Trends Biotech.* **12,** 114–117.
34. Kirkwood, T. B. L. and Tydeman, M. S. (1984) Design and analysis of accelerated degradation tests for the stability of biological standards II. A flexible computer program for data analysis. *J. Biol. Stand.* **12,** 207–214.
35. Tydeman, M. S. and Kirkwood, T. B. L. (l984) Design and analysis of accelerated degradation tests for the stability of biological standards I. Properties of maximum likelihood estimators. *J. Biol. Stand.* **12,** 195–206.
36. Nash, R. A. (1987) A new linear model for stability prediction. *Drug Dev. Ind. Pharm.* **13,** 487–499.

CHAPTER 33

Electroelution of Proteins from Polyacrylamide Gels

Michael J. Dunn

1. Introduction

The ability of polyacrylamide gel electrophoresis (PAGE) techniques to resolve the individual components of complex mixtures of several thousand proteins has resulted in this group of methods being one of the most widely used laboratory techniques in biochemistry. Until relatively recently, these techniques were used almost exclusively as analytical tools to characterize proteins in terms of their size, charge, relative hydrophobicity, and abundance. However, the development of a battery of highly sensitive techniques of microchemical characterization, including N-terminal and internal protein sequencing, amino acid compositional analysis, peptide profiling, and mass spectrometry *(1,2)*, has resulted in PAGE often being the method of choice for the micropreparative purification of proteins for subsequent chemical analysis.

A major obstacle to successful chemical characterization is efficient recovery of the separated proteins from the polyacrylamide gel, since most procedures are not compatible with the presence of the gel matrix. The most popular approach to this problem is currently the use of Western electroblotting techniques in which the proteins, after separation by PAGE, are transferred ("blotted") onto the surface of inert membranes (e.g., polyvinylidene [PVDF]), which are compatible with the reagents and solvents used in chemical characterization procedures (*see* Chapter 34).

An alternative approach that has been extensively used is electroelution. In this technique, protein zones are localized after electrophoresis by staining with Coomassie brilliant blue R-250. Protein-containing gel pieces are then excised and placed in an electroelution chamber, where the proteins are transferred in an electric field from the gel into solution and concentrated over a dialysis membrane with an appropriate mol-wt cutoff. Although this method can have a high efficiency (>90%) of protein recovery, it suffers from a number of disadvantages, including:

1. Relatively slow isolation of a small number of samples;
2. Contamination of the eluted protein with SDS, salts, and other impurities;
3. Peptide chain cleavage during staining or elution; and
4. Chemical modification during staining or elution, which can lead to N-terminal blockage *(3)*.

In the simplest system for electroelution, gel slices containing the protein of interest are mixed with a small volume of polyacrylamide or agarose gel, and cast into a glass tube (such as that used for rod gel electrophoresis). A small sack made from dialysis tubing and containing electrophoretic buffer is then fastened with elastic bands over the end of the tube. The tube is then inserted into a rod gel electrophoresis apparatus and the protein eluted into the dialysis sack, from which it can subsequently be recovered. Electroelution is generally more efficient if 0.1% (w/v) SDS is added to the elution buffer, since it helps to maintain proteins in solution and provides them with a high negative charge density, thereby increasing their electrophoretic mobility.

More sophisticated variants of this approach have been described *(3)*, and commercial equipment is now available that allows several different gel segments to be eluted into a series of elution chambers. In this chapter, the use of one such commercial device, the Biometra Elucon (Biometra Maidstone, Kent, UK), to recover proteins separated by SDS-PAGE is described.

2. Materials

Prepare all solutions freshly from analytical grade reagents and dissolve in deionized water.

1. Fixing and destaining solution: 450 mL methanol, 100 mL acetic acid, made up to 1 L in water.
2. Staining solution: 0.2 g Coomassie brilliant blue R-250 made up to 100 mL in water.

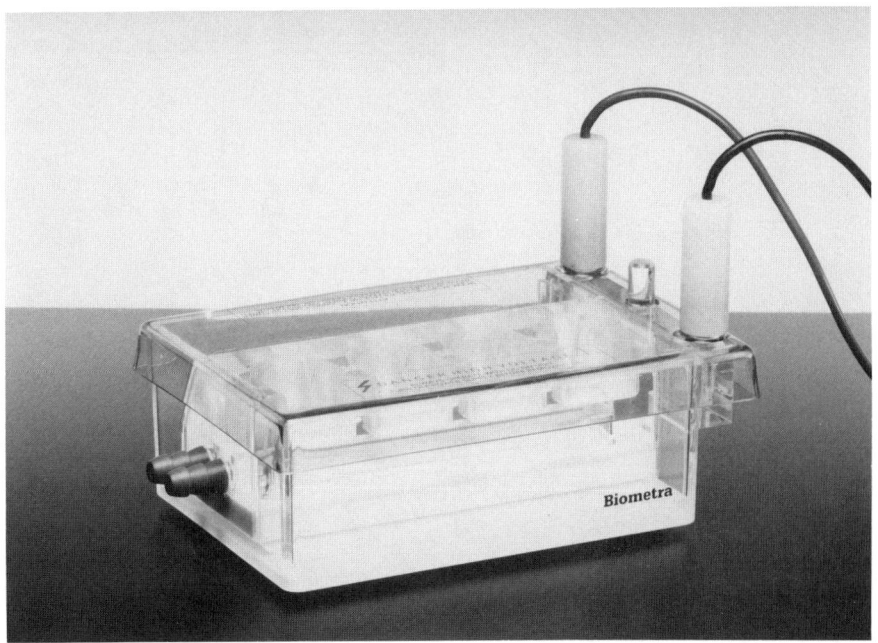

Fig. 1. Biometra Elucon apparatus for the simultaneous electroelution of proteins from up to four different polyacrylamide gel samples. Photograph courtesy of Biometra.

3. Electroelution apparatus: Biometra Elucon apparatus for the simultaneous elution of up to four samples (Fig. 1).
4. Dialysis membranes: Precut disks of dialysis membrane, exclusion limit 2 kDa (Biometra).
5. Elution buffer: For subsequent chemical characterization, it is preferable to use a volatile, basic buffer, such as 50 mM NH_4HCO_3 or 50 mM N-ethylmorpholine acetate, pH 8.5, containing 0.1% (w/v) SDS (*see* Note 1).

3. Methods
3.1. Gel Staining

1. Following separation of the proteins by SDS-PAGE, place the gel in fixing solution for at least 1 h at room temperature with gentle agitation.
2. Place the gel in staining solution for at 1 h at room temperature with gentle agitation.
3. Transfer the gel to destaining solution, and agitate gently at room temperature. After about 24 h and with several changes of destaining solution, the

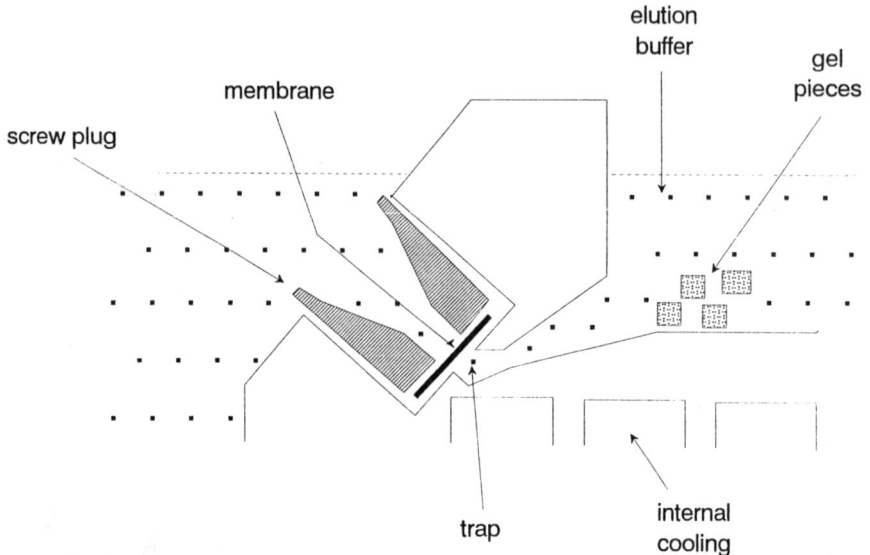

Fig. 2. Diagrammatic cross-section through an elution channel of the Elucon apparatus.

gel background will become colorless, leaving the separated proteins as blue-colored bands.

3.2. Electroelution

1. Connect the Elucon to a recirculating, cooling water bath set to 20°C.
2. Soak the dialysis membrane disks in elution buffer.
3. Fill the Elucon chamber up to 4 cm with elution buffer.
4. Place each prewetted dialysis membrane disk on top of a screw plug (*see* Fig. 2), and fix the screw with the aid of a spatula into the elution channel. Seal any elution channels not being used for electroelution with a rubber sealing disk in place of a dialysis membrane disk.
5. Fill each side of the elution channel with elution buffer, avoiding entrapment of air bubbles on either side of the fixed membrane.
6. Excize the stained protein bands of interest, and cut each band into small (3–5 mm) pieces.
7. Place the gel pieces in front of each elution channel (*see* Fig. 2).
8. Connect the Elucon to an electrophoresis power supply, and elute the proteins at 200 V constant for the appropriate time (*see* Note 2).
9. The protein is eluted out of the gel pieces and is concentrated in front of the dialysis membrane in the trap (*see* Fig. 2).

10. At the completion of electroelution, reverse the polarity of the Elucon for 1 min to release any protein that has adsorbed to the surface of the dialysis membrane.
11. Disconnect the Elucon from the power supply.
12. Insert a micropipet into the elution channel, and remove approx 100 μL of protein-enriched solution from the trap in front of the dialysis membrane (*see* Note 3).

3.3. Protein Precipitation

1. Lyophilize the electroeluted protein sample, or dry using a rotary vacuum evaporator.
2. Add 50 μL of water.
3. Add 450 μL of ice-cold acetone acidified to a final concentration of 1 mM HCl and incubate for 3 h at –20°C.
4. Pellet the precipitated protein using a microcentrifuge.
5. Wash the pellet three times with 100 μL ice-cold acetone.
6. Air-dry the pellet, which can then be redissolved in the appropriate solvent for the method of chemical characterization to be employed (*see* Note 4).

4. Notes

1. Electroelution is generally more efficient in the presence of 0.1% SDS, since it helps to maintain proteins in solution and provides them with a high negative charge density, thereby increasing their electrophoretic mobility. It can be reduced or omitted in the case of some highly soluble and highly charged proteins.
2. The time taken for electroelution is dependent on the molecular weight of the protein to be eluted, ranging from <1 h for proteins up to 50 kDa to 5 h for proteins of 150 kDa (*see* Fig. 3). Proteins larger than 200 kDa should be eluted overnight, but care must then be taken that the apparatus is effectively cooled (*see* Section 3.2., step 1) to avoid damage to the apparatus by overheating.
3. The efficiency of protein recovery by electroelution should be between 70 and 90%.
4. After electroelution, the protein-enriched solution also contains SDS, Coomassie Brilliant Blue stain, buffer salts and other contaminants. It has proven possible to apply electroeluted proteins directly onto sample targets for matrix-assisted laser desorption/ionization mass spectrometry (MALDI-MS) *(4),* where volatile buffer salts are removed by air-drying. However, for most high-sensitivity chemical characterization techniques, it is essential that the protein be separated from the impurities by precipitation before being redissolved for subsequent analysis.

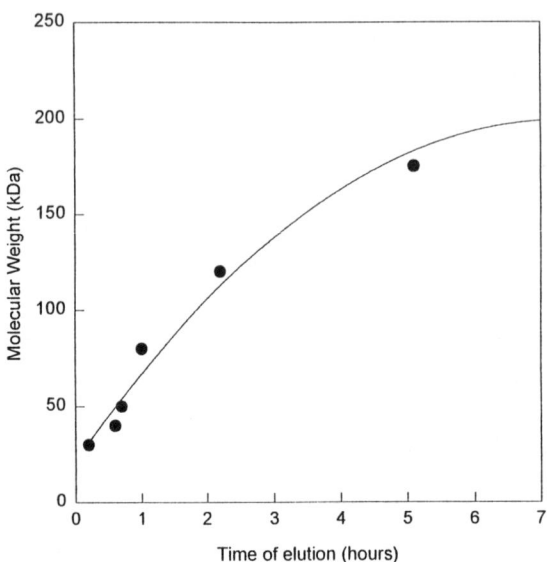

Fig. 3. Electroelution time of marker proteins in the Elucon chamber (data from Biometra). Buffer: 0.05M Tris-glycine, pH 8.6, 0.05% SDS. Electroelution conditions: 200 V, 20 mA at 20°C with cooling.

Acknowledgment

I am grateful to Peter Wolstenholme of Biometra for providing the photographic illustration of the Elucon apparatus.

References

1. Patterson, S. D. (1994) From electrophoretically separated protein to identification: strategies for sequence and mass analysis. *Anal. Biochem.* **221,** 1–15.
2. Kellner, R., Lottspeich, F., and Meyer, H. E. (eds.) (1994) *Microcharacterization of Proteins.* VCH Verlagsgesellschaft, Weinheim, Germany.
3. Aebersold, R. (1991) High sensitivity sequence analysis of proteins separated by polyacrylamide gel electrophoresis, in *Advances in Electrophoresis,* vol. 4 (Chrambach, A., Dunn, M. J., and Radola, B. J., eds.), VCH Verlagsgesellschaft, Weinheim, Germany, pp. 81–168.
4. Haebel, S., Jensen, C., Andersen, S. O. and Roepstorff, P. (1995) Isoforms of a cuticular protein from larvae of the meal beetle, *Tenebrio molitor,* studied by mass spectrometry in combination with Edman degradation and two-dimensional polyacrylamide gel electrophoresis. *Prot. Sci.* **4,** 394–404.

CHAPTER 34

Electroblotting of Proteins from Polyacrylamide Gels

Michael J. Dunn

1. Introduction

The ability of polyacrylamide gel electrophoresis (PAGE) techniques to resolve individual proteins of complex mixtures has resulted in this group of methods being indispensable to the protein biochemist. Although PAGE procedures can provide characterization of proteins in terms of their charge, size, relative hydrophobicity, and abundance, they provide no direct clues to the identity or function of the separated components.

A powerful approach to this problem was provided by the development of protein blotting techniques, based on those developed by Southern (1) for the analysis of electrophoretic separations of DNA. In these procedures the proteins, after separation by electrophoresis, are transferred ("blotted") onto the surface of an inert membrane, such as nitrocellulose. When immobilized on the surface of a membrane, the proteins are readily accessible to interaction with probes, such as antibodies or other ligands specific for the protein(s) being analyzed.

More recently, blotting has become the method of choice for the micropreparative purification of proteins for subsequent chemical characterization. In this approach, proteins separated by 1-D or 2-D PAGE are blotted onto an appropriate membrane support, the total protein pattern visualized using a total protein stain, and the protein band or spot of interest is excised. The protein, while still on the surface of the inert membrane support, can then be subjected to a battery of sensitive microchemical characterization techniques, including N-terminal and internal amino acid sequencing, amino acid compositional analysis, peptide profiling, and mass spectrometry (2,3).

The most popular method for the transfer of electrophoretically separated proteins to membranes is the application of an electric field perpendicular to the plane of the gel. This technique of electrophoretic transfer, first described by Towbin et al. (4), is known as Western blotting. Two types of apparatus are in routine use for electroblotting. In the first approach (known as "tank" blotting), the sandwich assembly of gel and blotting membrane is placed vertically between two platinum wire electrode arrays contained in a tank filled with blotting buffer (5). The disadvantages of this technique are that:

1. A large volume of blotting buffer must be used;
2. Efficient cooling must be provided if high current settings are employed to facilitate rapid transfer; and
3. The field strength applied (V/cm) is limited by the relatively large interelectrode distance.

In the second type of procedure (known as "semidry" blotting), the gel-blotting membrane assembly is sandwiched between two horizontal plate electrodes, typically made of graphite (6). The advantages of this method are that:

1. Relatively small volumes of transfer buffer are used;
2. Special cooling is not usually required, although the apparatus can be run in a cold room if necessary; and
3. A relatively high field strength (V/cm) is applied owing to the short interelectrode distance resulting in faster transfer times.

In the following sections, both tank and semidry electroblotting methods for recovering proteins separated by gel electrophoresis either for immunoprobing or for subsequent microchemical characterization are described. In addition, total protein staining procedures compatible with immunoprobing and microchemical techniques are discussed.

2. Materials

2.1. Electroblotting

Prepare all buffers from analytical-grade reagents and dissolve in deionized water. The solutions should be stored at 4°C and are stable for up to 3 mo.

Electroblotting of Proteins

1. Blotting buffers are selected empirically to give the best transfer of the protein(s) under investigation. The following compositions are commonly used:
   a. For immunoprobing of proteins with pIs between pH 4.0 and 7.0: Dissolve 2.42 g Tris base and 11.26 g glycine, and make up to 1 L (*see* Note 1). The pH of the buffer is 8.3 and should not require adjustment.
   b. For chemical characterization of proteins with pIs between pH 4.0 and 7.0 (*see* Note 2): Dissolve 6.06 g Tris base and 3.09 g boric acid, and make up to 1 L (*see* Note 1). Adjust the solution to pH 8.5 with $10N$ sodium hydroxide *(7)*.
   c. For proteins with pIs between pH 6.0 and 10.0: Dissolve 2.21 g 3-(cyclohexyl-amino)-1-propanesulfonic acid (CAPS) and make up to 1 L (*see* Note 1). Adjust the solution to pH 11.0 with $10N$ sodium hydroxide *(8)*.
2. Filter paper: Whatman 3MM (Maidstone, UK) filter paper cut to the size of the gel to be blotted.
3. Transfer membrane:
   a. For immunoprobing: Hybond-C Super (Amersham International, Aylesbury, UK) cut to the size of the gel to be blotted (*see* Note 3).
   b. For chemical characterization: FluoroTrans (Pall, Havant, UK) cut to the size of the gel to be blotted (*see* Note 4).
4. Electroblotting equipment: A number of commercial companies produce electroblotting apparatus and associated power supplies. For tank electroblotting the TE 42 Transphor Unit from Hoefer Scientific Instruments (Newcastle-under-Lyme, UK) is used, whereas for semidry electroblotting, the NovaBlot apparatus from Pharmacia Biotechnology (St. Ablans, UK) is used.
5. Rocking platform.
6. Plastic boxes for gel incubations.

2.2. Protein Staining

2.2.1. Immunoprobing

1. Instaview Nitrocellulose staining kit (BDH, Poole, UK) (*see* Note 5). The kit comprises three components:
   a. 50X stain concentrate.
   b. 100X enhancer concentrate.
   c. 10X destain concentrate.

2.2.2. Chemical Characterization

1. Destain: 450 mL methanol, 100 mL acetic acid made up to 1 L in deionized water.
2. Stain: 0.2 g Coomassie brilliant blue R-250 made up to 100 mL in destain.

3. Methods
3.1. Electroblotting
3.1.1. Semidry Blotting

1. Following separation of the proteins by gel electrophoresis, place the gel in equilibration buffer, and gently agitate for 30 min at room temperature (*see* Note 6).
2. Wet the lower (anode) plate of the electroblotting apparatus with deionized water.
3. Stack six sheets of filter paper wetted with blotting buffer on the anode plate, and roll with a glass tube to remove any air bubbles.
4. Place the prewetted transfer membrane (*see* Note 7) on top of the filter papers, and remove any air bubbles with the glass tube.
5. Place the equilibrated gel on top of the blotting membrane, and ensure that no air bubbles are trapped.
6. Apply a further six sheets of wetted filter paper on top of the gel, and roll with the glass tube.
7. Wet the upper (cathode) plate with deionized water, and place on top of the blotting sandwich.
8. Connect the blotter to power supply, and transfer at 0.8 mA/cm^2 of gel area (*see* Note 8) for 1 h at room temperature (*see* Note 9).

3.1.2. Tank Blotting

1. Following separation of the proteins by gel electrophoresis, place the gel in equilibration buffer, and gently agitate for 30 min at room temperature (*see* Note 6).
2. Place the anode side of the blotting cassette in a dish of blotting buffer.
3. Submerge a sponge pad taking care to displace any trapped air, and place on top of the anodic side of the blotting cassette.
4. Place two pieces of filter paper onto the sponge pad, and roll with a glass tube to ensure air bubbles are removed.
5. Place the prewetted transfer membrane (*see* Note 7) on top of the filter papers, and remove any air bubbles with the glass tube.
6. Place the equilibrated gel on top of the blotting membrane, and ensure that no air bubbles are trapped.
7. Place a sponge pad into the blotting buffer taking care to remove any trapped air bubbles, and then place on top of the gel.
8. Place the cathodic side of the blotting cassette on top of the sponge, and clip to the anode side of the cassette.
9. Remove the assembled cassette from the dish, and place into the blotting tank filled with transfer buffer.
10. Connect to the power supply, and transfer for 6 h (1.5-mm thick gels) at 500 mA at 10°C (*see* Note 9).

3.2. Protein Staining

3.2.1. Immunoprobing

1. Prepare a working staining solution by the addition of 1 mL of 100X enhancer concentrate and 2 mL 50X stain concentrate to 97 mL deionized water. Mix well.
2. Place the membrane to be stained in the staining solution, and agitate for 2–5 min.
3. While the membrane is staining, prepare a working solution of the enhancing solution by addition of 1 mL 100X enhancer concentrate to 99 mL deionized water, and mix well.
4. Pour off the staining solution from the membrane, and replace with the working enhancing solution. Agitate until a clear background is obtained (usually <5 min) (*see* Note 10).
5. At this stage, the membrane may be dried and stored, or destained (*see* step 6 and Note 11).
6. Prepare a working destaining solution by addition of 10 mL 10X destain concentrate to 90 mL deionized water.
7. Place the membrane in the destain solution, and agitate until all the color has been removed (typically 5–10 min).

3.2.2. Chemical Characterization

1. Remove the blotting membrane from the sandwich assembly.
2. Place the membrane into a dish containing the Coomassie blue staining solution for 2 min, and agitate gently on the rocking platform.
3. Place the membrane into destaining solution, and agitate for 10–15 min (or until the background is pale).
4. Wash the membrane with deionized water, place on filter paper, and allow to air-dry.
5. Place the membrane into a clean plastic bag, and seal until required for further analysis. The membrane can be stored in this state at room temperature for extended periods without any apparent adverse effects on subsequent chemical characterization.
6. An example of a membrane stained by this method is shown in Fig. 1.

4. Notes

1. Methanol (10–20% [v/v]) is often added to transfer buffers, since it removes SDS from protein–SDS complexes and increases the affinity of binding of proteins to nitrocellulose. However, methanol acts as a fixative and reduces the efficiency of protein elution, so that extended transfer times must be used. This effect is worse for high-mol-wt proteins, so that methanol is best avoided if proteins >100 kDa are to be transferred.

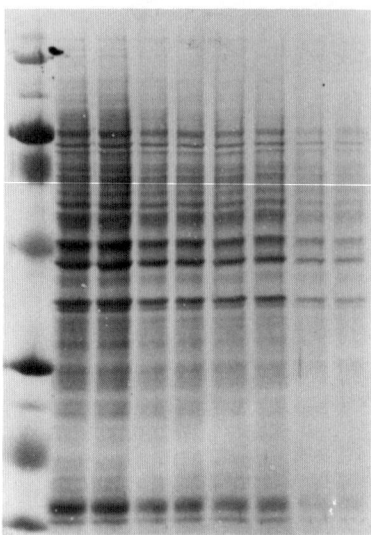

Fig. 1. PVDF Western electroblot transfer of human endothelial cell proteins separated by 10%T SDS-PAGE stained with Coomassie brilliant blue R-250. Different amounts of protein were applied to each lane: (1) 40 µg, (2) 20 µg, (3) 10 µg, (4) 5 µg, (M) mol-wt marker proteins.

2. The use of transfer buffers containing glycine or other amino acids must be avoided for proteins to be subjected to microchemical characterization.
3. Nitrocellulose is the most popular support for electroblotting, since it is compatible with most general protein stains, it is relatively inexpensive, and has a high protein binding capacity (249 µg/cm^2 [9]). Hybond-C Super is a 0.45-µm pore size supported nitrocellulose membrane, which is more robust than an unsupported matrix. Nitrocellulose membranes of smaller pore size are available (0.1 and 0.2 µm) and can give better retention of small proteins (<1500 kDa). Polyvinylidene difluoride (PVDF) membranes, which have a high mechanical strength and a binding capacity similar to that of nitrocellulose (172 µg/cm^2 [9]), are compatible with most immunoblotting protocols.
4. Nitrocellulose membranes are not compatible with the reagents and organic solvents used in automated protein sequencing. A range of PVDF-based (FluoroTrans, Pall; ProBlott, Applied Biosystems [Warrington, UK]; Immobilon-P and Immobilon-CD, Millipore [Watford, UK]; Westran, Schleicher and Schuell [Dassel, Germany]; Trans-Blot, Bio-Rad [Hercules,

CA]), glass fiber-based (Galssybond, Biometra [Maidstone, UK]; PCGM-1, Janssen Life Sciences [Geel, Belgium]), and polypropylene-based (Selex 20, Schleicher and Schuell) membranes have been developed to overcome this problem *(2,10,11)*. PVDF-based membranes have a higher protein binding capacity, and result in better average repetitive and initial sequencing yields *(7)*. Nitrocellulose can be used as a support in applications, such as internal amino acid sequence analysis and peptide profiling, where the protein band or spot is subjected to proteolytic digestion prior to characterization of the released peptides.

5. A variety of methods providing different levels of sensitivity have been described for the detection of total protein patterns on nitrocellulose and PVDF membranes following electroblotting *(12)*. However, most of these methods (including Coomassie brilliant blue R-250, Amido Black 10B, India ink, colloidal gold) are incompatible with subsequent immunoprobing. Instaview Nitrocellulose is a rapid and sensitive (10 ng/band) method for visualization of protein patterns on nitrocellulose or PVDF membranes, and the membrane can be rapidly destained without loss of immunoreactivity. This allows a second probing of the membrane after initial protein detection, either with specific activity stains or antibody detection methods.

6. Gels are equilibrated in blotting buffer to remove excess SDS and other reagents that might interfere with subsequent analysis (e.g., glycine). This step also minimizes swelling effects during protein transfer. Equilibration may result in diffusion of zones and reduced transfer efficiencies of high-mol-wt proteins. It is important to optimize the equilibration time for the protein(s) of interest.

7. Nitrocellulose membranes can be wetted with blotting buffer, but PVDF-based membranes must first be wetted with methanol prior to wetting with the buffer.

8. The maximum mA/cm^2 of gel area quoted applies to the apparatus that was used. This should be established from the manual for the particular equipment available.

9. Blotting times need to be optimized for the particular proteins of interest and according to gel thickness. Larger proteins usually need a longer transfer time, whereas smaller proteins require less time. Proteins will also take longer to be transferred efficiently from thicker gels. The transfer time cannot be extended indefinitely (>3 h) using the semidry technique, since the small amount of buffer used will evaporate. If tank blotting is used, the transfer time can be extended almost indefinitely (>24 h) providing that the temperature is controlled.

10. With PVDF membranes, a blue background staining may still be observed.

11. The destaining step completely destains the protein bands and the background of the blot, allowing the subsequent use of specific methods of probing the membrane using normal protocols.

References

1. Southern, E. (1975) Detection of specific sequences among DNA fragments separated by gel electrophoresis. *J. Mol. Biol.* **98,** 503–517.
2. Patterson, S. D. (1994) From electrophoretically separated protein to identification: strategies for sequence and mass analysis. *Anal. Biochem.* **221,** 1–15.
3. Kellner, R., Lottspeich, F., and Meyer, H. E. (eds.) (1994) *Microcharacterization of Proteins.* VCH Verlagsgesellchaft, Weinheim, Germany.
4. Towbin, H., Staehelin, T., and Gordon, G. (1979) Electrophoretic transfer of proteins from polyacrylamide gels to nitrocellulose sheets: procedure and some applications. *Proc. Natl. Acad. Sci. USA* **76,** 4350–4354.
5. Gooderham, K. (1984) Transfer techniques in protein blotting, in *Methods in Molecular Biology, Vol. 1: Proteins* (Walker, J. M., ed.), Humana, Clifton, NJ, pp. 165–178.
6. Peferoen, M. (1988) Blotting with plate electrodes, in *Methods in Molecular Biology, Vol. 3: New Protein Techniques* (Walker, J. M., ed.), Humana, Clifton, NJ, pp. 395–402.
7. Baker, C. S, Dunn, M. J., and Yacoub, M. H. (1991) Evaluation of membranes used for electroblotting of proteins for direct automated microsequencing. *Electrophoresis* **12,** 342–348.
8. Matsudaira, P. (1987) Sequence from picomole quantities of proteins electroblotted onto polyvinylidene difluoride membranes. *J. Biol. Chem.* **262,** 10,035–10,038.
9. Pluskal, M. G., Przekop, M. B., Kavonian, M. R., Vecoli, C., and Hicks, D. A. (1986) Immobilon PVDF transfer membrane: a new membrane substrate for Western blotting of proteins. *Bio Techniques* **4,** 272–283.
10. Aebersold, R. (1991) High sensitivity sequence analysis of proteins separated by polyacrylamide gel electrophoresis, in *Advances in Electrophoresis,* vol. 4 (Chrambach, A., Dunn, M. J., and Radola, B. J., eds.), VCH Verlagsgesellschaft, Weinheim, Germany, pp. 81–168.
11. Eckerskorn, C. (1994) Blotting membranes as the interface between electrophoresis and protein chemistry, in *Microcharacterization of Proteins* (Kellner, R., Lottspeich, F., and Meyer, H. E., eds.), VCH Verlagsgesellchaft, Weinheim, Germany, pp. 75–89.
12. Dunn, M. J. (1993) *Gel Electrophoresis: Proteins.* Bios Scientific Publishers, Oxford, UK.

CHAPTER 35

High-Performance Electrophoresis Chromatography

Serge Desnoyers, Sylvie Bourassa, and Guy G. Poirier

1. Introduction

Protein purification is important because it gives essential data on an enzyme, like its catalytic activity, its interactions with other proteins or DNA, amino acid quantitation, N-terminal microsequencing, and other information. These analyses will provide important data for designing oligodeoxynucleotide probes for gene cloning and peptide synthesis for antibody production. Purified proteins can be obtained by several methods: gel-filtration chromatography, ion-exchange chromatography, immunoaffinity chromatography, immunoprecipitation, and salt precipitation *(1)*.

Micropreparative electrophoresis is a relatively new protein purification method *(2–4)* that gives fast results (usually a few hours) and yields enough purified protein (≤200 µg) to allow microsequencing either directly from the elution chamber *(5)* or by centrifugation on polyvinylenedifluoride (PVDF) membrane *(6)*. Micropreparative electrophoresis purification is based on the principle of relative mobility of proteins in sodium dodecyl sulfate-polyacrylamide gel electrophoresis (SDS-PAGE).

Originally described by Ornstein *(7)* and subsequently modified by Laemmli *(8)*, SDS-PAGE is still the most reliable analytical system for proteins. However, it can also be used as a preparative system. This system is based on the effect of an electrical field causing charged molecules to move toward the electrode of opposite polarity. The mobility of each molecule decreases as it interacts with the surrounding matrix, in the present case, polyacrylamide gel. The gel acts as a molecular sieve. The amphoteric nature of proteins implies that they will migrate toward the electrode opposite their net charge. In SDS-PAGE, the net charge of each protein is the same because the proteins are solubilized in an ionic detergent, SDS. The anion dodecyl sulfate gives each molecule of protein a negative charge. Proteins are thus separated according to their molecular weight, and small proteins move faster than larger ones. The criterion of purity is to obtain a single band on SDS-PAGE, on which is based the principle of protein purification by micropreparative electrophoresis. Instead of stopping the run when the front of migration reaches the bottom of the gel, electrophoresis is continued until each zone is eluted from the gel. If eluate is fractionated, each zone can be collected as a purified one.

Several devices have been described for micropreparative electrophoresis *(2,9),* and there are now commercialized apparatuses, such as the high-performance electrophoresis chromatography system (HPEC™) 230A from Applied Biosystems Inc. (Foster City, CA), ELFE from Genofit Inc. (Grand Lancy, Switzerland), Prep-Cell 491 from Bio-Rad (Hercules, CA), and 1100PG from BRL (Gaithersburg, MD). This chapter describes a method originally described for a slab gel apparatus *(10),* but applied on a commercialized micropreparative electrophoresis apparatus from Applied Biosystems, model 230A HPEC *(11).* The HPEC apparatus is provided with an on-line UV detector, an elution module, and a fraction collector. Proteins are chromatographed in a polyacrylamide gel matrix under the influence of an electrical field. The gel column stands between the cathodic module and the anodic module. Buffers are constantly renewed by a pressurized tubule system, eliminating heat build-up and ion depletion.

The protein sample is loaded onto the top of the gel at the cathodic end. Once current is applied, the proteins are eletrophoresed under thermostated conditions. Since each separated zone is eluted from the bottom of the gel, proteins are swept through the detector by a flow of elution buffer and collected by a fraction collector. The proteins so purified can be analyzed further.

Most of the known buffer systems existing for slab gel can be applied for HPEC purification, with slight modifications *(12).* The inconvenience with the Laemmli buffer system Tris-glycine-SDS is that the sample, once purified, needs to be handled for desalting *(6)* to allow microsequencing of the purified protein, which implies a loss of material. We describe here a micropreparative electrophoretic system buffer that is completely compatible with the microsequencing of protein permitting one-step protein purification and direct microsequencing. This buffer increases the resolution of the different proteins and is faster to recover the proteins because its use makes the blotting or the desalting step unnecessary. Virtually 100% of the proteins eluted from the HPEC is recovered.

2. Materials

It is recommended that reagents of the highest grade of purity possible or of electrophoresis grades be used.

2.1. Apparatus

Separation system HPEC 230A (Applied Biosystems):

1. Gas cylinder and regulator.
2. Gel tubes.
3. Column filters (Zytex® membrane).
4. Dialysis membrane.
5. Forceps.
6. Razor blade.
7. Pasteur pipet.
8. 0.5-mL Eppendorf tube with ventilation holes.

2.2. Stock Solutions and Buffers

1. Acrylamide/bis-acrylamide concentrate solution (30%T, 2.7%C): Dissolve 29.2 g of acrylamide and 0.8 g of bis-acrylamide in 80 mL of deionized (DI) water. Stir and once completely dissolved make up to 100 mL with DI water. Filter using a 0.22-µm filter. Store at 4°C in a brown bottle for 1 mo (see Note 2).
2. Gel buffer (10X): $0.6M$ 2-dimethylamionethanol (DME) solution titrated with HCl at pH 8.3. Add 16 mL of DME (straight from the commercial bottle) to 70 mL of DI water. Adjust pH to 8.3 with concentrate HCl. Make up to 100 mL with DI water. Store at 4°C in a brown bottle for 6–8 wk.
3. 10% Ammonium persulfate (APS): Add 50 mg of APS to 0.5 mL DI water. Prepare fresh before each use.
4. Upper buffer (1X): 120 mM DME-HCl, 150 mM boric acid, 0.1% SDS (see Note 1). Add 24 mL of DME (straight from the commercial bottle) to 1.5 L of DI water. Dissolve 9.27 g of boric acid. Stir and once dissolved make up to 1.98 L with DI water, filter using a 0.22-µm filter, and sparge with helium for 20 min. Add 20 mL of 10% SDS, stir gently. The pH should be close to 9.3.
5. Lower and elution buffers (1X): 150 mM boric acid-DME, pH 8.3. Dissolve 13.9 g of boric acid in 2.8 L of DI water and adjust pH to 8.3 with a few

Table 1
Gel Formulation for 10 mL of Final Mixture

Solutions	7.5%T	10%T	15%T
Acrylamide-bis, mL	2.44	3.33	4.88
10X gel buffer, mL	1	1	1
Water, mL	6.44	5.6	4
10% APS, μL	70	70	70
TEMED, μL	7	7	7

drops of DME (usually around 4 mL). Make up to 3 L with DI water, filter using a 0.22-μm filter, and sparge with helium for 20 min.

6. Sample buffer (2X): $0.06M$ DME-HCl, 6% SDS, 20% glycerol, 0.4% 2-mercaptoethanol. Combine 1 mL $0.6M$ DME-HCl, 3 mL 20% SDS, 4 mL 50% glycerol, and 40 μL 2-mercaptoethanol. Mix all reagents in water to a final volume of 10 mL.

3. Methods

3.1. Tube Gel Preparation

1. Gels are poured into glass tubes, which must have been thoroughly cleaned with detergent, such as Alconox® overnight.
2. Rinse with DI water and dry.

3.2. Dialysis Membrane

This is an important component because it is placed between the elution block and the lower block of the HPEC230A, preventing proteins from going into the lower buffer tubule. Boil the membrane in DI water for 30 min. Handle the membrane with forceps. Store at 4°C in a 50:50 water:methanol solution.

3.3. Gel Preparation (see Table 1)

1. Put a plastic cap at one end of the tube, and a plastic collar at the other end. Tubes must be maintained in a vertical position.
2. Prepare the required gel volume, excluding APS and TEMED, according to the number of desired gel columns to be used (one gel column = 1 mL).
3. Deaerate the gel solution for 15 min under vacuum.
4. Add catalyst (APS and TEMED), and gently swirl the solution (*see* Table 1).
5. Fill tubes using a 9-in. Pasteur pipet.
6. Remove any air bubbles.
7. Add a layer of water-saturated butanol over the gel solution.
8. Allow to polymerize for 1 h (*see* Note 3).

9. Remove the collar, and cut the protruding portion of the gel with a razor blade while leaving 2 mm of gel protruding for storage.
10. Seal the end with a plastic cap and Parafilm™. Put in a small container (Falcon 50-mL conical tube with 1 or 2 mL of water). Gels can be conserved for 2 wk in such conditions.

3.4. Buffer Installation and Circulation

1. Use buffers described in Section 2.2.
2. Fill appropriate bottles with upper, lower, and elution buffers.
3. Adjust pressure to 1–2 psi in the tubule system to allow buffers to circulate (*see* Note 4).

3.5. Gel Installation

1. Carefully remove cap and collar from the gel tube by cutting the plastic with a razor blade. Be careful not to touch or move the gel and thus introduce an air pocket.
2. With razor blade, cut the protruding gel at the top and bottom of the gel tube, which must be as even as possible. Cut the gel under water to avoid introduction of an air pocket.
3. Wet the bottom and top ends of the gel tube with a drop of upper and lower buffer, respectively. Place a Zytex® membrane on each end (they will stay in place because of the surface tension).
4. Place the gel tube between the upper and lower electrode assembly. Be sure there is no leak of running buffers.
5. Purge the tubule system with buffer to make sure there is no air bubble in the system, especially in the elution line, which goes directly through the UV detector.
6. Elution buffer flow rate should be 15–20 µL/min for good analysis and collection of the sample protein.

3.6. Sample Preparation

The sample must be dissolved in the 2X sample buffer and in the smallest volume possible (10 µL) even if the apparatus allows a sample volume of 90 and 200 µL *(13)* (*see* Note 5).

1. If the sample is solid, dissolve the proteins in 5–10 µL of DI water and mix an equal volume of 2X sample buffer. Heat at 100°C for 2 min. Place on ice for 5 min, and spin at 13,000g for 2 min (*see* Note 6).
2. For sample of <25 µL, mix an equal volume of 2X sample buffer. Heat at 100°C for 2 min. Place on ice for 5 min and spin at 13,000g for 2 min.
3. For a sample of >25 µL, concentrate the sample before solubilization as a liquid (*see* Note 7).

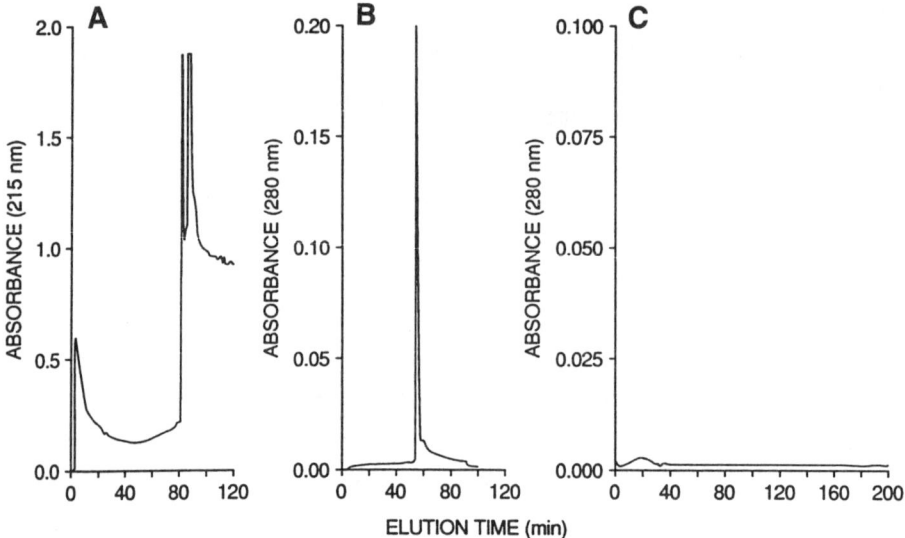

Fig. 1. Prerunning in Tris-glycine-SDS compared with DME buffers. The Tris-glycine-SDS buffer shows a typical shift in the optical density when the glycine ions replace the chloride ions (**A**). There is no shift in the optical density with the DME buffer system, but removal of UV-absorbing matter is still visible (**B**). The baseline in such a system is very stable (**C**).

3.7. Prerunning

Each new gel used for micropreparative electrophoresis must be prerun to remove UV-absorbing contaminant. This is done by applying a current of 1 mA for 1–2 h. As depicted in Fig. 1, it is possible to see the peaks of absorption of these contaminants in both of the systems, i.e., Tris-glycine-SDS and DME-borate. The shift in the baseline level is caused by glycinate ions for the Tris-glycine-SDS system (Fig. 1A), but in DME-borate system, there is no shift because of the presence of DME in both the lower and elution buffers (Fig. 1B; *see* Note 8).

3.8. Electrophoresis

Electrophoresis is conducted at 1 mA/gel after loading the sample at the top of the gel column. Eluted fractions are collected in a fraction collector in 0.5-mL conical tubes with ventilation holes (Fig. 2). Sample proteins can be further analyzed by microsequencing (Fig. 3; *see* Note 9).

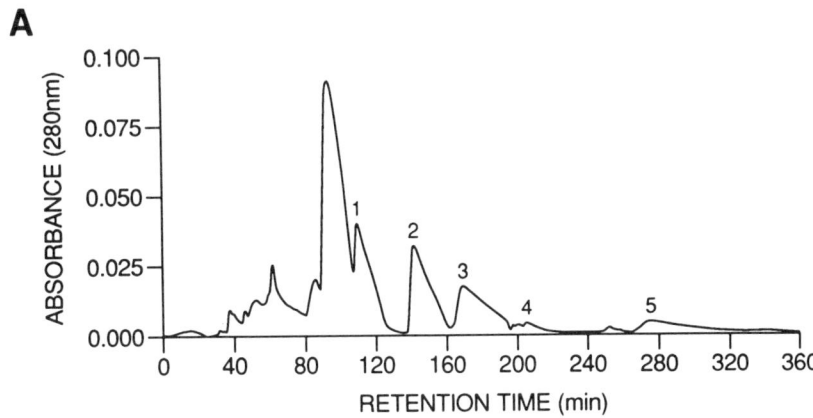
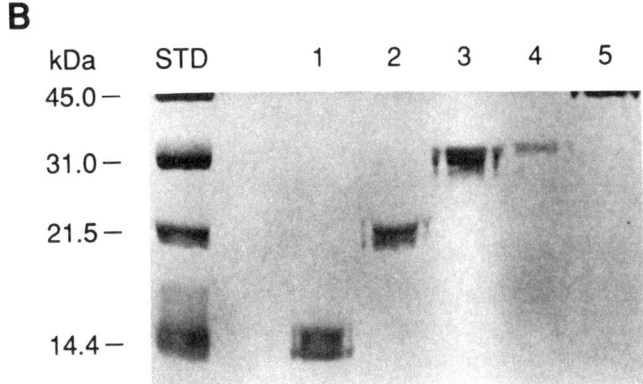

Fig. 2. Separation of four different proteins using a 2.5 × 50-mm polyacrylamide column (15%T) and DME buffer. Twenty micrograms of protein were loaded onto the column, and electrophoresis was performed for 360 min at 1 mA (**A**). An aliquot of each peak was analyzed on a mini-SDS-PAGE and silver-stained (**B**). 1, lysozyme; 2, soybean trypsin inhibitor; 3 and 4, carbonic anhydrase; 5, ovalbumin.

4. Notes

1. The quality of SDS is important because an impure C_{12} SDS source can cause an electrophoretic pattern to change *(13)*. Bio-Rad SDS has been proven to be of the highest quality.
2. Acrylamide is a neurotoxic agent. Wear protective gloves and mask when handling the powder and the liquid. To dispose of old liquid acrylamide solution, add an excess of catalyst (APS and TEMED), and let polymerize. Discard the solid product. Handle bis-acrylamide with the same precaution as acrylamide, even though there are no data on its toxicity.

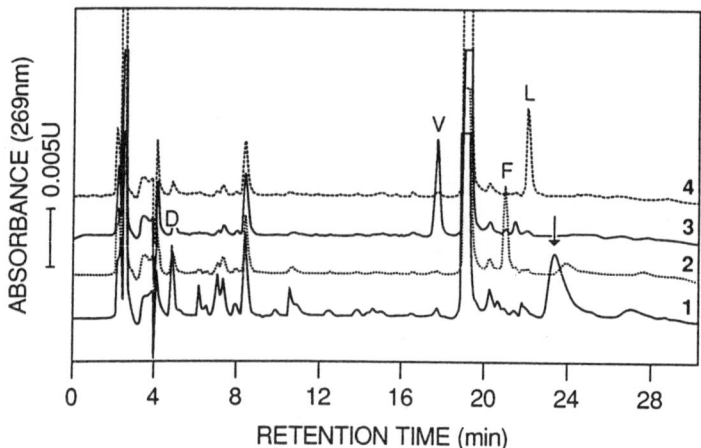

Fig. 3. Microsequencing of HPEC-DME buffer-eluted soybean trypsin inhibitor. The protein, after elution from the HPEC, was directly applied to a polybrene membrane and sequenced. The cycles are numbered in the figure. The first cycle shows a contaminant peak at 23 min (arrow), which is out of the range of separation of the amino acids. Microsequencing was performed on an ABI Model 473A.

3. Polymerization of polyacrylamide is an exothermic process, and control of heat production might be important in a high percentage gel because of the possibility of bubble formation in the gel. Polymerization should then be conducted at 4°C.
4. Gas flows from the gas tank to pressurize the buffer bottles. Upper buffer flows across the cathodic electrode and out to the waste bottle. Lower buffer flows across the anodic electrode and out to waste. Elution buffer flows between the anodic end of the gel and the lower buffer.
5. Samples must be loaded in small volumes because there is no stacking gel in the HPEC system. Glycine can be eliminated and a single buffer (DME) used in place of Tris/glycine. DME contains no primary amino groups and is therefore compatible with microsequencing, unlike Tris.
6. Centrifugation is important to remove particulate insoluble material from the sample. If particulate material is loaded on the gel, it may dissolve during the run and cause a "ghost" peak to appear.
7. Protein samples can be concentrated by ethanol or trichloroacetic acid precipitation or lyophilization. Precipitated proteins are usually difficult to resolubilize and require longer heat treatment. If a sample cannot be concentrated sufficiently, it can be dialyzed against 1X sample buffer without 2-mercaptoethanol, which has to be added just prior to heating. Another pos-

sibility is to achieve the composition of 1X sample buffer within the sample using concentrated gel buffer, SDS, 2-mercaptoethanol, and glycerol.

A minimum of protein must be loaded on the gel column to obtain maximum recovery. Recovery of more than 95% is achieved with 50 µg or more of protein. Loading of less protein could yield a recovery of as low as 25% after an HPEC run *(14)*.

8. Prerunning the gel column in the HPEC apparatus is necessary to establish a continuous system of ions in the gel and is a useful diagnostic tool. In the standard Tris-glycine-SDS system, the prerun showed (Fig. 1A) two peaks eluted between 80 and 100 min of the run. These two peaks are gel contaminants and must be removed before the run. As also shown, the baseline value increased compared with the initial value at the beginning of the run. This increase in baseline value was caused by glycine ions replacing chloride ions in the gel. Thus, the prerun is necessary to establish a continuous system of ions. Glycine and SDS are not recommended for Model 230A. Glycine generated high background absorbance at low wavelength and interferes with microsequencing. The SDS caused foaming to occur and disturbed the baseline. The DME buffer system is better used at 280 nm for detection, because it has a strong absorption at 215 nm, and baseline stability is thus affected.

9. To demonstrate the compatibility of the elution buffer with the microsequencing, soybean trypsin inhibitor collected after micropreparative electrophoresis was directly applied to a polybrene membrane and sequenced. The first four amino acid cycles are shown in Fig. 3: Asp (D), Phe (F), Val (V), and Leu (L). There is a single additional peak in the first cycle (arrowed), which presumably represented a contaminant. This peak did not interfere with the separation and identification of any authentic amino acids in the cycle. Since DME contains no primary amine that would be derivatized, there is little probability that these contaminant peaks come from DME. With a repetitive yield of 95.7% and an initial yield of 95.6 pmol, which is 10% of that applied on HPEC and almost 100% of that recovered from HPEC, this yield compared very well with sequencing results from soluble proteins. Similar results were obtained by sequencing lysozyme separated under the same conditions (data not shown). Although sample recovery on the HPEC is better in the Tris system (25%, data not shown), the separation is far better in DME-borate (Fig. 2 and ref. *14*) because the soybean trypsin inhibitor is completely separated from lysozyme, which is not the case in the Tris system.

Acknowledgments

This work was supported by a grant from the Medical Research Council of Canada (grant no. 12344). The authors thank Michael H. P. West for his contribution to the method described in this chapter, and Van Luu The for making the HPEC available for our research program. Figures are reprinted with permission from ref. 5, *Anal. Biochem.*, Academic, San Diego, CA.

References

1. Walker, J. M. (ed.) (1987) *Methods in Molecular Biology, Vol. 3: New Protein Techniques,* Humana, Clifton, NJ.
2. Chrambach, A. and Nguyen, N. Y. (1979) Preparative electrophoresis, isotachophoresis and electrofocusing on acrylamide gel, in *Electrokinetic Separation Methods* (Righetti, P. J., van Oss, C. J., and Vanderhoff, J. W., eds.), Elsevier/North Holland, Amsterdam, pp. 337–367.
3. Sheer, D. G., Yamane, D. K., Hawke, D. H., and Yuan, P.-M. (1990) The use of micropreparative electrophoresis of protein/peptide isolations for primary structure determination. *Biotechniques* **9,** 486–495.
4. Sheer, D. (1990) Sample centrifugation onto membranes for sequencing. *Anal. Biochem.* **187,** 76–83.
5. Desnoyers, S., Bourassa, S., West, M. H. P., and Poirier, G. G. (1994) One-step protein purification using micropreparative electrophoresis fully compatible with protein microsequencing. *Anal. Biochem.* **221,** 418–420.
6. Sheer, D. G. (1989) The use of PVDF membranes in sample recovery following an HPEC™ isolation bioseparation. *User Bulletin No. 4,* Applied Biosystems Inc., Foster City, CA.
7. Ornstein, L. (1964) Disc electrophoresis-I: background and theory. *Ann. NY Acad. Sci.* **121,** 321–349.
8. Laemmli, U. K. (1970) Cleavage of structural protein during the assembly of the head of bacteriophage T4. *Nature* **227,** 680–685.
9. Baumann, M. and Lauraeus, M. (1993) Purification of membrane proteins using a micropreparative gel electrophoresis apparatus: purification of subunits of the integral membrane protein *Bacillus subtilis* aa3-type quinol oxidase for low level amino acid sequence analysis. *Anal. Biochem.* **214,** 142–148.
10. Wu, R. S., Stedman, J. D., West, M. H. P., Pantazis, P., and Bonner, W. M. (1982) Discontinuous agarose electrophoretic system for the recovery of stained proteins from polyacrylamide gels. *Anal. Biochem.* **124,** 264–271.
11. Sheer, D. and Kochersperger, M. (1990) Separation and characterization of proteins in the range of 1 to 200 kDa with HPEC, in *Current Research in Protein Chemistry* (Villafranca, J., ed.), Academic, New York, pp. 245–262.
12. Sheer, D. G. (1990) The Tris-glycine-SDS system for HPEC™ applications. *User Bulletin No. 1,* Applied Biosystems Inc., Foster City, CA
13. Margulies, M. M. and Tiffany, H. L. (1984) Importance of sodium dodecyl sulfate source to electrophoretic separations of thylakoid polypeptide. *Anal. Biochem.* **136,** 309–313.
14. (1991) Installation and use of the expanded volume (200 µL) sample block on the model 230A. *User Bulletin No. 9,* Applied Biosystems Inc., Foster City, CA.

CHAPTER 36

Practical Column Chromatography

Shawn Doonan

1. Introduction

Many of the most powerful methods of protein fractionation involve the use of column chromatography. Earlier chapters in this volume describe specific chromatographic techniques, but the focus is mainly on the particular application rather than on the generalities of how to do column chromatography in practice. This final chapter is intended to fill in some of the practical gaps in terms of what materials and equipment are needed and how they should be used.

There is no doubt that for convenience, and to some extent for quality of the results obtained, it is ideal to set up a laboratory with a completely automated set of fractionation equipment and to use commercial chromatography columns and ancillary equipment. This is not always realistic. For example, if the intention is to do a one-off purification for a particular purpose, then the expense involved in setting up a dedicated laboratory will not be justified. It is, therefore, important to realize that much can be achieved with a minimum of equipment and that lack of more sophisticated facilities should not deter one from attempting protein purification by chromatographic methods; in principle, all that is needed is a column and some plastic tubing! In what follows, these low-cost options are dealt with in addition to a survey of available commercial equipment. Detailed descriptions of how to set up a particular piece of equipment are always provided by the supplier and are not given here.

There are several suppliers of fractionation equipment, but reference will be made mainly to Pharmacia Biotech (Milton Keynes, UK), since this is the equipment with which the author is familiar; catalogs from other suppliers (Bio-Rad [Hercules, CA], Watson-Marlow [Falmouth, UK], Whatman [Maidstone, UK], Amicon [Beverley, MA], Pierce [Rockford, IL]) should be consulted for alternatives. Approximate prices for equipment are given to provide an idea of the cost of setting up a laboratory. No reference is made to specialist equipment for FPLC or for HPLC, since these topics are covered in Chapters 26 and 27, respectively. Similarly, Fig. 2 in Chapter 25 gives a diagram of the basic arrangement for column chromatography, and this is not repeated here.

2. Equipment and Materials
2.1. Columns

For most applications columns should have a length-to-diameter ratio in the range 5/1–20/1 (*see* Note 1). Pharmacia Biotech markets two main series of columns for general laboratory-scale work. The XK series is suitable for all applications, except for use with organic solvents (where the SR series is required). Internal diameters of 16, 26, and 50 mm are available in each case with four tube lengths in the range 20–100 cm. Each column is supplied with a variable-flow adaptor that allows complete enclosure of the matrix bed, so that sample and buffer application is directly onto the bed with no dead space. The adaptors are continuously adjustable up to 18 cm, so that a 40-cm tube can have an enclosed matrix bed anywhere in the range 22–40 cm. Flow adaptors can be used at both ends of the column, giving greater flexibility in bed height, although normally a fixed-end piece is used at the bottom of the column. By appropriate choice of column dimensions and with the use of adaptors, any bed volume up to 1880 mL can be obtained. The columns come with thermostatic jackets for temperature control by circulation of coolant from an external source. Prices are in the range of $165 for the XK 16/20 to $330 for the XK 50/100.

The more basic C range of columns come with diameters of 10, 16, and 26 mm with lengths in the range 10–100 cm (40-cm maximum for the 10-mm diameter column). They are supplied without flow adaptors and without thermostatic jackets, although these are available as optional extras, and in this basic format are about 50% of the cost of the corresponding XK column. If purchased without the thermostatic jacket, then the column must be run in a cold room if temperature control is required. Packing reservoirs, which effectively extend the length of the column, are available for both the XK and the C series.

Practical Column Chromatography

A cheap alternative is provided by standard laboratory chromatography columns. These are essentially glass tubes drawn out to a taper at one end and with a sintered glass disk (usually of porosity P160; pore size 100–160 μm) sealed in above the taper. For example, BDH (Poole, UK) markets these with diameters of 10, 20, and 30 mm, and 30-cm long at a cost of about $15 each. Larger sizes can be made by purchasing Buchner filter funnels with diameters in the range 30–95 mm (porosity P160), and then getting a professional glass blower to join a piece of glass tube of the same diameter and of the desired length onto the funnel. A 70 mm × 50 cm column (bed volume 2 L) made in this way is very useful for large-scale work and easier to handle than commercial columns of a similar size. The major problem with them is the dead space under the sinter in which substantial mixing can occur, but for initial relatively crude separations, this is not important. At the other end of the range for very small-scale work, a Pasteur pipet or small syringe plugged with glass wool makes a perfectly adequate column.

For most purification schedules, a range of column sizes will be required. A good selection from the Pharmacia Biotech range would be one each of C 10/20, C 16/40, and C 26/40 with a flow adaptor in each case (or the corresponding XK columns if a cold room is not available) plus one or two large homemade columns. The Pharmacia Biotech columns can, however, be substituted by cheap glass columns if funds are limited (*see* Note 2).

2.2. Pumps

Columns can be operated under gravity flow with control exercised by the height of the solvent reservoir above the column and/or by a screw clip on the tubing at the column outlet. Ideally, however, a peristaltic pump should be used. Pharmacia Biotech markets a basic single-channel pump (P-1) at about $1300. This gives a continuous range of flow rates from 0.6–500 mL/h, which is adequate for most purposes (*see* Note 3). This range is achieved by using tubing (silicone rubber is standard) of three different internal diameters (1.0, 2.1, and 3.1 mm) with two continuous ranges of roller speeds differing by a factor of 10. Pumping can be in either direction. The pump can be controlled from Pharmacia Biotech fraction collectors or from a chromatography controller. It can also be safely used in a cold room, which is essential for many applications. More expensive piston pumps that give greater flow rates and can exert moderately high pressures are available (e.g., the HiLoad P-50 from

Pharmacia Biotech, maximum rate 50 mL/min, cost $3300), but there are few situations in which their use is essential.

2.3. Fraction Collectors

Although it is obviously possible, it is very tedious to collect fractions from a column by hand, so a simple fraction collector is an important item of equipment (*see* Note 4). The basic RediFrac from Pharmacia Biotech has a standard rack for 95 test tubes of 10–18-mm diameter; accessory racks for 45 tubes of 28-mm diameter or for 175 tubes of 12-mm diameter are also available. The cost is about $1500 with extra racks at about $80 each. Collection can be on the basis of time or of drop count, and there is a facility for marking tube changes on an attached recorder. Shut-down can be programmed and linked to switching off the peristaltic pump, so that flow through the column is terminated. The fraction collector can be used in a cold room. The GradiFrac collector has the same basic features, but with additional capabilities for control of ancillary equipment (pumps, monitors, switch valves) and for analysis of chromatograms; it costs about $3000. At the top of the range ($4300) is the SuperFrac collector, which has very sophisticated control functions, including a peak slope detection system, which allows for accurate collection of poorly resolved peaks. It is also easily adaptable for the collection of large-volume fractions. These features might be in demand in a laboratory that is dedicated to protein purification, but they are luxuries that can be dispensed with, and for most applications the basic RediFrac is perfectly adequate. Certainly for an active laboratory, two cheaper fraction collectors would be much better use of resources than one top-of-the-range model.

2.4. Monitors

Fractions from a column can be analyzed for protein content by manual absorption measurements at an appropriate wavelength (generally 280 nm; *see* Note 5), but it is more convenient to use a monitor attached to a chart recorder for continuous display of the elution profile. The UV-1 monitor uses a mercury discharge lamp with filters to select wavelengths of 254, 280, or 405 nm. It has a sensitivity range of 0.01–2 absorbance units for full-scale deflection and provides a 0–10 mV signal for connection to standard chart recorders. The cost of the basic instrument with flow cell and two filters is about $5000. One problem with this monitor is that it should not be used in a cold room. This restriction does not apply to the Uvicord S II. The latter also provides a greater range of wavelengths (including 206 and 226 nm, which are useful for detecting proteins at very low levels) and has the capability of connection to a peak integrator. The cost is higher at about $6300 (with three filters), but the extra expense is justified by the extra performance. Either machine requires, of course, a chart recorder for display of the monitor signal. Pharmacia Biotech offers single-channel (REC 101, about $2300) and dual-channel (REC 102, about $2800) machines, both of which are fully compatible with their other equipment. The latter is to be preferred if it is intended to monitor more than one property of the column effluent.

In addition to absorption monitors, Pharmacia Biotech also produces flow-though conductivity and pH monitors (both cost about $2200). They are useful for monitoring gradients (particularly the former, since salt gradients are much more commonly used than are pH gradients), but neither can be considered as essential pieces of equipment.

2.5. Gradient Makers

Many chromatographic procedures require elution of adsorbed protein with a gradient usually of increasing salt concentration. Pharmacia Biotech markets a gradient mixer (GM-1, cost about $800) that produces linear gradients of between 40 and 600 mL. These are produced by having two reservoirs each of 300-mL capacity and of equal diameter connected at the base, such that as solvent is withdrawn from the reservoir originally containing start buffer, the limit buffer flows in to maintain equal levels in the reservoirs (*see* Note 6). Mixing is provided by a motor-driven paddle.

Such devices can also be easily made by a competent glass blower. All that is required is two flat-bottomed glass cylinders, one (for the limit buffer) with a single outlet near the base, and the other (for the start buffer) with two such outlets. The two reservoirs are connected by a sort piece of flexible tubing that can be closed off with a screw clip. Buffer is drawn off from the reservoir initially containing start buffer through plastic tubing connected to its second outlet. Mixing in this reservoir is provided by a magnetic stirrer and pellet. It is useful to have a range of reservoirs available with volumes in the range 100 mL to 2 L.

It is even easier to construct a system with two measuring cylinders connected by a syphon of narrow-bore plastic tube buffer being withdrawn from the cylinder originally containing start buffer through a second plastic tube. With this arrangement, it is essential to ensure that the plastic tubings do not slip out during gradient operation. This can be done by passing the plastic tubings through lengths of glass tube and fixing the latter in place with tape to the tops of the measuring cylinders. Beakers can be used instead of measuring cylinders, but they should be tall and narrow. Otherwise small differences in level of the beakers will cause a large movement of buffer from one to the other.

Yet another possibility is to use a peristaltic pump to deliver limit buffer into a mixing vessel from which solvent is withdrawn onto the column by a second pump. If the flow rate of the pump delivering limit buffer is one-half of the rate at which solvent is removed onto the column, then the gradient will be linear as with the above devices. The problem with this method, of course, is that it ties up a second peristaltic pump.

2.6. Valves

It is convenient, but by no means essential, to include three- or four-way valves in the solvent stream before and after a column. These can be used for sample application and for diversion of the effluent stream to waste, respectively. Pharmacia Biotech markets manual valves (LV-3 and LV-4) at about $170. If the chromatography system is to be automatically controlled, then the valves must be solenoid-driven (e.g., the PSV-50 at about $330). The latter can be controlled by the fraction collector in an automated system.

2.7. Tubing

It is important to have available a supply of flexible plastic tubings for transporting solvents from one part of the chromatographic system to another. These are marketed by most suppliers of laboratory materials, but again a convenient range is available from Pharmacia Biotech. Standard polyethylene tubing with an internal diameter of 1 mm and external diameter of 1.8 mm is generally useful. Polyvinyl chloride tubings with larger internal diameters (1.6–4.0 mm) are useful when faster flow rates are required. A supply of flexible Microperpex (silicone rubber) tubings with internal diameters of 1.3, 2.7, and 4.0 mm is very useful for making joins between lengths of capillary tubings or joining these to glass tubing. For example, a 1-cm length of the 1.3-mm Microperpex tubing makes a good air-tight connection between lengths of the standard 1.8-mm polyethylene tubing.

3. Methods

3.1. Packing Columns

1. Suspend the appropriate amount of matrix (*see* Note 7) in equilibration buffer. Check the pH, and if it is far removed from the target value, then adjust it by careful addition of the acidic or basic component of the buffer.
2. Resuspend the matrix by gentle stirring, and allow to settle until a firm bed has formed. If there is fine material in suspension, then remove this by aspiration with a water vacuum pump or by syphoning. Repeat this step until no more fines remain (*see* Note 8).
3. Choose a column of such a size that the matrix bed will fill it sufficiently for the adaptor to reach the top of the bed (if using a commercial column with adaptor) or to about 80% of its height. Clamp the column vertically in a position where it is protected from drafts or sources of radiant heat (convection currents owing to these will cause uneven column packing). Ideally, the column should be packed at the temperature at which it is to be run.
4. Seal off the outlet tube of the column, and pour in some equilibration buffer. Open the outlet, and allow buffer to flow through to remove air from the bottom net or sinter and from the outlet tubing. Clamp off the outlet again.
5. Suspend the settled matrix in about an equal volume of buffer, and pour the slurry into the chromatography column being careful not to trap any air bubbles (pouring it down a glass rod touching the side of the column can be helpful). If using a column packing extension or reservoir, the remainder of the slurry can also be poured in.
6. Open the bottom tube of the column, and allow buffer to flow through at a rate about 20% greater than that at which it is intended to run the column (*see* Note 9); flow rate can be controlled either with a peristaltic pump or by adjustment to the pressure on a screw clip attached to the outlet tubing.
7. As packing proceeds, a discontinuity will be seen between the packed bed, and the suspension of matrix above it. If using a packing extension then allow packing to continue until this discontinuity has reached the desired height, and remove any remaining suspension from the column. If not using a packing extension, then as clear liquid forms above the suspension, remove it by aspiration and replace it with fresh suspension until the bed has reached the desired height (*see* Note 10). Stop flow through the column.
8. If using a column with a flow adaptor, then layer buffer on top of the matrix bed, being careful not to disturb it, and fill the column to the top. Slide the upper adaptor across the top of the column so that no air is trapped between the net, and the liquid meniscus, push the adaptor just into the column, and tighten the compression nut so that the sealing O-ring makes contact with the sides of the column. Push the adaptor slowly down the col-

umn until the net is in contact with the matrix bed; during this process, excess buffer will exit though the tubing of the flow adaptor, the end of which should be submerged in the buffer reservoir. When the adaptor is in place, then tighten the compression nut so that the O-ring makes a firm contact with the sides of the column. If using a column without an adaptor, then carefully layer buffer on the top of the matrix (the depth depending on the size of the column), and place a tightly fitting rubber bung with a glass tube passing through it in the top; the latter should have a length of capillary tubing passing through, such that the tubing inside the column ends somewhat above the matrix bed (and ideally, touches the wall of the column) and the other end is immersed in a buffer reservoir; the whole arrangement must be air-tight (*see* Note 11).

9. Flow equilibration buffer through the column either under the control of a peristaltic pump (*see* Note 12) or under gravity; in the latter case, flow rate can be controlled by a combination of adjusting the height of the buffer reservoir above the top of the column and adjustment of a screw clip on the outflow tubing. Generally two or three column volumes of buffer are sufficient for equilibration, but this should be confirmed by measuring the pH of the effluent and, if possible, the conductivity.

3.2. Sample Application

1. The sample should have been equilibrated in application buffer either by dialysis or by gel filtration (*see* Chapter 11). Except when using gel filtration (*see* Chapter 25), there is usually no limit on the volume of sample that can be applied (*see* Note 13); protein loading will, of course, depend on the size of the column (*see* Note 7).
2. If using a flow adaptor and peristaltic pump, stop flow through the column by switching off the pump (*see* Note 14). Transfer the inlet tubing to the sample container, and restart the flow when sample will be sucked onto the column. Be careful not to introduce air bubbles; if there is a bubble at the end of the inlet tubing when it is transferred to the sample container, then briefly reversing the direction of the peristaltic pump with the tubing immersed in the sample will get rid of it. Stop the pump, transfer the inlet tube to the buffer reservoir, and restart the flow.
3. If not using a flow adaptor, then remove the rubber bung from the top of the column, and continue buffer flow until the meniscus is flush with the top of the bed. Stop the flow. Layer sample carefully on top of the bed using a pipet or a syringe, taking great care not to disturb the surface; allowing the sample to flow down the wall of the tube helps. Restart the flow. If the volume of sample is large, then replace the bung in the column, and place the inlet tube in the sample container; sample will be sucked onto the column by vacuum. Run the sample completely on to the matrix bed (*see*

3.3. Column Development

1. Wash the column with starting buffer until no more protein is eluted (*see* Note 16). The flow-through peak can often be collected as a single fraction (*see* Note 17).
2. Elute remaining protein by application of one or more step changes in eluent or by application of a gradient (*see* Note 18). If using a column with flow adaptor, then be careful not to introduce air bubbles into the line when switching from the equilibration buffer reservoir to the gradient maker (*see* Section 3.2., step 2). If using a standard laboratory column, then the gradient can be applied by simply transferring the column inlet tube to the outlet of the gradient maker; some mixing between the gradient, and the liquid layer over the matrix bed is unavoidable. Make sure that the mixing paddle or the magnetic stirrer is switched on.
3. Collect appropriate size fractions (*see* Note 19).
4. Combine fractions using appropriate criteria (*see* Note 20).
5. Regenerate and store the column as necessary (*see* Note 21).

4. Notes

1. This is the optimum range for applications such as ion-exchange and other types of adsorption chromatography. Short, fat columns give better flow rates and are less prone to development of high back-pressures. They are, however, less easy to pack evenly. Poor packing can lead to distortion of bands of material passing through the column, which in turn can give poor resolution if two bands are eluting close together. Long, thin columns (up to 100/1 length/diameter) are appropriate for true partition chromatography applications, such as gel filtration.
2. The great advantage of commercial columns is in the design of the end pieces, i.e., the parts in contact with the matrix bed. These have very low liquid volumes, so the possibility of mixing (and of dilution) of eluted materials is minimized. This is important in the most demanding applications, e.g., in gel filtration or in adsorption chromatography where poorly resolved peaks are eluted by application of a gradient. With standard laboratory columns, it is impossible to avoid some dead volume and consequent mixing in the space under the sintered support. In addition, there is always a liquid layer over the matrix bed, which interferes with application of gradients. That being said, there are few occasions where the refinements offered by commercial columns are essential (gel filtration being one such), and it is perfectly feasible to carry out protein purification pro-

cedures using unsophisticated homemade columns. (Note that if money is no object, Pharmacia Biotech markets many of its chromatography matrices as prepacked columns. These offer consistency of behavior as well as convenience, but of course, at a price. This company's catalog should be consulted for a full list of columns available.)
3. Very large columns (e.g., 1-L bed volume) of ion-exchange materials can be run at flow rates of up to 1–2 L/h. It is perfectly feasible to run these under gravity.
4. When using large columns at high flow rates, particularly at the beginning of a purification procedure where the separation required may be relatively crude, the fractions can be quite large and may conveniently be taken by hand. For example, with a 7 × 50-cm column of CM cellulose (vol 2 L) developed with a 4-L salt gradient at a flow rate of 2 L/h, it might be adequate to collect 20 fractions of 200 mL each over the 2-h period of the chromatographic run; this is more easily done manually than by adapting a laboratory-scale fraction collector to cope with the large-volume fractions.
5. Nearly all proteins absorb light at 280 nm owing to their content of the aromatic amino acids tryptophan and tyrosine. A very useful rule of thumb is that in most cases, an absorption of 1 in a 1-cm cell corresponds roughly to a protein concentration of 1 mg/mL (or $A_{280nm}^{1\%} = 10$), although for proteins devoid of tryptophan, this may be wrong by a factor of 4. However, a plot of A_{280nm} against volume provides a useful elution profile for a column. The absorbance at 254 nm is generally about one-half of that at 280 nm, and hence gives lower sensitivity; it is used only because there is a strong line in the mercury emission spectrum at this wavelength. For very low protein concentrations, absorbance at 226 or 206 nm can be used, since these wavelengths are in the region of absorption by the peptide bond, and the absorbance is very high. The problem is that many buffers absorb at these wavelengths (particularly at 206 nm), thus limiting the usefulness of the detectors. Even with detection at 280 nm, the possibility of buffer absorption should be considered if the buffering species is aromatic or contains conjugated double bonds. A low-level and constant buffer absorption can usually be blanked out, but problems will arise if a gradient of increasing buffer concentration is used.
6. The shape of the gradient obtained depends of the cross-sectional areas of the two reservoirs. It can be shown (1) that:

$$C = C_1 - (C_1 - C_i)(1 - v)^{A_1/A_2} \qquad (1)$$

where C is the concentration after a fraction v of the total gradient has been withdrawn, C_1 and C_i are the limit and initial concentrations, respectively, and A_1 and A_2 are the cross-sectional areas of the mixing vessel and of the reservoir for the higher concentration solution, respectively. In the simplest case where $C_i = 0$, and the cross-sectional areas are equal, this reduces to:

$$C = C_1 \cdot v \qquad (2)$$

i.e., a linear gradient. Concave or convex gradients can be obtained by appropriate choice of the ratio A_1/A_2, but this is rarely done. For example, if the protein of interest elutes early in a particular gradient, then better resolution could in principle be obtained either by using a concave gradient or, more simply, a linear gradient with a lower final concentration. If the protein of interest elutes late in the gradient, then improved results could be obtained with a convex gradient or, more simply, by using an initial concentration (C_i in Eq. 1) >0. It should be noted that the method of making gradients with two interconnected vessels works by maintaining the heights of the solutions in the vessels equal, and hence, with vessels of the same cross-sectional areas, this means starting with equal weights of the two solutions. To a first approximation, this can be taken as equality of volumes, but if the densities are markedly different (e.g., 0 and $1M$ NaCl solutions), then some of the strong solution will move into the mixing vessel as soon as connection is made; this is only likely to be important if the protein of interest elutes very early in the gradient.

7. The amount required will depend on the type of chromatography to be done, on the capacity of the matrix (*see* previous chapters on specific techniques), and on the state of purity of the sample. To some extent, amounts have to be determined by trial, but there are some principles and practices that can be useful in deciding the approach to take.

   In ion-exchange chromatography, for example, the protein binding capacities of the matrices vary in the range 10–30 mg/mL of column bed; more precise values are given in suppliers' literature. If the chosen protocol is for the protein of interest to be retained on the column and then eluted by a single step change in the elution buffer, then the whole of the binding capacity of the column may be used since no further separation can be achieved once the protein is displaced from the matrix. If the intention is to do gradient elution, however, then only a fraction (~20–30%) of the binding capacity of the column should be used so that the remaining matrix is available for ion exchange as the displaced material passes down the column. Similarly, if conditions are chosen so that the protein of interest passes through the column without retention, then the whole capacity of the matrix can be used to bind impurities.

   An estimate of the amount of matrix required can be obtained in a trial using batch adsorption. For example, take 10 aliquots of the equilibrated protein solution, and add increasing quantities of equilibrated matrix to each. Mix gently for 10 min, and then sediment the matrix by centrifugation. Determine the amount of the protein of interest remaining in the supernatant in each case. This will allow an estimate of the minimum amount of matrix required to bind all of the protein of interest, and hence, the amount to used in practice can be chosen on the basis of the outlined considerations. If the objective is to bind impurities, but not the target protein, then a similar experiment can be used to determine the amount of matrix that gives maximum removal of protein from solution without loss of target protein. It is important to bear in mind that the ability of a matrix to bind a particular protein will depend on the state of purity of the sample. Hence if conditions for adsorption, and the binding capacity of a matrix for a particular protein, have been determined using a partially purified sample, then the capacity may be considerably less if a cruder preparation is used. In these circumstances, it may well happen that more strongly binding impurities saturate the column and displace the protein of interest, so that it emerges without adsorption to the column contrary to expectations. Conversely, if a protein is partially purified using a particular ion-exchange material followed by rechromatography on the same material under the same conditions, then a smaller amount of matrix will be required for the second step, since much of the impurities will already have been removed.

There is a temptation to "play it safe" and use much larger columns than are actually required; this should be resisted. The consequences will be reduced yield owing to nonspecific adsorption of protein on the matrix and unnecessary dilution of the active fraction. There will also be a cost penalty, which can be severe with the more expensive matrices.

8. Care should always be taken to remove fines. If this is not done, then the fine particles will pack into the interstices between matrix particles and block flow of solvent through the column. If this happens, then the only recourse is to repack the column.
9. If the flow rate of the column is too great, then equilibration of protein between the matrix and the solvent may not be achieved; in addition, compaction of the matrix bed may occur. On the other hand, low flow rates will lead to peak spreading by diffusion with consequent dilution and loss of resolution, and possibly to loss of activity because of the extended time of the procedure. Flow rates are generally in the range 0.1–0.5 cm/min (i.e., 0.1–0.5 cm^3/cm^2 of cross-sectional area/min).
10. It is important not to allow all the matrix to settle, and, then replace the supernatant liquid with more matrix suspension, and so on. Because the coarser particles settle faster, this would result in a series of discontinuities between bands of fine and coarse material in the packed bed with the consequence of inferior chromatographic performance, and the possibility of poor flow rates.

11. With this arrangement, as liquid is removed from above the matrix bed, the partial vacuum produced will cause buffer to be sucked out of the reservoir and into the column; there needs to be sufficient buffer above the matrix bed before flow is started to ensure that the syphoning starts before the bed goes dry. The tube carrying the buffer should extend to close to the top of the bed, so that incoming buffer does not "bomb" the bed and disrupt the surface. The arrangement must, of course, be air-tight. This can be arranged by pushing one end of a piece of flexible plastic tubing onto the glass tube, and then inserting one or more short lengths of tightly fitting plastic tubings into the other end until the bore of the innermost insert is somewhat less than the external diameter of the capillary tubing to be used to connect to the reservoir. The capillary can then be pushed through this last insert by a sufficient distance to end just above the matrix bed. If very small columns are being used with rubber bungs that are too small to bore easily, then a hypodermic needle can be substituted for the glass tube.

12. If a flow adaptor is being used, then the peristaltic pump can be placed before or after the column (except during packing, of course, when it must be after). One disadvantage of placing it after the column is that some mixing of the eluent will occur owing to the volume of the pump tubing and to the pumping action; this can cause loss of resolution. Another problem arises if it is attempted to pump liquid through the column at a rate greater than its natural flow rate when air bubbles will form in the outlet tubing. Nevertheless, it is common practice to leave the peristaltic pump positioned after the column after packing is complete and when the column is run.

13. Large-volumes of dilute protein solution do not usually pose a problem, particularly if the protein of interest binds to the column. Indeed, adsorption of the protein onto a column from a large-volume of solution followed by stepwise or gradient elution provides concentration as well as purification.

14. Having a three-way valve in the buffer line (before the peristaltic pump), with the tubing from the third port dipped into the sample reservoir, is useful here, since simply turning the valve will direct sample rather than buffer onto the column. If the sample tubing contains air, then reversing the direction of flow of the pump temporarily will displace the air by buffer from the column; reversing it again will then load the sample onto the column.

15. Some problems can arise at this stage. Particularly if the protein solution is concentrated, and the column diameter is large, imperfections in packing sometimes result in the protein solution channeling or slipping between the matrix bed and the walls of the column; this is easy to see if the solu-

tion is colored. The only remedy is to stop flow through the column, stir up the top section of the bed (to below the point of visible channeling), and allow it to settle again before applying the rest of the sample. Another problem can arise with crude protein samples that either have not been properly clarified before application to the column or from which protein precipitates (owing to the particular pH, and so forth) during application. The particulate matter will form a layer on top of the matrix bed; this will slow down or stop buffer flow, or may lead to channeling. The situation may be recoverable by stirring the top of the bed to disperse the layer of particulate matter, but subsequent running of the column is unlikely to be ideal. This emphasizes the desirability of using properly clarified protein solutions for chromatography. When flow adaptors are being used, particulate matter can block the applicator net; the only solution if this occurs is to dismantle the adaptor and clean or replace the net.

16. This can be judged by the trace on a recorder returning to base line or by manual measurements of A_{280nm}. This should occur with a volume somewhat larger than the liquid volume of the column, unless material is partially adsorbed, in which case the volume may be considerably greater; conditions should be chosen, if possible, to avoid the latter situation.

17. This is obviously the case if the protein of interest is retained on the matrix (although it is always advisable to retain the unadsorbed fraction until it is confirmed that the target protein has indeed bound—failure to equilibrate the matrix or the sample properly or overloading the column can lead to unexpected behavior, and it is easier to rechromatograph the fraction than to start again from the beginning). Note that diversion of the flow-through fraction to a collecting reservoir is one of the functions that can be programmed into more sophisticated fraction collectors equipped with solenoid valves.

If conditions have been chosen such that the protein of interest is not adsorbed onto the matrix, then the breakthrough material can usually still be collected in a single fraction, since it should have a uniform composition. One circumstance where this is not so is if either the target protein or some of the impurities are retarded, but not retained by the column. The target protein will then be enriched in either the back part or the front part of the breakthrough peak, and individual fractions will need to be taken for optimal purification. Similarly, if the column is overloaded so that the protein of interest, which was intended to be retained, is displaced by more tightly binding contaminants (*see* Note 7), then the target protein will be concentrated in the back part of the breakthrough peak, and potential purification will be lost if individual fractions are not collected. This situation should be suspected if the breakthrough peak is markedly asymmetric with increasing protein content in the back half.

18. The total volume of the gradient will generally be in the range of four to eight times the volume of the matrix bed. The most commonly used salt for gradients is NaCl. In the absence of previous information on the salt concentration required to elute a particular protein, a gradient of 0–1.0M should be tried. From this trial, the salt concentration at the maximum of the eluted protein peak can be calculated from the equation in Note 6. This value plus about 20% can then be used as the limit of the gradient for subsequent experiments. If the target protein elutes at relatively high salt concentration, then the gradient can be started at a value >0. For example, if the protein elutes at 0.6M NaCl, then a gradient from 0.4–0.75M would probably give good results. It is important not to have the protein eluting at the limit of the gradient because tailing will then occur.

    To set up the gradient, pour equal volumes (*see* Note 6) of the start buffer, and the start buffer containing salt, into the mixing chamber and into the second reservoir, respectively, making sure that the connection between the vessels is closed off and that the vessels are at the same height. If it is intended to use a nonlinear gradient with vessels of unequal cross-section (*see* Note 6), then the vessels must be filled to the same level. At the time when the gradient maker is connected to the column, open the connection between the vessels, and switch on the stirring motor. If the gradient device consists of two cylinders or beakers that need to be connected by a syphon tube, then fill them to the desired level, insert the tubing into one of the vessels, suck liquid up it with a syringe, sqeeze the tube so that liquid cannot escape, and place the end below the surface of the liquid in the second vessel.

19. It is not usually worthwhile collecting a large number of small fractions—this simply increases the amount of analysis to be done. Fraction volumes should be between 1/50th and 1/20th of the total elution volume of the column, depending on the degree of resolution expected. Take larger rather than smaller volumes, unless there is a good reason to do otherwise.

20. Chromatography at early stages of a purification is unlikely to result in the protein of interest being obtained in a symmetric, well-resolved protein peak. Combination of fractions is therefore likely to be a payoff between yield and purification. The first requirement is to construct a protein profile (if not using an automatic monitor) by plotting $A_{280\,nm}$ against elution volume or fraction number. Next, the fractions containing the target protein must be established using a quantitative assay, and the activity profile superimposed on the protein profile. From this, it will be possible to see how to combine the fractions in such a way as to optimize both purification and yield. For example, it is usually worth rejecting fractions at the beginning and/or end of the peak of active material if their contents of

contaminating protein are very high. It is also very important to determine the total yield of the target protein, since a particular chromatographic procedure may result in substantial purification, but may still be unacceptable if it results in a low overall yield. The importance of having a quantitative assay for the protein of interest is difficult to overestimate, particularly for the analysis of fractions from chromatographic procedures. In addition, the assay should be as rapid as possible, since large numbers of fractions may have to be analyzed. Time devoted to developing a rapid quantitative assay will be well spent if a substantial amount of purification work is to be undertaken. A good example of what can be done with a little ingenuity is given in ref. 2.

21. With most matrices, regeneration can be done by washing with one or two column volumes of a solution of high salt concentration (~$1M$), followed by a similar volume of equilibration buffer; manufacturers' instructions should be followed if different from the above. The column should then be stored in equilibration buffer, containing an antibacterial agent, such as sodium azide (0.2% [w/v]). Matrices used to fractionate crude protein mixtures may retain some protein and pigmented material after such a treatment. This can usually be removed by washing with $0.1M$ NaOH (if the matrix is stable under these conditions) before re-equilibration. As a last resort, the discolored matrix can be removed from the top of the column and replaced. Some matrices (particularly those based on Sephadex) shrink considerably when subjected to solutions of high ionic strength, and the bed may not return to its original dimensions on re-equilibration. If this occurs, then the column should be repacked.

References

1. Bock, R. M. and Ling, N.-S. (1954) Devices for gradient elution in chromatography. *Anal. Chem.* **26,** 1543–1546.
2. MacGregor, S. E. and Walker, J. M. (1994) A microtiter plate assay for the detection of inhibitors of the Na^+, K^+-ATPase. *Appl. Biochem. Biotechnol.* **49,** 135–141.

Index

A

Accelerated degradation testing, 335, 339, 350–352, 354, 355
Acetone, fractional precipitation with, 139
Acetonitrile, 286
Activity,
    loss of, 83, 84, 332
    measurements, 12, 13
    preservation of, 2, 3, 27
Additives, stabilizing, 344–348
Adsorption coefficient, 278
Affigel 10, 189–191
Affigel 15, 181, 189
Affinity,
    chromatography, *see* Chromatography, affinity
    elution, 173
    precipitation, 10, 217–236
        metal ion, 159, 222
    tag, 158
Agarose, carbonyldiimidazole-activated, 179–181
Aggregation,
    of particulate matter, 22, 140
    of proteins, 33, 340, 346
Amido black 10B, 369
Amino acid,
    compositional analysis, 364
    sequencing, 1, 4, 364
Ammonium sulfate, 345
    concentration with, 97, 98
    fractional precipitation with, 25, 74, 137–139
    percentage saturation, 137
    storage of proteins in, 348
    trial precipitation with, 141
Animal tissues, extraction of proteins from, 5, 17–22
Antibodies, 37, 157
    concentration of, 132, 133
Antigens, 157, 158
Antimicrobials, 342
Antioxidants, 77, 78
Arrhenius equation, 350–352, 354
Ascorbate, 72
Aspartate aminotransferase, purification of, 152
Aspartic proteinases, 87, 88, 343
Assays, 395, 396
Automated systems, 190, 381, 382

B

Bacterial contamination, prevention of, 342, 343
Band spreading, 262
Bed volume of columns, 261, 387
Binding,
    nonspecific, 188
    proteins, nucleic acid, 157
    specific, 10
Bioassay, 13
Biogel, 258, 261
Biological affinities, 157
Bis-ATP, 227, 228, 233–235
Bis-ligands, 217
Bis-NAD$^+$, 218–234

Blotting,
　semidry, 364, 366
　tank, 364, 366
Blue Dextran, 169
Blue Sepharose CL-6B, 171, 173, 174
Breakthrough curves, 161, 162
Browning reactions, 71
Buffer,
　compounds, 106
　exchange, 2, 103–113, 265
　　by ultrafiltration, 127, 128
　values, 104, 105
Buffering capacity, 104, 105
Buffers,
　amphoteric, 251
　changing, 103–113
　chromatofocusing, 252
　fast protein liquid
　　chromatography, 273
　"Good," 105
　high-performance electrophoresis
　　chromatography, 373, 374
　homogenization, 20
　hydroxyapatite chromatography, 212
　making, 103–113
　size-exclusion chromatography, 260
Bulking agents, 332, 333

C

C_{18} columns, 286
C_8 columns, 286
Ca^{2+}-ATPase,
　assay of, 299–302
　extraction of, 299, 301
Calibration of size-exclusion
　columns, 263
Calsequestrin,
　assay of calcium binding to, 299,
　　301, 302
　extraction of, 300, 301

Carboxymethyl group, 146, 147
Carrier ampholytes, 240, 241
Centrifugation, 19
　precautions in use of, 20, 21, 99, 142
Cereal seed proteins,
　extraction of, 26
Chaikoff press, 50
Chaotropic agents, 33, 151, 160, 163, 164
CHAPS, 295–297
Charge, separation by, 8, 9
Chelators, 342
Chemical characterization of proteins, 363, 365, 367
Chloroplasts, 57
　assay of intactness, 62
　assay of photosynthetic activity, 62, 63
　assay of purity, 63, 64
　isolation,
　　from barley, 61
　　from pea leaves, 61, 62
　　from wheat, 60, 61
Chromatofocusing, 9, 249–254, 269
Chromatography,
　affinity, 2, 10, 157–168, 217
　detergent-exchange, 309
　dye-ligand affinity, 10, 169–175
　fast protein liquid (FPLC), 8, 269–275
　gel filtration, 269
　high-performance electrophoresis, 371–380
　hydrophobic interaction, 151–155, 269
　hydroxyapatite, 211–215
　immobilized metal ion affinity, 197–208
　　high performance, 205–207
　immunoaffinity, 2, 10, 187–195
　ion-exchange, 2, 8, 145–150, 317, 318

Index

lectin affinity 10, 177–183, 313, 315–318
practical column, 381–396
reversed-phase, 2, 163, 269, 277–291
  columns for, 279–281
size-exclusion, 9, 255–267
  matrices for, 258, 260, 261
Cibacron Blue, 158, 169–171
Cleland's reagent, 347
Coenzyme A synthase, purification of, 173, 174
Cold denaturation, 340, 348
Colloidal gold, 369
Column chromatography, practical, 381–396
Columns, 382, 383
  sizes of, 382
Concanavalin A, 178
  immobilization of, 180, 181
Concentration,
  by ammonium sulfate precipitation, 95–102
  by forced dialysis, 95–102
  by ultrafiltration, 115–134
Conformational changes, 346
Coomassie brilliant blue R-250, 241, 358, 365, 369
Critical micelle concentration (CMC), 295–297, 306
Crosslinking agents, 260, 261
Cyanogen bromide, 160
Cysteine proteinases, 85, 87, 343

D

3,4-DCI, see 3,4-Dichloroisocoumarin
Dead volume, 290, 382, 383
Deamidation, 340
Decay, first-order, 341
Denaturants, 85
Denaturation, 82, 163, 277

Denaturing agents, use in extraction of proteins, 33
Desalting, 148, 257, 265
Detergents, 163, 295–297, 305
  removal of, 130, 305–311
Development of columns, 389
Dextran, 145
  crosslinked, 260
  granulated, 239
Dialysis, 307, 309, 310, 317, 319, 320
  changing buffers by, 109, 110
  equilibrium in, 100, 101
  removal of ammonium sulfate by, 143, 144
  tubing, 97–101, 107, 109, 143
3,4-Dichloroisocoumarin (3,4-DCI), 87, 89
Diethlydithiocarbamate (DIECA), 72
Diethylaminoethyl,
  group, 147
  cellulose, 146–149
Differential centrifugation, 49–51
Digitonin, 295–297
Distribution, subcellular, 18
Documentation of purification, 12, 13
Donnan equilibrium, 111
Dounce homogenizer, 52
Drying for storage, 349, 350
DTT, see Cleland's reagent
Dye-ligand affinity chromatography, see Chromatography, dye-ligand affinity

E

E-64, 87, 89
E. coli, enzymatic lysis of, 34
EDTA, 20, 88, 89, 99
Egg surface membrane complex, isolation of, 315, 317, 318
Electroblotting, 363–370
Electroelution, 357–362

Electrophoresis, micropreparative, 371
β-Elimination, 340
Elution profile, 390
Endopeptidases, 81, 343
    assay of, 88, 89
Equilibration of ion-exchange resins, 148
Equipment,
    column chromatography, 382–386
    fast protein liquid chromatography, 270
    high-performance electrophoresis chromatography, 372
    lyophilization, 324–327
    reversed-phase chromatography, 285–287
    size-exclusion chromatography, 257–259
    ultrafiltration, 119–122
1-Ethyl-3-(3-dimethylaminopropyl)-carbodiimide, 224
Example purification, 13–15
Exopeptidases, 81, 343
Extracellular proteins, fungal, 39, 40
    electrophoresis of, 42, 43
Extracts, concentration of, 95–102

F

Fast protein liquid chromatography, see Chromatography, fast protein liquid
Filtration of protein solutions, 342, 343, 352, 353
    sterile, 341
Fines in chromatographic materials, 392
FITC-casein substrate, 88–91
Flow adaptors, 162, 382, 383, 387, 388
Flow rate, 162, 261, 262, 392
Flowthrough fraction, 165

Fluorescein isothiocyanate, see FITC
Forced dialysis, 97–99
FPLC, see Chromatography, fast protein liquid
Fraction collectors, 384
Fractional precipitation, bulk purification by, 2, 135–144
Freeze dryer,
    filling and loading, 328, 329
    operation, 327–331
    shutting down, 330–331
    simple, 331
    start-up, 328
    termination of run, 330
Freeze-concentration, 332
Freeze-drying, see Lyophilization
Freezing of protein solutions, 348, 349
Freezing temperature, 329
French press, 36
Fumarase, separation of isoenzymes of, 212, 213
Fungi,
    cell breakage, 45, 46
    cell walls of, 39
    extraction of proteins from, 4, 39–46
    filamentous, 39
    growth,
        media, 40, 41
        phase, 39, 44, 45
    starter cultures, 44

G

Gel filtration,
    changing buffers by, 109, 110
    chromatography, see Chromatography, size-exclusion
Glutamate dehydrogenase, purification of, 231–233
Glutathione S-transferases, purification of, 249–251

Glycerol,
  stabilization by, 345, 346
  storage in, 348
Glycoconjugates, 177
Glycoproteins, 166, 177
Golgi apparatus, 54
Gradient, 152, 390, 391, 395
  elution, 150, 284, 389
  makers, 385, 386
Guanidinium hydrochloride, 33
Glycation, 340

H

Heat denaturation,
  use in purification,
    11, 12, 22
Henderson-Hasselbalch equation,
  103, 104
High-performance electrophoresis
  chromatography, see Chromatography, high-performance electrophoresis
Hofmeister series, 153, 345
Homogenates, 49
  preparation of, 17, 52
Homogenization, 50
Hormones, 157
HPLC, reversed-phase, see
  Chromatography,
  reversed-phase
Hybridomas, 132, 158
Hydrophobic,
  adsorption chromatography,
    308–310
  interaction, 151, 152, 278, 279
    chromatography, see
      Chromatography,
      hydrophobic interaction
Hydrophobicity, 157
Hydroxyapatite chromatography,
  see Chromatography,
  hydroxyapatite

I

IDA, see Iminodiacetate
IMAC, see Chromatography,
  immobilized metal ion affinity
Iminodiacetate (IDA), 198–200
Immobilized,
  chelating groups, 197–199
  metal ion affinity chromatography,
    see Chromatography,
    immobilized metal ion
    affinity
Immunoaffinity chromatography,
  see Chromatography,
    immunoaffinity
Immunoassay, 13
Immunoglobulins, purification of,
  164, 165, 179
Immunoprobing of proteins, 365, 367
Inactivation, irreversible, 341
Inclusion bodies, 31–36
India ink, 369
Inhibitors,
  stock solution, 89
  working cocktails, 90
Ion-exchange chromatography,
  see Chromatography,
  ion-exchange
Isoelectric,
  focusing, 2, 9, 239–247
  point, 157, 239, 249

L

Lactate dehydrogenase,
  purification of, 218, 219
  separation of isoenzymes of,
    233, 234
Lectin affinity chromatography, see
  Chromatography, lectin affinity
Lectins, 158
  used in affinity chromatography,
    178

Ligands, 159
  biotinylated, 167
  competing, 163
  group specific, 157, 158
  leakage, 188, 189
  mono-specific, 157
  specific, 346, 347
Locking-on effect, 221
Long-term stability, 340
Low-temperature storage, 348, 349
Lubrol PX, 295–297
Lyophilization, 317, 319, 320, 323–336, 346, 349, 350
Lysosomes, 52, 54
Lysozyme, lysis of *E. coli* with, 34

M

Maillard reaction, 340
Mantin Gaulin press, 36
Marker enzymes, 54, 55
Mass spectrometry, 364
Matrices, 159, 160
  activation, 160
  affinity, 159, 160
  capacity of, 161, 391, 392
  ion-exchange, 72, 145–147
  polyacrylamide, 266
  regeneration of, 149, 164, 396
  streptavidin-activated, 167
Matrix-assisted laser desorption ionization mass spectrometry (MALDI-MS), 206, 361
Membrane proteins, 6, 257, 266, 293
  extraction of, 293–303, 313
  extrinsic, 293
  integral, 293, 295–297, 313
  intrinsic, 293
  peripheral, 293–295, 313
  purification of, 313–321
  solubilization, 297
Membranes,
  biological, 293

  semipermeable, 115
MemSep™ cartridges, 146
Metalloproteinases, 85, 88, 343
Metarhizium, intracellular proteins from, 41, 42
Methyl sulfonate group, 147
Microheterogeneity, 84
Microsequencing, 1, 371
Microsomes, 52
Microtiter plates, 327
Mitochondria, 18, 52, 57
  assay of activity, 65
  isolation,
    from pea leaves, 64, 65
    from rat liver, 52, 54
Mobile phase, 278, 281–284
Molecular weight, 255
  cutoff, 100, 101, 116, 122–125
  standards, 264
Monitors, 384, 385
Mono Q exchanger, 271–273
Mono S exchanger, 271–273
Monoclonal antibodies, 4, 158, 187–189
  immobilization, 188–192
  purification, 194, 195
Mycellium, extraction of proteins from, 42

N

N_2,N_2'-adipodihydrazido-bis-(N^6-carbonylmethyl)NAD$^+$, *see* Bis-NAD$^+$
N_2,N_2'-adipodihydrazido-bis-(N^6-carbonylmethyl-ATP), *see* Bis-ATP
Nitocellulose, 368
  staining (Instaview), 365
Nitrate reductase,
  purification of, 24

Index

O

Octylglucoside, 295–297, 315
Osborne fractionation, 24–26
Osmolytes, 342, 345, 346
Overexpression of recombinant proteins, 31
Oxidase inhibitors, 77, 78
Oxidation, 82, 340
   prevention of, 20, 347

P

Packing columns, 387, 388
PAGE, see Polyacrylamide gel electrophoresis
Particulate matter, removal, 21, 22
Partition coefficient, 256
Partitioning, 255, 262, 263, 277
PEG, see Polyethylene glycol
Phenyl Sepharose CL-4B, 153
Pepstatin, 88, 89
Peptide,
   bond cleavage, 340
   profiling, 364
pH gradient, 239, 249
   measurement of, 244–246
1,10-Phenanthroline, 88, 89
Phenolic compounds, 69–79
Phenylmethylsulfonyl fluoride, 23, 28, 33, 78, 79, 87, 89, 161, 342
Phosphofructokinase, purification of from bovine heart, 222, 233–235
pK_a values, 106, 110
Plant cell walls, 57
Plant tissues,
   extraction,
      of enzymes from, 4, 69–79
      of proteins from, 23–29
Plate height, theoretical, 263, 313
PMSF, see Phenylmethylsulfonyl fluoride

Polyacrylamide gel electrophoresis, 163, 357, 363,
Polybrene membranes, 379
Polybuffer exchangers, 249–252
Polybuffers, 250–252
Polyclar AT, see Polyvinylpolypyrrolidone
Polyclonal,
   antibodies, 1, 4, 167
   antiserum, 164
Polyethylene glycol,
   fractionation with, 140
   as a stabilizer of proteins, 344
Polyphenoloxidases (PPO), 71
Polyphenols, 4, 69–79
   adsorbents for, 75–77
Polyvinylidene difluoride (PVDF) membranes, 357, 368, 371
Polyvinylpolypyrrolidone, 72–77
Polyvinylpyrrolidone, 25, 72
Preservatives, 324
Primary drying, 323, 329, 333, 334
Product formulation, 324, 349
Protein A,
   affinity matrices, 163
   Sepharose 4B, 164, 165
   from *Staphylococcus aureus*, 158
Protein,
   blotting, 363
   extraction,
      animal tissues, 17–22
      bacteria, 31–37
      fungi, 39–46
      plant tissues, 23–30, 69–80
   precipitation,
      organic solvents, 135, 136
      salts, 135, 136
Protein–matrix interaction, 266
Proteinases, inhibition of, 78, 79, 86–88, 161, 302, 320, 342–344, 353
Proteolysis, avoidance of, 3, 81–93, 343, 344

Proteolytic attack,
    recognition of, 82–84
Protoplasts, 57, 58,
    isolation from,
        barley, 61
        wheat, 60
Pumps, 383, 384
Purification factor, 14, 15
Purity, 81, 82, 163, 167
    need for, 3, 4
PVP, *see* Polyvinylpyrrolidone
PVPP, *see* Polyvinylpolypyrrolidone
Pyrogens, 264

Q

Quantity, 1, 2
Quaternary,
    aminoethyl group, 147
    ammonium group, 147
Quinones, 71

R

Receptor for sperm from sea urchin
        eggs, 314–320
    solubilization of, 315–318
Receptors, 157
Recombinant,
    proteins, 6, 12, 31
        extraction from bacteria, 31–37
    DNA technology, 31
Recovery,
    maximizing, 125, 126, 167
    of protein from gels, 129, 130
Red Sepharose, 171, 173, 174
Reducing agents, 342, 347
Refolding of proteins, 36
Rehydration of dry proteins, 335, 349
Relative centrifugal force, 51
Residual moisture, 334
Resolution, 263
Resolving power, 2

Reversed-phase chromatography, *see*
    Chromatography, reversed-
    phase
Ribulose bisphosphate carboxylase,
    purification of, 24, 25, 73, 74
RUBISCO, *see* Ribulose
    bisphosphate carboxylase

S

Salt removal,
    by dialysis, 105–107
    by gel filtration, 105–107
    by ultrafiltration, 127, 128
Salting out, 345
Salts, structure-forming, 152
Sarcoplasmic reticulum, 298–302
Scale of purification, 1, 2
Schedule of purification, 1
SDS-PAGE, 3, 9, 27, 35, 85, 88, 165,
        166, 174, 250, 319, 359, 371
Secondary drying, 323, 329, 334, 335
Seed storage proteins, 24
Sephacryl, 258
Sephadex, 145, 258
    G-25 for changing buffers, 109,
        110
Sepharose, 258
Serine proteinases, 86, 87, 343
Shelf life, 324, 335, 339
Size, separation by, 9, 10
Size-exclusion chromatography,
    see Chromatography,
    size-exclusion
Skeletal muscle, 298, 300
Sodium cholate, 295–297
Solubilization of recombinant
    proteins, 35, 36
Solubility, 7, 8, 135
Sonication, 36
Source of protein, choice of, 4–6
Spacers in affinity chromatography
    matrices, 160

Index

Stability,
  analysis of, 350–352
  conformational, 350
  kinetic, 350
  long term, 336
Standard buffers, 107
Stationary phases, 278, 280
  for IMAC, 201–203
Stepwise elution, 150
Storage,
  containers, 341
  of matrices, 164
  of proteins, 264, 265, 339–355
Strategies for purification, 1–14
Subcellular fractionation, 5
  of animal tissues, 49–56
  of plant tissues, 57–68
Substrates, 346, 347
Sucrose,
  concentration with, 348
  as protein stabilizer, 344–346
Sulfobetain 14, 295–297
Sulphopropyl group, 147
Superdex, 258, 261
Superose, calibration
  of columns, 275

T

Tannins, 70
TED, see Tris(carboxymethyl)ethylenediamine
Theoretical plates, 266
Thiol-disulfide exchange, 340, 347
Tissue distribution of proteins, 18
Triazine dyes, 169–171, 222
Trichloroacetic acid, 85, 88
Tris(carboxymethyl)ethylenediamine, 198, 199
Triton X-100, 264, 295–297
Tubing, chromatography, 386

U

Ultrafiltration, 9, 317, 319, 320, 348, 353
  centrifugal devices for, 120–122
  fractionation by, 130
  purification and concentration by, 115–134
  solute retention, 123–125
  static devices for, 122
  stirred cells for, 119, 120
Unfolding, reversible, 340
Urea, 33

V

Valves, chromatography, 386
Vials,
  pharmaceutical, 327
  sealing under vacuum, 327, 330
Visking tubing, see Dialysis tubing
Volume, size-exclusion chromatography,
  elution, 255, 256
  total, 255, 256
  void, 255, 256

W

Waring blender, 18, 50
Western,
  blotting 163,
  electroblotting, 357
Wheat germ agglutinin, 315, 318

Y

Yeast alcohol dehydrogenase,
  purification of, 218, 219, 230, 231
Yeasts, 39,
  breakage of, 46
Yield, 12, 13, 15, 326, 335

Z

Zone spreading, 266
Zymography, 88, 91, 92

STAFFORD LIBRARY
COLUMBIA COLLEGE
1001 ROGERS STREET
COLUMBIA, MO 65216